Mycosynthesis of Nanomaterials
Perspectives and Challenges

Editors

Mahendra Rai
Department of Biotechnology
SGB Amravati University
Amravati, Maharashtra, India

Patrycja Golińska
Department of Microbiology
Nicolaus Copernicus University
Torun, Poland

CRC Press
Taylor & Francis Group
Boca Raton London New York

CRC Press is an imprint of the
Taylor & Francis Group, an **informa** business

A SCIENCE PUBLISHERS BOOK

Cover illustrations reproduced by kind courtesy of Elanit Rubalcava Avila and Yossef Rubalcava Avila, Chihuahua City, Mexico.

First edition published 2023
by CRC Press
6000 Broken Sound Parkway NW, Suite 300, Boca Raton, FL 33487-2742

and by CRC Press
4 Park Square, Milton Park, Abingdon, Oxon, OX14 4RN

© 2023 Mahendra Rai and Patrycja Golińska

CRC Press is an imprint of Taylor & Francis Group, LLC

Reasonable efforts have been made to publish reliable data and information, but the author and publisher cannot assume responsibility for the validity of all materials or the consequences of their use. The authors and publishers have attempted to trace the copyright holders of all material reproduced in this publication and apologize to copyright holders if permission to publish in this form has not been obtained. If any copyright material has not been acknowledged please write and let us know so we may rectify in any future reprint.

Except as permitted under U.S. Copyright Law, no part of this book may be reprinted, reproduced, transmitted, or utilized in any form by any electronic, mechanical, or other means, now known or hereafter invented, including photocopying, microfilming, and recording, or in any information storage or retrieval system, without written permission from the publishers.

For permission to photocopy or use material electronically from this work, access www.copyright.com or contact the Copyright Clearance Center, Inc. (CCC), 222 Rosewood Drive, Danvers, MA 01923, 978-750-8400. For works that are not available on CCC please contact mpkbookspermissions@tandf.co.uk

Trademark notice: Product or corporate names may be trademarks or registered trademarks and are used only for identification and explanation without intent to infringe.

Library of Congress Cataloging-in-Publication Data (applied for)

ISBN: 978-1-032-35547-4 (hbk)
ISBN: 978-1-032-35549-8 (pbk)
ISBN: 978-1-003-32738-7 (ebk)

DOI: 10.1201/9781003327387

Typeset in Times New Roman
by Radiant Productions

Preface

Since the inception of the concept of green chemistry by Paul Anastas in the nineties, there has been a great awareness among the researchers for the synthesis of chemical products and processes to minimize or remove the application or production of harmful products in order to create the sustainable future. This green approach also applies to nanotechnology, as a consequence, a new branch of nanotechnology is emerging as green nanotechnology. In light of this, the synthesis of nanoparticles by microbes such as fungi, bacteria, algae, and plants is referred to as biosynthesis, bio-inspired synthesis or green synthesis. The synthesis of nanoparticles from the fungi is absolutely a green process known as Mycosynthesis.

The fungi are ubiquitous in nature and are our friends and foes. On one hand, they are responsible for several beneficial products including life-saving antibiotics while on the other hand, they cause several diseases to humans and animals. More recently, fungi are exploited for the synthesis of different types of nanoparticles. The enzymes and proteins secreted by the fungi are responsible for the reduction of metal ions into the nanoparticles. Such synthesis is eco-friendly, economically viable, and can be carried out at ambient temperature without the use of high temperature, pressure, or toxic chemicals.

The proposed book on Mycosynthesis of nanomaterials deliberate the role of different fungi in the fabrication of inorganic and organic nanoparticles; the characterization of the nanoparticles and also discusses the enigmatic mechanistic aspect of the synthesis process. The book has been divided into three sections: Section I incorporates general topics and different types of nanoparticles synthesized by fungi; Section II deals with characterization techniques for mycosynthesised nanoparticles and the mechanism involved in synthesis of different nanoparticles. Section III—the last section critically analyses the toxicity to the ecosystem and humans.

The book is highly interdisciplinary and delivers the reading material for postgraduate and research students of fungal biology, microbiology, chemistry, nanotechnology, biotechnology, and green nanotechnology. It will also enrich academicians and researchers of pharma industries those interested in eco-friendly and sustainable technologies.

We thank all the contributors for their continuous support in providing the chapter and also for revision.

We also thankfully acknowledge the financial support rendered by the Polish National Agency for Academic Exchange (NAWA) (Project No. PPN/ULM/2019/1/00117/U/00001) to visit the Department of Microbiology, Nicolaus Copernicus University, Toruń, Poland.

Mahendra Rai, India
Patrycja Golińska, Poland

Contents

Preface	iii

Section I: Different Nanoparticles Synthesized by Fungi

1. Using Fungi to Produce Metal/Metalloid Nanomaterials 3
Nikhil Pradhan and *Raymond J. Turner*

2. Yeast as a Cell Factory: Biological Synthesis of Selenium Nanoparticles 27
Farnoush Asghari-Paskiabi, Mehdi Razzaghi-Abyaneh and *Mohammad Imani*

3. *Penicillium* species as an Innovative Microbial Platform for Bioengineering of Biologically Active Nanomaterials 50
Hamed Barabadi, Kamyar Jounaki, Elahe Pishgahzadeh, Hamed Morad and *Hossein Vahidi*

4. Biosynthesis of Iron Oxide Nanoparticles Using Fungi 81
Amanpreet K. Sidhu and *Priya Kaushal*

5. Mycosynthesis of Chitosan Nanoparticles 99
Mayuri Napagoda and *Sanjeeva Witharana*

6. Fungi-Mediated Fabrication of Copper Nanoparticles and Copper Oxide Nanoparticles, Physical Characterization and Antimicrobial Activity 112
Ishita Saha, Parimal Karmakar and *Debalina Bhattacharya*

7. Mycogenic Synthesis of Silver Nanoparticles and its Optimization 126
Joanna Trzcińska-Wencel, Magdalena Wypij and *Patrycja Golińska*

8. Biosynthesis of Gold Nanoparticles by Fungi: Progress, Challenges and Applications 146
Adriano Brandelli and *Flávio Fonseca Veras*

9. Biosynthesis of Zinc Oxide Nanoparticles and Major Applications 172
Shrutika Chaudhary, Saurabh Shivalkar and *Amaresh Kumar Sahoo*

vi   *Mycosynthesis of Nanomaterials: Perspectives and Challenges*

10. Fungi-Mediated Synthesis of Carbon-Based Nanomaterials   193
Pramod U. Ingle, Kunal Banode, Suchitra Mishra, Aniket K. Gade
and *Mahendra Rai*

11. Fungi-Based Synthesis of Nanoparticles and its Large-Scale   215
Production Possibilities
Nadun H. Madanayake and *Nadeesh M. Adassooriya*

**Section II: Characterization Techniques of Mycosynthesised Nanoparticles
and Mechanism of Synthesis**

12. Techniques for Characterization of Biologically Synthesized   233
Nanoparticles by Fungi
Pramod Ingle, Kapil Kamble, Patrycja Golinska, Mahendra Rai and
Aniket Gade

13. Mechanism of Synthesis of Metal Nanoparticles by Fungi   254
Nelson Durán, Marcelo B. De Jesus, Ljubica Tasic, Wagner J. Fávaro
and *Gerson Nakazato*

Section III: Toxicity of Nanoparticles to Human and Environment

14. Toxicity of Mycosynthesised Nanoparticles   283
Indarchand Gupta, Arun Ghuge, Hanna Dahm and *Mahendra Rai*

Index   297

Editors' Biography   299

Section I
Different Nanoparticles Synthesized by Fungi

CHAPTER 1

Using Fungi to Produce Metal/Metalloid Nanomaterials

Nikhil Pradhan and *Raymond J. Turner**

Introduction

Nanoscience is an emerging and rapidly growing field that focuses on material with at least one dimension between 1–100 nm in size (Baig et al. 2021). Material within this size is known as nanomaterial (NM) and comes in a variety of structures, shapes, and forms from nanoparticles (NP), nanorods (NR), nanosheets (NS), and a subtype of NP referred to as quantum dots (QD) amongst many others (Baig et al. 2021). Material with size dimensions in this nano range offers unique quantum properties compared to their larger counterparts particularly due to their surface area to volume ratios (Wu et al. 2020; Baig et al. 2021). Increased photonics via plasmons, enhanced antimicrobial activity, and conductivity are few properties that NM exhibit making them incredibly useful in medical fields for sterilization or bioimaging purposes, as well as modern computer chipsets and solar cells (Ealias and Saravanakumar 2017; Wu et al. 2020; Yaqoob et al. 2020; Aflori 2021; Baig et al. 2021).

With a wide array of useful applications for the modern world, significant focus has been placed on producing NM of various compositions. Compositions of nanomaterial can be of almost any element or mixtures, but carbon and silica-based are quite common (Maiti et al. 2019; Bernardos et al. 2019). A unique carbon-based NM is a lipid-based particle that is widely studied and tends to find uses in drug delivery (Mukherjee et al. 2009) and in applications such as the Covid-19 RNA vaccines (Buschmann et al. 2021). Metal-based compositions are of interest due to their plasmon, electrical, magnetic, and catalytic properties (Yaqoob et al. 2020). Many metals such as silver (Ag), gold (Au), zinc (Zn), and copper (Cu) have been used to create single element NM as well as mixed element NM.

Department of Biological Sciences, University of Calgary, 2500 University Dr. NW, Calgary, Alberta, T2N 1N4, Canada.
* Corresponding author: turnerr@ucalgary.ca

The composition of the NM is extremely important as the material will have a direct effect on the type of properties the NM can exhibit, while additionally, the size and shape of a particle directly influence those specific properties (Baig et al. 2021). For example, Au NM can be fluorescent, and a larger Au NP will fluoresce closer to the 800 nm region of light, while a smaller particle displays fluorescence around the 300 nm range (Hu et al. 2020). Producing a spherical Au NP allows fluorescence at one specific wavelength, while a cubic NP may fluoresce with two different emission wavelengths due to the edges of the particle (Zhong 2009). There is a wide range of chemical and physical methods to produce differently shaped particles of various metals, and this is typically the main route of synthesis for NM. However, these are not without flaws/drawbacks and as such different routes of synthesis are constantly being explored (de Oliveria et al. 2020).

Chemical syntheses are efficient, highly tunable, and controllable, and they can be easily scaled up as necessary (de Oliveria et al. 2020). However, chemical processes are expensive due to the cost of the materials needed to reduce the metal of choice down to form NM. They can also involve highly toxic reducing chemicals such as sodium borohydride, hydroxylamine, N, N-dimethylformamide, and a variety of acids all of which lead to disposal issues and can have negative effects on the environment (Rahimi and Doostmohammadi 2019). Physical methods tend to be very energy-demanding as they use heat, pressure, and lasers, which all add various complexities and energy costs to the NM synthesis (Rahimi and Doostmohammadi 2019).

Eco-friendlier routes have been explored for well over a decade, due to their sustainability advantages. Most eco-friendly approaches use plant material, wastes, and extracts with reasonable success, yet microorganisms also can meet the eco-friendly criteria. In this chapter, we explore the use of fungi, as there has been a surprising amount of success in synthesizing not only metal and metalloid NPs but also reduced graphene oxide using edible and medicinal mushrooms (Owaid 2020).

Biogenically produced Nanomaterial

The issues around chemical and physical production approaches have been the driving factor in the search for green or eco-friendly synthesis solutions (Abdul Salam et al. 2014). Many microbial species and organisms have developed mechanisms to tolerate metals through the reduction of a harsh metal ion into a less harmful one, typically precipitating to elemental forms (Lemire et al. 2013). Often, this reduction results in the formation of NM, at least at the early stages. Thus, this metal reduction property of the reducing biochemistry of biological systems has been explored over the past 15 years towards a biotechnology approach for NM synthesis.

An important feature of nanomaterials is that they need a coating or cap on its surface to stabilize them and adjust their properties depending on their intended use (Javed et al. 2020). Chemically, this coat composition is chosen by the chemist and is typical of a singular molecule composition (Javed et al. 2020). Biological species can produce this coat themselves. The biogenic coat originates from the biomolecules present as part of the metabolism of the organism involved and their present physiological state (Guilger-Casagrande et al. 2021). Thus, biogenically produced

NM may contain a unique or a variety of biomolecules (Guilger-Casagrande et al. 2021). This has the further advantage that such diverse biomolecular caps will further adjust the properties of the produced NM as well as significantly enhance their stability (Piacenza et al. 2018a). Many biogenically produced NM have been shown to display similar or even superior properties to chemical or physically produced NM because of their biogenic coat/cap and comes at a lower cost and reduced impact on the environment.

There are a wide range of organisms explored for nanomaterial production including plants and plant extracts (Rahman et al. 2020; Nande et al. 2021), Lichen (Rattan et al. 2021), algae (Mukherjee et al. 2021), fungi, and yeast. For general reviews on the power of bacteria and microbes for the biogenic synthesis of nanoparticles see Dhanker et al. (2021), Koul et al. (2021), or Kapoor et al. (2021). Here in this chapter, we will focus specifically on overviewing the use of fungi for metal nanomaterial production (for reviews see: Sastry et al. (2003); Dhillon et al. (2012); Moghaddam et al. (2015); Saxena et al. (2014); Yadav et al. (2015); Bhardwaj et al. (2020); Adebayo et al. (2021); Sudheer et al. (2022). The early work using fungi in nanoscience led to the introduction of the term 'myconanotechnology' (Rai et al. 2009).

Research into the plant and microbial syntheses of NM has shown us that different species can be better at producing certain types of NM (Iravani 2011; Ghosh et al. 2021). Some strains display higher success with certain types of metals over others and will produce different shapes due to the biochemistry of the given species. As such, expanding biogenic NM possibilities even further has led to research into using fungi as a vessel to produce NM. The difference in fungi metabolism and pathways between bacteria and plants offers unique methods to produce NM and can result in differences in NM composition (Keller 2018; Li et al. 2021). The field of fungi-based NM synthesis or myconanotechnology is an emerging and exciting one. In this chapter, we will explore how fungi can be harnessed to produce NM, the species of fungi that have been shown to produce NM, as well as the types of NM that have been produced, and the differences between them and other biologically produced NM.

How do fungi produce nanomaterials?

Fungi face metal exposure in nature, as they have in part evolved with the planets' geochemistry (Cervantes and Gutierrez-Corona 1994; Acosta-Rodríguez et al. 2018; Oladipo et al. 2018). Like all organisms, some metal ions are vital for cellular function that must be taken up in moderation, while others pose threats to cellular components and can harm the fungi (Robinson et al. 2021). As such cells must have methods to import some metals as well as deal with harmful metals either by reducing uptake or actively exporting them, but also alterations of a site of interaction to prevent a metal-binding to it, or overproduction of proteins being damaged or for repair and finally by modifying the metal to a less harmful form (Lemire et al. 2013). Metals enter fungi through transporters as metal-organic chelates, metal salts, oxides, and sulfites, which can pose a threat to cellular machinery such as enzymes and proteins as the metal ions can bind and disrupt folding and function (Ghannoum and Rice 1999).

This leads to a bit of a conundrum for a cell that protects itself by reducing metal ions to elemental forming NMs. This then gives metal NPs inside the cell and the metal NPs can also be harmful to fungi as they release metal ions through redox decomposition that can disrupt the biochemistry of a cell in the same way (Gudikandula and Charya Maringanti 2016; Jamdagni et al. 2018). In fact, many chemogenic metal-based nanoparticles are toxic to beneficial soil fungi (Ameen et al. 2021). However, NPs tend to release the metal ions at a slower rate which can buy time for a fungus to export or secrete the NP (Cheong et al. 2020). Yet overall, fungi appear to be more resistant to most metals and metalloid ions than bacteria (Priyadarshini et al. 2021). An exciting aspect of this metal/metalloid ion bioconversion to NPs is in the field of bioremediation of metal-contaminated environments (Kumari and Singh 2016; He et al. 2017; Mandeep and Shukla 2020; Kumar and Dwivedi 2021).

Fungi produce NM by reducing metal ions via enzymes which enable that reduction process (Siddiqi and Husen 2016). This process can occur via fungi importing the metal ions and carrying out the reduction process intracellularly (Siddiqi and Husen 2016). Within this process, the metal precursors are added to a mycelial culture where it can then be taken into the fungi and is dependent on the type of metal (Gericke and Pinches 2006; Prabhu and Poulose 2012; Husen and Siddiqi 2014). In this process, the NM is produced intracellular, and to recover the NM, the fungi must be broken up to extract the NM from the fungi (Sneha et al. 2010). A variety of cell lysis and processing methods are available that are routine to various biotechnology industries such as combining chemical treatment, sonication, homogenization, centrifugation, gel filtration, dialysis, and filtration (Guilger-Casagrande et al. 2019). This extraction process allows for all fungi as well as residual components to be removed and leaves the NM coated with the biomolecules needed to maintain the stability of the particle (Guilger-Casagrande et al. 2019). While intracellular synthesis is effective, it is typically not preferred as the extraction process involves more steps and is thus more time-consuming, making it less ideal.

Fungi can also produce NM extracellularly, and this is the more favored route of synthesis as it is easier to separate the fungi from the NM at the end. Extracellular synthesis occurs if the fungi can secrete key enzymes that can partake in the reduction process (Ahmad et al. 2003). This has been observed by taking fungal filtrates and detecting enzymes present indicating that fungi are capable of this secretion (Ahmad et al. 2003). As an example, biomolecules such as NADPH/NADH dependent reductases and NADH-dependent nitrate reductase enzymes have been suggested to play a role in Ag NP synthesis (Gudikandula et al. 2017). The addition of the metal salt to the fungal culture leads to the secreted enzymes reducing it to its less harmful elemental form enabling NP synthesis much like the intracellular method. Following the formation of the particles, they can then be extracted via simple cell removal by differential centrifugation and/or filtration making it a much simpler extraction process (Tyagi et al. 2019).

Production of NM using fungi is possible and effective, however, the in-depth mechanisms are yet to be explored in detail in most systems. Most current work focuses on the production of the particles and characterization of them to explore their functionality and usefulness. The current propositions suggest the general methods NM is believed to be produced either intracellularly and/or extracellularly. A large portion of studies are focused on Ag and Au NP production, as are most biogenic approaches, however, this leaves gaps in understanding the mechanism behind other metal NM production. Thus, there is significant bias in our understanding. However as the field grows, it would be expected that a greater understanding of the biochemical processes involved in the synthesis of biogenic metal and metalloid NMs will be identified, and a detailed understanding of fungi NM synthesis will be obtained.

The importance of the nanomaterial cap

Fungal-based NP synthesis is one of many approaches to NM synthesis, along with traditional chemical syntheses, plant material mediated syntheses and bacterial mediated syntheses. While fungi fall under the category of biological syntheses, the differences in their metabolism and molecular makeup provide a uniqueness that sets them apart, in terms of how they produce NM as well as the stability of the produced particles (Wisecaver et al. 2014; Calvo and Cary 2015; Moghaddam et al. 2015). Beyond the composition of NM, the coating molecules or the cap play a large role in how NP will function. These molecules help to stabilize a particle via electrostatic, steric or electrostatic interactions which prevent particles from aggregating and instead remain as individual particles in solution (Piacenza et al. 2018a; 2019). Chemically, this cap comprises of a single type of molecule such as surfactants, small ligands, polymers, dendrimers, cyclodextrins and polysaccharides (Javed et al. 2020). The molecules are used are typically amphiphilic and are comprised of a polar head group and a non-polar hydrocarbon tail enabling them to be soluble in various solutions and increasing their functionality (Gulati et al. 2018).

Biogenic NM do not require the addition of the cap molecules to the synthesis as the organism producing the particles will provide them (Chowdhury et al. 2014). The biomolecules that are used to reduce and form the NM are capable of binding to the metal NP produced and stabilize it (Chowdhury et al. 2014). They largely comprise of various proteins, secreted amino acids, and cell wall enzymes (Husseiny et al. 2015; Bulgarini et al. 2021). Transmission electron microscopy can be used to visualize this cap as organic material will appear as a lighter coloured thin layer surrounding the outside of the produced NP (Elgorban et al. 2016). Characterization of the NM via UV-Visible absorption, Energy Dispersive X-Ray Analysis (EDX), FTIR and/or electrophoresis techniques can give more insight to composition of the cap (Devi and Joshi 2015; Elgorban et al. 2016). An example is shown in Figure 1.

Silver NPs were produced by *Aspergillus niger* were evaluated by SDS-polyacrylamide gel electrophoresis and showed molecular weight bands between 50–116 kDa relating to the proteins needed to synthesize and stabilize the NP (Chowdhury et al. 2014). The NP were then separated, and the cap was chemically removed and then reanalyzed via electrophoresis, showing a band at 85 kDa is likely a protein required to keep the NP stable (Chowdhury et al. 2014). EDX analysis of the extracted Ag NPs produced by *Aspergillus versicolor* showed the presence of oxygen and carbon indicating organic molecules were coating the surface of the Ag

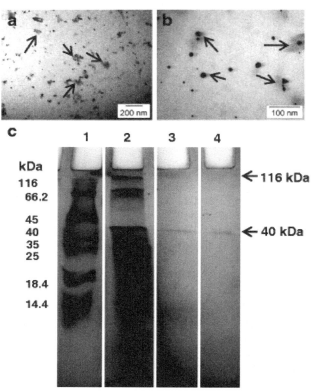

Figure 1. Protein cap analysis of Au NPs produced by *Tricholoma crissum*. **(a, b)** TEM images showing capping protein layer (arrows) around AuNPs. **(c)** SDS-PAGE of extracellular protein secreted from *T. crassum* and protein associated with nanoparticles. Lane 1, molecular size marker. Lane 2, total extracellular protein. Lane 3, nanoparticles loaded without boiling show a faint protein band bound to the AuNPs at 116 kDa mark. There is also a band of detached protein at 40 kDa mark. Lane 4, nanoparticles after boiling with 1% SDS loading buffer showing disappearance of the 116 kDa band and a distinct 40 kDa band. Arrow indicates 40 kDa. Figure from Basu et al. (2018) [Creative Commons Licence - http://creativecommons.org/licenses/by/4.0/].

particles (Elgorban et al. 2016). UV-Visible spectroscopy analysis of Ag particles produced from *Aspergillus tamarii*, *Aspergillus niger*, and *Penicillium ochrochloron* revealed an absorption peak at 280 nm that was caused by the presence of tryptophan and tyrosine amino acids attached to the particles (Devi and Joshi 2015). Gold NPs were found to be protein-coated after the particle removal from the coat yielded a band at 40 kDa that matched a band found within the separate fungal filtrate (Basu et al. 2018).

As seen in Table 1, *Fusarium* sp. is the most common fungus for the synthesis of nanoparticles. *Fusarium* sp. has been shown to be able to produce a wide range of metal/metalloid NMs. The proposed mechanism is via an extracellular secreted NAD(P)-dependent nitrate reductase coupled to a quinone (Durán et al. 2005; 2011). It is also suggested that this enzyme becomes part of the cap of the NM, yet other enzymes and proteins may also be present (see review by Rai et al. 2021) see also Figure 1.

The specifics of the coats of various biogenic NP have not been studied and produced from any biological system, as such the exact compositions and inner workings remain unknown. Looking towards NM produced by plants and bacteria shows variations in the effectiveness of the NM produced dependent on the species used. An example from bacteria showed quite a diversity of lipid : protein: carbohydrate ratios for Se NPs produced from a variety of selenium-resistant environment isolates (Bulgarini et al. 2021). Selenium NP produced *E. coli, Staphylococcus aureus, Pseudomonas aeruginosa* and *Rhodococcus aetherivorvans* are compositionally equivalent and all display antimicrobial properties, however, the effectiveness of the particles as an antimicrobial agent differs due to the coating molecules (Medina Cruz et al. 2018). Variations in the metabolism of the organism used enable different proteins, enzymes, and amino acids to be secreted and act as a cap for the NP. With the large variety of fungal species and the differences in fungi life cycles compared to other biological species, this same effect should be seen and thus should be explored for the organisms studied to date. Currently biogenic NMs offer the advantage of being inherently biocompatible making them ideal for biomedical uses such as drug delivery when compared to a chemically produced equivalent (Sharma et al. 2015). Further research and expanded knowledge of the composition of fungal NP caps could better allow for NM to be produced with a targeted application focus.

Controlling synthesis

A large goal of NM production is to scale up processes that can produce monodisperse populations of a specific size and shape (Khan et al. 2019). As size and shape define the properties displayed by a particle, a heavily diverse NM population will prevent one from seeing consistent results in any application. Therefore, fine-tuning the relevant synthesis parameters can allow for consistently monodisperse populations of NM to be produced. Biogenic NM synthesis is affected by a variety of conditions such as pH, temperature, pressure, culture media, light, the quantity of biomass, initial metal precursor as well as the concentration used, and exposure period (Durán et al. 2011; Dahoumane et al. 2017; Guilger-Casagrande et al. 2019; Kapoor et al. 2021). The mechanisms for the synthesis of NM using fungi have not been studied in-depth, however, some studies have noticed certain patterns when altering certain conditions. These studies focus on the effect of modifying parameters to produce Ag NPs; however, it does give perspective to how the parameters can affect NP synthesis in fungi.

Myconanotechnology displays the ability to produce NM at temperatures lower than 100°C in comparison to the > 300°C needed for some chemical syntheses (Karakoti et al. 2012; Sreenivasa Kumar et al. 2017). Often the temperature chosen will be the optimal temperature to allow the fungi to grow and survive. Variations in temperature have shown to increase protein secretion, therefore, improving particle production rate, with very high temperatures and very low temperatures reducing synthesis rates (Azmath et al. 2016; Abdel Rahim et al. 2017). Certain fungi have been used to produce particles at high temperatures, however, increasing temperatures greater than 80°C can lead to the denaturation of most eukaryotic proteins (Birla et al. 2013). If the proteins are the catalytic components, synthesis will stop. Also, if key proteins are vital for providing the cap of the NP, their alteration leads to less stable particles that tend to aggregate and increase in size losing functionality (Birla et al. 2013). Temperature affects the rate of the NP synthesis, and this can affect the size of the particles produced (Abdel Rahim et al. 2017; Shahzad et al. 2019).

Fungi are able to produce particles under atmospheric pressure making the synthesis of differently shaped particles simple when compared to physical routes of synthesis (Tran et al. 2013). Metal precursor concentration and pH are factors that must be considered. An organism's tolerance level will need to be accounted for when considering what metal concentration levels to choose (Suresh et al. 2010). Likewise, pH must be considered to ensure optimal fungal growth and as well as it affects the size of the particles produced and the rate of production (Amaladhas et al. 2012). The addition of many metals or metalloid salts can dramatically change pH. In studies producing Ag NM, increasing pH has been shown to alter the nitrate reductase enzymes that play a role in reducing $AgNO_3$ (Husseiny et al. 2015). This in turn alters the morphology of the produced particles with some studies showing reduced synthesis time and smaller size distributions.

Culture media, light exposure, the quantity of biomass, and exposure time are also factors that must be considered. The media used to culture the fungi will alter the physiological fitness, growth rate and priority metabolism of the fungi and can alter the size or shape of the NM that the fungi will produce (Basu et al. 2015). Additionally, metabolites in the media can alter the metal precursor, either through chemistry or as a chelate (metallophore) (Kuzyk et al. 2021). The changing metal speciation can affect the bioavailability that can affect the resulting NM yield and/or properties. Light exposure is another factor affecting the fungal species used to synthesize the particles, as adjusting intensity and wavelength can optimize the growth of the fungus (Birla et al. 2013; Gudikandula et al. 2017).

The above points to a wide variety of experimental production parameters that can be modulated by using fungi compared to chemistry and physical production space. The outcome of changes in any parameter can be either advantageous or destructive to the target NM functionality. At this point in the field there is little predictive knowledge available and thus requires considerable iterative empirical work and a bit of luck to find the optimum production conditions.

Types of nanomaterial produced

Nanomaterials are produced in a variety of compositions, each with their unique properties and applications. Different fungal species have been utilized to produce NM with some species more studied than others. Here we will explore some of the more popular types of fungi produced metal/metalloid NM to see their function as well as look at the fungal species that have successfully been used to produce them. Each section provides reasons to produce these metal NMs produced by various fungi species. Table 1 summarizes the different materials produced by different species of fungi beyond what is discussed here.

Silver NM is among one of the most popular due to the properties that it has. Silver as a metal displays strong antimicrobial activity and this effect is amplified when looking at Ag NP (Sim et al. 2018; Anees Ahmad et al. 2020). This is hypothesized

Table 1. Examples of Fungi produced Nanomaterials.

Organism	Metal Metalloid	Production	Size (nm)	Characteristics Comments	References
	Silver				
Anthroderma fulvum	Ag	cell filtrate	Ave=15	spherical Abs=420 Antimicrobial	Xue et al. 2016
Aspergillus clavatus	Ag	extracellular	10–25	spherical & hexagonal Antimicrobial	Verma et al. 2009
Aspergillus foetidus	Ag	cell filtrate (proteins)	Ave=105	spherical and clustered Abs=425 Antimicrobial	Roy et al. 2013
Aspergillus fumigatus	Ag	extracellular cell filtrate	5–25	nanocrystalline Abs = 420	Bhainsa and D'Souza 2006
Aspergillus niger	Ag	cell filtrate (proteins)	Ave=20	Abs=420 Antimicrobial	Gade et al. 2008
Colletotrichum sp.	Ag	cell free extract (peptides, biomolecules)	5–60	myriad Abs = 420 Antimicrobial	Azmath et al. 2016.
Coriolus versicolor	Ag	extra/intercellular (proteins) {surface SH}	25–75	spherical, Abs=440/430	Sanghi and Verma 2009
Cryphonectria sp.	Ag	cell filtrate	30–70	polydisperse Abs=440 Antimicrobial	Dar et al. 2013
Duddingtonia flagans	*Ag*	cell filtrate (protein=chitinase}	30–409	quasi-spherical Abs=406-414	Costa Silva et al. 2017
Epicoccum nigrum	Ag	extracellular	1–22	spherical, Abs=424 Antimicrobial	Qian et al. 2013
Fusarium acuminatum	Ag	cell extract	Ave=13 5–40	Abs=420 Antimicrobial	Ingle et al. 2008
Fusarium oxysporum	Ag	cell filtrate {Nitrate dependent Reductase & quinones}	20–50	Abs=415-420	Durán et al. 2005

Table 1 contd. ...

...*Table 1 contd.*

Organism	Metal Metalloid	Production	Size (nm)	Characteristics Comments	References
Fusarium semitectum	Ag	cell filtrate (proteins) {NADH reductase}	10–60	spherical Abs=378, 420	Basavaraja et al. 2008
Guignardia mangiferae	Ag	extracellular/filtrate Abs=417Antimicrobial	5–30	spherical	Balakumaran et al. 2015
Helminthosporium tetramera	Ag	cell filtrate	17–33	spherical to oval Abs=400 Antimicrobial	Shelar and Chavan 2014
Penicillium oxalicum	Ag	extracellular (proteins)	10–40	spherical Abs=420-450 Antimicrobial	Rose et al. 2019
Pestalotia sp.	Ag	cell filtrate (proteins)	10–40	spherical, polydisperse Abs=415 Antimicrobial	Raheman et al. 2011
Phoma glomerata	Ag	cell filtrate (proteins)	60–80	spherical Abs= 440 Antimicrobial	Birla et al. 2009
Pycnoporus sanguineus	Ag	in mycelium	52.8–103.3	spherical Abs=420	Chan and Mat 2013
Trichoderma reesei	Ag	extracellular (proteins)	5–50	Abs=414-420 EM=490	Vahabi et al. 2011
Trichoderma viride	Ag	cell filtrate (proteins)	2–4	spherical Abs=405, EM=320-520	Fayaz et al. 2010
Alternaria sp.	Au	cell filtrate (peptides, proteins)	7–18	multiple shapes Abs= 535-550	Dhanasekar et al. 2015
Aspergillus nger	Au	cell filtrate	10–30	different structures Abs=530; insecticidal	Soni and Prakash 2012.
Pleurotus ostreatus	Au	laccase enzyme	22–39	Abs=550	El-Batal et al. 2014
Paraconiothyrium variabile	Au	laccase enzyme	71–266	spherical Abs=530	Faramarzi and Forootanfar 2011
Penicillium chrysogenum	Au	cell filtrate	5–100	spherical, triangle, rods Abs=532.	Sheikhloo and Salouti 2011
Neurospora crassa	AuAg	Intracellular	3–110	triangular nanoplates Abs=450, 520	Castro-Longoria et al. 2011

Species	Metal	Location	Size (nm)	Shape/Properties	Reference
Fusarium oxysporum	Pt	intracellular Cell filtrate	10–100 10–30	mixed square rectangles mixed mostly square	Riddin et al. 2006
Aspergillus sp.	Pb	intracellular Cell associated	1.8–5.8 5–10	spherical	Pavani et al. 2012
Phanerochaete chrysosporium	Se	intracellularly	30–400	spherical	Espinosa-Ortz et al. 2015
Trichoderma atroviride	Se	intra/extracellular (peptides)	60-123	spherical, Antimicrobial	Joshi et al. 2019
Aspergillus welwitschiae	Te	cell filtrate (proteins)	Ave=60	oval-sphere Abs=400 Antimicrobial	Abo Elsoud et al. 2018
Metal oxides					
Rhizopus oligosporus	Cu_2O, CuO	extracellular	Ave=23 aggregate	plate like to spherical no SPR signals	Pamungkas et al. 2021
Stereum hirsutum	Cu, CuO	mycelium free extract (Extracellular proteins)	5-20	spherical Abs=580-590 for CuNP Abs=590-800 for CuO	Cuevas et al. 2015
Aspergillus niger	ZnO	cell filtrate	Ave=20	hexagonal Abs=370, Antimicrobial	Mekky et al. 2021
Fusarium oxysporum (protein)	Bi_2O_2	cell filtrate	5–8	crystalline	Uddin et al. 2008
Fusarium oxysporum	Fe_3O_4	cell filtrate (protein)	20–50	quasi spherical	Bharde et al. 2006
Fusarium oxysporum	SiO_2 & TiO_2	extracellular (protein){~25kDa enzyme}	5–15	crystalline	Bansal et al. 2005
Fusarium oxysporum	ZrO_2	extracellular (protein){~25kDa enzyme}	3–11	quasi spherical	Bansal et al. 2004
Verticillium sp.	Fe_3O_4	cell filtrate (protein)	100–400	cuboid-octahedral	Bharde et al. 2006

Table 1 contd. ...

Organism	Metal Metalloid	Production	Size (nm)	Characteristics Comments	References
	Quantum dots				
Phanerochaete chrysosporium	CdS	cell surface (Cystine & proteins)	Ave= 2.58	cubic crystal Abs= 296-298; EM=458nm	Chen et al. 2014
Fusarium oxysporum	CdS [PbS ZnS, MoS$_2$]	Extracellular (proteins) {sulfate reductase}	5–20	various shapes Abs=450	Ahmad et al. 2003
Fusarium oxysporum	CdSe	intracellular {SOD catalase}	< 20	EM=410	Yamaguchi et al. 2016
Fusarium oxysporum	CdTe	extracellular	15–20	Abs=400-450 EM=475 Antimicrobial	Syed and Ahmad 2013

Note: This is not an exhaustive list. It is meant as an overview example taking examples of different organisms and or different NP produced.

Abs; absorbance maximum (nm), EM; Fluorescence emission maximum (nm), Ave; average, ND; not defined.

The cell filtrate refers to using a filter to remove the cells and thus using the secreted biomolecules from the fungi as the reactant. Whereas for cell extracts, the fungi cells are collected and lysed and cells removed and thus using the intracellular biomolecules as the reactant.

Information in parenthesis () under production includes information from FTIR or other experiments on the nature of the NP cap. Information in {} defines identified enzymes in the mechanism.

Antimicrobial: either or both antibacterial and fungicidal.

to be due to the release of Ag^+ ions by the NP and this release results in a more localized delivery method when compared to other Ag delivery methods (Zhai et al. 2016). Silver NPs have shown the ability to control pathogenic fungi and bacteria, viruses, larvicidal and insecticidal activities as well as the ability to combat cancer cells (Rodrigues et al. 2013; Sundaravadivelan and Padmanabhan 2013; Nagajyothi et al. 2014; Banu and Balasubramanian 2014; Pereira et al. 2014). This gives Ag NPs a role in the health industry, with the purpose of sterilizing medical equipment or to use as a coat on a bandage to help sterilize wounds (Prabhu and Poulose 2012; Burduşel et al. 2018). In addition, Ag particles also display unique optical properties that arise from the surface plasmon resonance (SPR) phenomenon (Elamawi et al. 2018). This lets smaller Ag particles absorb ~ 400 nm, where larger particles absorb light red-shifted towards the 800 nm range as well as scatter more light. As the size of particles can be controlled, Ag NPs have found uses in sensors and biodevices where they have been used to sense various analytes such as glucose or triacylglyceride (Varghese Alex et al. 2020; Tan et al. 2021). SPR also enables Ag NPs to be used as photocatalysts where it has been shown to efficiently photodegrade methyl red dye dependent on the concentration of particles used (Tamuly et al. 2013). Due to the versatility of Ag NPs, there is a heavy focus on the production of Ag NPs of various sizes and shapes using many different fungal strains.

Silver NM has had a lot of work done to find various fungal species that can produce them (Khan et al. 2018; Guilger-Casagrande and de Lima 2019). The primary routes of fungal synthesis for Ag NM are through an extracellular route and as such most use cell filtrates, which carry the proteins and enzymes from the fungi needed to synthesize Ag particles. As seen in Table 1, *Coriolus versicolor* stands out as producing particles intracellularly (Sanghi and Verma 2009). Silver uptake was enabled under alkaline conditions with a pH of 10 which also reduced the formation time of the NP from 72 to 1 hour (Sanghi and Verma 2009). This gave rise to spherical Ag NP between 25–75 nm in size. *Aspergillus* sp. and *Fusarium* sp. receive a lot of focus in Ag NP production with multiple species successfully synthesizing spherical Ag NP (Bhainsa and D'Souza S006). *Aspergillus clavatus* displayed the ability to synthesize hexagonal-shaped particles in addition to the spherical particles that displayed antimicrobial activity (Verma et al. 2009). The ability of this fungus to produce hexagonal particles demonstrates an advantage over typical physical approaches, as shapes differing from the standard spherical tend to be trickier to produce. All the fungi looked at produce relatively uniformly small size populations of particles with the exception being *Duddingtonia flagans* which produced a range between 30–409 nm.

Gold NPs are primarily known for their optical and electrical properties (Kong et al. 2017). These are dependent on the size and shape of the particle leading to focus being placed on size tunable Au NP syntheses. Particles of smaller size absorb and fluoresce around the 450 nm range and as size increases, so does the wavelength (Zheng et al. 2012). The optical properties give them applications in the medical fields for uses such as bio and molecular imaging (Dykman and Khlebtsov 2011; Si et al. 2021). Additionally, Au is a conductive material giving Au NPs unique size dependent electrical properties (Khalafalla et al. 2014). The uniqueness arises due to the large volume/surface area ratio which allows them to coat electrodes and improve conductivity (Young et al. 2016; Sousa et al. 2017). Gold NPs also are easily functionalized with various drugs, genes and targeting ligands letting them act in targeted drug delivery methods (Dykman and Khlebtsov 2011). Further use of Au NPs shows promise in photothermal cell damage as a method in tumor therapy and therapy of infectious diseases (Kennedy et al. 2011). Functionalization of Au NPs with antibodies or other various molecules can allow for targeting of cells allowing strong imaging of the tumor (El-Sayed et al. 2005).

Selenium is sometimes referred to a as a metalloid but is only as such in certain atomic forms such as crystalline NM. NPs are an increasingly popular NM due to the potential that comes from the wide array of properties. Firstly, they exhibit strong antimicrobial activity, targeting various pathogenic bacteria, fungi, and yeasts (Shahverdi et al. 2010; Beheshti et al. 2013; Filipović et al. 2021). In addition, these NPs show anticancer, antioxidant and antibiofilm properties (Yu et al. 2012; Forootanfar et al. 2014; Shakibaie et al. 2015). Selenium NM has been delivered orally and shows positive effects in helping with reproductive performance (Badade et al. 2011; Hosnedlova et al. 2018). With easy cellular uptake and the low cytotoxicity of Se NPs, they can be used to efficiently and safely deliver drugs (Maiyo and Singh 2017; Hosnedlova et al. 2018). Medical applications are only a portion of the uses for Se NM as they also display many photophysical properties such as high photoconductivity, piezoelectricity, thermoelectricity, and spectral sensitivity (Piacenza et al. 2018b). Additionally, they show strong potential for cellular imaging with strong fluorescence properties that can be tuned based on the size of the particles (Khalid et al. 2016).

Selenium NM has been produced extracellularly, intracellularly, and with the use of cell filtrates. *Trichoderma atroviride* produced spherical Se NM using an extracellular method through the secretion of reductive proteins and enzymes (Joshi et al. 2019). DLS analysis showed the particles were uniform in size ranging from 93.2–98.5 nm with zeta averages between negative 49.3–43.7 mV showing high stability (Joshi et al. 2019).

Interesting blend of chemical and biogenic approaches have also been used to generate unique Se NPs. Ascorbic acid synthesized Se NPs together with a polysaccharide purified from the fungi *Dictyophora indusiate* for the coat (Liao et al. 2015). These particles displayed anticancer properties showing antiproliferative activity, the ability to induce apoptosis, cell cycle arrest, and ROS overproduction leading to mitochondria dysfunction and the ability to activate the Fas-associated death domain protein (Liao et al. 2015). Antimicrobial and antibiofilm activity was seen from Se NP produced from *Penicillium chrysogenum* filtrate with the combination of gentamicin and gamma irradiation (El-Sayyad et al. 2019). Isolation of β-(1,3)-D-glucan (BFP) from *Auricularia auricula-judae* was done to produce Se NP embedded in the glycan triple helix structure (Jin et al. 2020). This was then tested as a therapeutic cure for acute myeloid leukemia, where the BFP-Se complex inhibited the proliferation of the leukemia cells with tests being done on isolated blood and bone marrow human cells and mice (Jin et al. 2020).

Metal oxides are a wide range of NMs including compositions such as CuO, ZnO, SnO_2, Al_2O_3, MgO, ZrO_2, AgO, TiO_2, CeO_2, Fe_2O_3, and many more (Chavali and Nikolova 2019). They carry strong electronic, magnetic, and chemical properties that alter in functionality based on the size of the produced particles. Fe_2O_3 is among the most popular due to its magnetic properties giving it large potential in magnetic storage devices and as a magnetic resonance imaging contrast agent (Lang et al. 2007). Metal oxides with stronger conductance such as TiO_2 show strong potential in sensors, optoelectronics, and photovoltaics (Franke et al. 2006). CuO, MgO, Al_2O_3, CeO_2 find uses as redox catalysts in differing situations from gas sensing, and biomedical scenarios to material chemistry (Chavali and Nikolova 2019). Along with Ag NPs, TiO_2 and ZnO may be the most commercialized and are found in thousands of nano-enhanced products from sunscreen, cosmetics, paints, preservatives, and processed food (Moloi et al. 2021).

Quantum dots (QD) are a subcategory of NPs that range between 1.5–10 nm (Cotta 2020). Their small size gives them unique photophysical properties that are seen as extremely strong fluorescence and conductivity (Bera et al. 2010). These properties are modifiable as the size of the QD changes allowing them to be produced and tuned to precision. QD can be composed of a large variety of metals with the most popular ones containing cadmium such as CdSe and CdTe (Mo et al. 2017). CdSe QDs are seen being used primarily for their optical properties, with strong tunable fluorescence (Ozkan 2004). Applications such as bioimaging, modern electronic displays, and sensors benefit greatly from the strong fluorescence these particles can produce when compared to conventional fluorescent dyes (Ozkan 2004). CdTe QDs are being looked at as a coating for solar cells as they have a band gap that is optimal for this purpose, as well as a high optical absorption coefficient (Yang and Zhong 2016). Cd QDs also show promise in drug delivery and cancer therapy however, Cd^{2+} ions are toxic and Cd-based QD facilitate the slow release of these ions, making it harder to implement them into medical applications safely (Yang et al. 2012; Mo et al. 2017).

Conclusion

Fungi-produced nanomaterial has garnered a lot of interest recently as the need for eco-friendly nanomaterial production has increased. Many other biological species such as plants (and plant extracts) and bacteria have also been used to produce a variety of NM compositions (Iravani 2011; Ghosh et al. 2021). However, the field of myconanotechnology is growing, and the understanding of the mechanisms behind how this production occurs has also grown. Techniques have been improved to allow modifications to parameters in synthesis to allow the modification of the size and shape of produced NPs. The surface coating of the NPs produced by fungi offers better biocompatibility in a number of applications. Additionally, biogenic nanomaterial coats are comprised of a variety of biochemicals that gives unique functionality compared to chemically synthesized equivalents. The field is barely older than 15 years and perhaps still in the growing stages and lags behind the use of bacteria for biogenic NM production. Thus, one still sees a large focus placed on producing particles of various compositions with a wide array of fungi, which has proven to be successful. In the future, placing more focus on understanding the specifics behind the mechanism for fungi-produced NPs will enable better control of the particles that are produced. This has been seen using the Fungi secreted enzyme laccase, that is

18 *Mycosynthesis of Nanomaterials: Perspectives and Challenges*

allowing for the use of cell-free systems. As well as studying the composition of the coat and understanding the effect certain biochemicals have in the properties we see can lead to a better selection of fungal species to create NPs for specific applications. The field is promising and as the understanding increases, fungi will continue to be a strong candidate for producing eco-friendly NPs on a large scale.

References

Abdel Rahim, K., Mahmoud, S.Y., Ali, A.M., Almaary, K.S., Mustafa, A.E.Z.M.A., and Husseiny, S.M. 2017. Extracellular biosynthesis of silver nanoparticles using *Rhizopus stolonifera*. Saudi J. Biol. Sci. 24: 208–216.

Abdul Salam, H., Sivaraj, R., and Venckatesh, R. 2014. Green synthesis and characterization of zinc oxide nanoparticles from *Ocimum basilicum* L. var. purpurascens Benth.-Lamiaceae leaf extract. Mater Lett. 131: 16–18.

Abo Elsoud, M.M., Al-Hagar, O.E.A., Abdelkhalek, E.S., and Sidkey, N.M. 2018. Synthesis and investigations on tellurium myconanoparticles. Biotechnol. Rep. 18: e00247.

Acosta-Rodríguez, I., Cardenás-González, J.F., Pérez, A.S.R., Oviedo, J.T., and Martínez-Juárez, V.M. 2018. Bioremoval of different heavy metals by the resistant fungal strain *Aspergillus niger*. Bioinorg Chem. Appl. 2: 3457196.

Adebayo, E.A., Azeez, M.A., Alao, M.B., Oke, A.M., and Aina, D.A. 2021. Fungi as veritable tool in current advances in nanobiotechnology. Heliyon 7: e08480.

Aflori M. 2021. Smart nanomaterials for biomedical applications—A review. Nanomater. 11: 1–33.

Ahmad, A., Mukherjee, P., Senapati, S., Mandal, D., Khan, M.I., Kumar, R., and Sastry, M. 2003. Extracellular biosynthesis of silver nanoparticles using the fungus *Fusarium oxysporum*. Coll. Surf. B: Biointerfaces 28: 313–318.

Amaladhas, T.P., Sivagami, S., Devi, T.A., Ananthi, N., and Velammal, S.P. 2012. Biogenic synthesis of silver nanoparticles by leaf extract of *Cassia angustifolia*. Adv. Nat. Sci. Nanosci Nanotech. 3: 045006.

Ameen, F., Alsamhary, K., Alabdullatif, J.A., and Nadhari, S. 2021. A review on metal-based nanoparticles and their toxicity to beneficial soil bacteria and fungi. Ecotox. Environ. Saf. 213: 112027.

Anees Ahmad, S., Sachi Das, S., Khatoon, A., Tahir Ansari, M., Afzal, M., Saquib Hasnain, M., and Kumar Nayak, A. 2020. Bactericidal activity of silver nanoparticles: A mechanistic review. Mat. Sci. Ener. Technol. 3: 756–769.

Azmath, P., Baker, S., Rakshith, D., and Satish, S. 2016. Mycosynthesis of silver nanoparticles bearing antibacterial activity. Saudi Pharm. J. 24: 140–146.

Badade, Z.G., More, K., and Narshetty, J. 2011. Oxidative stress adversely affects spermatogenesis in male infertility. Biomed. Res. 22: 323–328.

Baig, N., Kammakakam, I., Falath, W., and Kammakakam, I. 2021. Nanomaterials: a review of synthesis methods, properties, recent progress, and challenges. Mater. Adv. 2: 1821–1871.

Bansal, V., Rautaray, D., Bharde, A., Ahire, K., Sanyal, A., Ahmad, A., and Sastry, M. 2005. Fungus-mediated biosynthesis of silica and titania particles. J. Mater. Chem. 15:2583–2589.

Bansal, V., Rautaray, D., Ahmad, A., and Sastry, M. 2004. Biosynthesis of zirconia nanoparticles using the fungus *Fusarium oxysporum*. J. Mater Chem. 14:3303–3305

Basavaraja, S., Balaji, S.D., Lagashetty, A., Rajasab, A.H., and Venkataraman, A. 2008. Extracellular biosynthesis of silver nanoparticles using the fungus *Fusarium semitectum*. Mater Res. Bull. 43(5): 1164–1170.

Basu, A., Ray, S., Chowdhury, S., Sarkar, A., Mandal, D.P., Bhattacharjee, S., and Kundu, S. 2018. Evaluating the antimicrobial, apoptotic, and cancer cell gene delivery properties of protein-capped gold nanoparticles synthesized from the edible mycorrhizal fungus *Tricholoma crassum*. Nanoscale Res. Lett. 13: 154.

Basu, S., Bose, C., Ojha, N., Das, N., Das, J., Pal, M., and Khurana, S. 2015. Evolution of bacterial and fungal growth media. Bioinformation. 11: 182.

Balakumaran, M.D., Ramachandran, R., and Kalaicheilvan, P.T. 2015. Exploitation of endophytic fungus, *Guignardia mangiferae* for extracellular synthesis of silver nanoparticles and their *in vitro* biological activities. Microbiol. Res. 178: 9–17.

Banu, A.N. and Balasubramanian, C. 2014. Optimization and synthesis of silver nanoparticles using *Isaria fumosorosea* against human vector mosquitoes. Parasitol. Res. 113: 3843–3851.

Beheshti, N., Soflaei, S., Shakibaie, M., Yazdi, M.H., Ghaffarifar, F., Dalimi, A., and Shahverdi, A.R. 2013. Efficacy of biogenic selenium nanoparticles against *Leishmania major*: In vitro and *in vivo* studies. J. Trace Elem. Med. Biol. 27: 203–207.

Bera, D., Qian, L., Tseng, T.K., and Holloway, P.H. 2010. Quantum dots and their multimodal applications: A review. Materials. 3: 2260.

Bernardos, A., Piacenza, E., Sancenón, F., Hamidi, M., Maleki, A., Turner, R.J., and Martínez-Máñez, R. 2019. Mesoporous silica-based materials with bactericidal properties. Small. 15: 1900669.

Bhainsa. K.C., and D'Souza, S.F. 2006. Extracellular biosynthesis of silver nanoparticles using the fungus Aspergillus fumigatus. Coll. Surf. B Biointerfaces. 47(2): 160–4.

Bharde, A., Rautaray, D., Bansal, V., Ahmad, A., Sarkar, I., Yusuf, S.M., Sanyal, M., and Sastry, M. 2006. Extracellular biosynthesis of magnetite using fungi. Small 2: 135–141

Bhardwaj, K., Sharma, A., Tejwan, N., Bhardwaj, S., Bhardwaj, P., Nepovimova, E., Shami, A., Kalia, A., Kumar, A., Abd-Elsalam, K.A., and Kuča, K. 2020. *Pleurotus* macrofungi-assisted nanoparticle synthesis and its potential applications: A review. J. Fungi (Basel). 6(4): 351.

Birla, S.S., Tiwari, V.V., Gade, A.K., Ingle, A.P., Yadav, A.P., and Rai, M.K. 2009. Fabrication of silver nanoparticles by *Phoma glomerata* and its combined effect against *Escherichia coli*, *Pseudomonas aeruginosa* and *Staphylococcus aureus*. Lett. Appl. Micro. 48(2): 173–179.

Birla, S.S., Gaikwad, S.C., Gade, A.K., and Rai, M.K. 2013. Rapid synthesis of silver nanoparticles from *Fusarium oxysporum* by optimizing physico-cultural conditions. Sci. World J.

Bulgarini, A., Lampis, S., Turner, R.J., and Vallini, G. 2021. Biomolecular composition of capping layer and stability of biogenic selenium nanoparticles synthesized by five bacterial species. Micro. Biotech. 14: 198–212.

Burduşel, A.C., Gherasim, O., Grumezescu, A.M., Mogoantă, L., Ficai, A., and Andronescu, E. 2018. Biomedical applications of silver nanoparticles: an up-to-date overview. Nanomater 8.

Buschmann, M.D., Carrasco, M.J., Alishetty, S., Paige, M., Alameh, M.G., and Weissman, D. 2021. Nanomaterial delivery systems for mRNA vaccines. Vaccines 9(1): 65.

Calvo, A.M. and Cary, J.W. 2015. Association of fungal secondary metabolism and sclerotial biology. Front Micro. 6: 62.

Castro-Longoria, E., Vilchis-Nestor, A.R., and Avalos-Borja, M. 2011. Biosynthesis of silver, gold and bimetallic nanoparticles using the filamentous fungus *Neurospora crassa*. Coll. Surf. B: Biointerfaces 83(1): 42–48

Cervantes, C., and Gutierrez-Corona, F. 1994. Copper resistance mechanisms in bacteria and fungi. FEMS Micro. Rev. 14: 121–137.

Chan, Y.S., and Mat, D.M. 2013. Biosynthesis and structural characterization of Ag nanoparticles from white rot fungi. Mater Sci. Eng. C. 33: 282–288.

Chavali, M.S., and Nikolova, M.P. 2019. Metal oxide nanoparticles and their applications in nanotechnology. SN Appl. Sci. 1:6. 1: 1–30.

Chen, G., Yi, B., Zeng, G., Niu, Q., Yan, M., Chen, A., Du, J., Huang, J., and Zhang, Q. 2014. Facile green extracellular biosynthesis of CdS quantum dots by white rot fungus *Phanerochaete chrysosporium*. Coll. Surf. B Biointerfaces. 117: 199–205.

Cheong, Y.K., Arce, M.P., Benito, A., Chen, D., Crisóstomo, N.L., Kerai, L.V., Rodríguez, G., Valverde, J.L., Vadalia, M., Cerpa-Naranjo, A., and Ren, G. 2020. Synergistic antifungal study of PEGylated graphene oxides and copper nanoparticles against *Candida albicans*. Nanomater. (Basel) 10(5): 819.

Chowdhury, S., Basu, A., and Kundu, S. 2014. Green synthesis of protein capped silver nanoparticles from phytopathogenic fungus *Macrophomina phaseolina* (Tassi) Goid with antimicrobial properties against multidrug-resistant bacteria. Nanoscale Res. Lett. 9: 1–11.

Costa Silva, L.P., Oliveira, J.P., Keijok, W.J., Silva, A.R., Aguiar, A.R., Guimaraes, M.C.C. et al. 2017. Extracellular biosynthesis of silver nanoparticles using the cell-free filtrate of nematophagus fungus *Duddingtonia flagans*. Int. J. Nanomed. 12: 6373–6381.

20 *Mycosynthesis of Nanomaterials: Perspectives and Challenges*

Cotta, M.C. 2020. Quantum dots and their applications: what lies ahead? Appl. Nano. Mater. 3: 4920–4924.

Cuevas, R., Durán, N., Diez, M.C., Tortella, G.R., and Rubilar, O. 2015. Extracellular biosynthesis of copper and copper oxide nanoparticles by *Stereum hirsutum*, a native white-rot fungus from Chilean forests. J. Nanomater. 16: 1–7.

Dahoumane, S.A., Jeffryes, C., Mechouet, M., and Agathos, S.N. 2017. Biosynthesis of inorganic nanoparticles: A fresh look at the control of shape, size and composition. Bioeng (Basel). 4(1): 14.

Dar, M.A., Ingle, A., and Rai, M. 2013. Enhanced antimicrobial activity of silver nanoparticles synthesized by *Cryphonectria* sp. evaluated singly and in combination with antibiotics. Nanomedicine. 9: 105–106.

Devi, L.S., and Joshi, S. 2015. Ultrastructures of silver nanoparticles biosynthesized using endophytic fungi. J. Microscopy Ultrastruct. 3: 29.

Dhanasekar, N.N., Ravindran Rahul, G., Badri Narayanan, K., Raman, G., and Sakthivel, N. 2015. Green chemistry approach for the synthesis of gold nanoparticles using the fungus *Alternaria* sp. J. Microbiol. Biotechnol. 25(7): 1129–1135.

Dhanker, R., Hussain, T., Tyagi, P., Singh, K.J., and Kamble S.S. 2021. The emerging trend of bio-engineering approaches for microbial nanomaterial synthesis and its applications. Front Microbiol. 2: 638003.

Dhillon, G.S., Brar, S.K., Kaur, S., and Verma, M. 2012. Green approach for nanoparticle biosynthesis by fungi: current trends and applications. Crit. Rev. Biotechnol. 32(1): 49–73.

Durán, N., Marcato, P.D., Alves, O.L. et al. 2005. Mechanistic aspects of biosynthesis of silver nanoparticles by several *Fusarium oxysporum* strains. J. Nanobiotechnol. 3: 8

Durán, N., Marcato, P.D., Durán, M., Yadav, A., Gade, A., and Rai, M. 2011. Mechanistic aspects in the biogenic synthesis of extracellular metal nanoparticles by peptides, bacteria, fungi, and plants. Appl Microbiol. Biotechnol. 90(5): 1609–1624.

Dykman, L.A., and Khlebtsov. N.G. 2011. Gold nanoparticles in biology and medicine: recent advances and prospects. Acta Naturae. 3: 34.

Ealias, A.M. and Saravanakumar, M.P. 2017. A review on the classification, characterisation, synthesis of nanoparticles and their application. *IOP Conference Series:* Mater. Sci. Eng. 263: 032019.

Elamawi, R.M., Al-Harbi, R.E., and Hendi, A.A. 2018. Biosynthesis and characterization of silver nanoparticles using *Trichoderma longibrachiatum* and their effect on phytopathogenic fungi. Egyp J. Biol. Pest Cont. 28: 1–11.

Elgorban, A.M., Am, E., Sm, A., Sm, S., Km, E., Ah, B., Sr. S., and Ma, M. 2016. Extracellular synthesis of silver nanoparticles using *Aspergillus versicolor* and evaluation of their activity on plant pathogenic fungi. Mycosphere. 7(6): 844–852.

El-Batal, A.I., Elkenawy, N.M., Yassin, A.S., and Amin, M.A. 2014. Laccase production by *Pleurotus ostreatus* and its application in synthesis of gold nanoparticles. Biotechnol. Rep. 5: 31–39.

El-Sayed, I.H., Huang, X., and El-Sayed, M.A. 2005. Surface plasmon resonance scattering and absorption of anti-EGFR antibody conjugated gold nanoparticles in cancer diagnostics: applications in oral cancer. Nano lett. 5: 829–834.

El-Sayyad, G.S., El-Bastawisy, H.S., Gobara, M., and El-Batal, A.I. 2019. Gentamicin-assisted mycogenic selenium nanoparticles synthesized under gamma irradiation for robust reluctance of resistant urinary tract infection-causing pathogens. Biol. Trace Elem. Res. 195: 323–342.

Espinosa-Ortz, E.J., Gonzalezgil, G., Saikaly, P.E., van Hullebusch, E.D., and Lens, P.N. 2015. Effects of selenium oxyanions on the white-rot fungus *Phanerochaete chrysosporium*. Appl. Microbiol. Biotechnol. 99: 2405–2418.

Faramarzi, M.A., and Forootanfar, H. 2011. Biosynthesis and characterization of gold nanoparticles produced by laccase from *Paraconiothyrium* variabile. Coll Surf B Biointerfaces. 87: 23–27.

Fayaz, M., Tiwary, C.S., Kalaichelvan, P.T., and Venkatesan, R. 2010. Blue orange light emission from biogenic synthesized silver nanoparticles using *Trichoderma viride*. Coll Surf. B. Biointerfaces 75: 175–178.

Filipović, N., Ušjak, D., Milenković, M.T., Zheng, K., Liverani, L., Boccaccini, A.R., and Stevanović, M.M. 2021. Comparative study of the antimicrobial activity of selenium nanoparticles with different surface chemistry and structure. Front Bioeng. Biotechnol. 8: 1591.

Using Fungi to Produce Metal/Metalloid Nanomaterials 21

Forootanfar, H., Adeli-Sardou, M., Nikkhoo, M., Mehrabani, M., Amir-Heidari, B., Shahverdi, A.R., and Shakibaie, M. 2014. Antioxidant and cytotoxic effect of biologically synthesized selenium nanoparticles in comparison to selenium dioxide. J. Trace Elem. Med. Biol. 28: 75–79.

Franke, M.E., Koplin, T.J., and Simon, U. 2006. Metal and metal oxide nanoparticles in chemiresistors: Does the nanoscale matter? Small. 2: 36–50.

Gade, A.K., Bonde, P.P., Ingle, A.P., Marcato, P.D., Duran, N. and Rai, M.K. 2008. Exploitation of *Aspergillus niger* for synthesis of silver nanoparticles. J. Biobased. Mater. Bioener. 2(3): 243–247.

Gericke, M., and Pinches, A. 2006. Microbial production of gold nanoparticles. Gold Bull. 39: 22–28.

Ghannoum, M.A., and Rice, L.B. 1999. Antifungal agents: mode of action, mechanisms of resistance, and correlation of these mechanisms with bacterial resistance. Clin. Microbiol. Rev. 12: 501–517.

Ghosh, S., Ahmad, R., Zeyaullah, M., and Khare, S.K. 2021. Microbial Nano-Factories: Synthesis and Biomedical Applications. Front Chem. 9: 626834.

Gudikandula,, K., and Charya Maringanti, S. 2016. Synthesis of silver nanoparticles by chemical and biological methods and their antimicrobial properties. J. Exper. Nanosci. 11: 714–721.

Gudikandula, K., Vadapally, P., and Singara Charya, M.A. 2017. Biogenic synthesis of silver nanoparticles from white rot fungi: Their characterization and antibacterial studies. OpenNano. 2: 64–78.

Guilger-Casagrande, M., Germano-Costa, T., Bilesky-José, N., Pasquoto-Stigliani, T., Carvalho, L., Fraceto, L.F., and Lima, R. de. 2021. Influence of the capping of biogenic silver nanoparticles on their toxicity and mechanism of action towards *Sclerotinia sclerotiorum*. J. Nanobiotechnol. 19: 1–18.

Guilger-Casagrande, M., and de Lima, R. 2019. Synthesis of silver nanoparticles mediated by fungi: A review. Front Bioeng. Biotechnol. 7: 287.

Gulati, S., Sachdeva, M., and Bhasin, K.K. 2018. Capping agents in nanoparticle synthesis: Surfactant and solvent system. AIP Conf. Proc. 1953: 030214.

He, K., Chen, G., Zeng, G., Huang, Z., Guo, Z., Huang, T., Peng, M., Shi, J., and Hu, L. 2017. Applications of white rot fungi in bioremediation with nanoparticles and biosynthesis of metallic nanoparticles. Appl. Microbiology Biotechnol. 101: 4853–4862.

Hosnedlova, B., Kepinska, M., Skalickova, S., Fernandez, C., Ruttkay-Nedecky, B., Peng, Q., Baron, M., Melcova, M., Opatrilova, R., Zidkova, J., Bjørklund, G., Sochor, J., and Kizek, R. 2018. Nano-selenium and its nanomedicine applications: A critical review. Intern. J. Nanomed. 13: 2107.

Hu, X., Zhang, Y., Ding, T., Liu, J., and Zhao, H. 2020. Multifunctional gold nanoparticles: a novel nanomaterial for various medical applications and biological activities. Front Bioeng. Biotechnol. 8: 990.

Husen, A., and Siddiqi, K.S. 2014. Plants and microbes assisted selenium nanoparticles: characterization and application. J. Nanobiotechnol. 12: 28.

Husseiny, S.M., Salah, T.A., and Anter, H.A. 2015. Biosynthesis of size-controlled silver nanoparticles by *Fusarium oxysporum*, their antibacterial and antitumor activities. Beni-Suef Univ. J. Basic. Appl. Sci. 4: 225–231.

Ingle, A., Gade, A., Pierrat, S., Sonnichsen, C., and Rai, M. 2008. Mycosynthesis of silver nanoparticles using the fungus *Fusarium acuminatum* and its activity against some human pathogenic bacteria. Curr. Nanosci. 4: 141–144.

Iravani, S. 2011. Green synthesis of metal nanoparticles using plants. Green Chem. 13: 2638–2650.

Jamdagni, P., Khatri, P., and Rana, J.S. 2018. Green synthesis of zinc oxide nanoparticles using flower extract of *Nyctanthes arbor-tristis* and their antifungal activity. J. King Saud Univ.—Science. 30: 168–175.

Javed, R., Zia, M., Naz, S., Aisida, S.O., Ain, N., and Ao, Q. 2020. Role of capping agents in the application of nanoparticles in biomedicine and environmental remediation: recent trends and future prospects. J. Nanobiotechnol. 18: 1–15.

Jin, Y., Cai, L., Yang, Q., Luo, Z., Liang, L., Liang, Y., Wu, B., Ding, L., Zhang, D., Xu, X., Zhang, L., and Zhou, F. 2020. Anti-leukemia activities of selenium nanoparticles embedded in nanotube consisted of triple-helix β-d-glucan. Carbo Poly. 240: 116329.

Joshi, S.M., Britto, S. de, Jogaiah, S., and Ito, S.I. 2019. Mycogenic selenium nanoparticles as potential new generation broad spectrum antifungal molecules. Biomolec. 9: 419.

22 *Mycosynthesis of Nanomaterials: Perspectives and Challenges*

Kapoor, R.T., Salvadori, M.R., Rafatullah, M., Siddiqui, M.R., Khan, M.A., and Alshareef, S.A. 2021. Exploration of microbial factories for synthesis of nanoparticles—a sustainable approach for bioremediation of environmental contaminants. Front Microbiol. 12: 1404.

Karakoti, A.S., Munusamy, P., Hostetler, K., Kodali, V., Kuchibhatla, S., Orr, G., Pounds, J.G., Teeguarden, J.G., Thrall, B.D., and Baer, D.R. 2012. Preparation and characterization challenges to understanding environmental and biological impacts of nanoparticles. Surf. Interf. Analys. 44: 882.

Keller, N.P. 2018. Fungal secondary metabolism: regulation, function and drug discovery. Nat. Rev. Microbiol. 17: 167–180.

Kennedy, L.C., Bickford, L.R., Lewinski, N.A., Coughlin, A.J., Hu, Y., Day, E.S., West, J.L., and Drezek, R.A. 2011. A new era for cancer treatment: gold-nanoparticle-mediated thermal therapies. Small. 7: 169–183.

Khalafalla, M.A.H., Mesli, A., Widattallah, H.M., Sellai, A., Al-Harthi, S., Al-Lawati, H.A.J., and Suliman, F.O. 2014. Size-dependent conductivity dispersion of gold nanoparticle colloids in a microchip: contactless measurements. J. Nanopart. Res. 16: 1–8.

Khalid, A., Tran, P.A., Norello, R., Simpson, D.A., O'Connor, A.J., and Tomljenovic-Hanic, S. 2016. Intrinsic fluorescence of selenium nanoparticles for cellular imaging applications. Nanoscale. 8: 3376–3385.

Khan, A.U., Malik, N., Khan, M., Cho, M.H., and Khan, M.M. 2018. Fungi-assisted silver nanoparticle synthesis and their applications. Bioproc. Biosyst. Eng. 41: 1–20.

Khan, I., Saeed, K., and Khan, I. 2019. Nanoparticles: Properties, applications and toxicities. Arab. J. Chem. 12: 908–931.

Kong, F.Y., Zhang, J.W., Li, R.F., Wang, Z.X., Wang, W.J., and Wang, W. 2017. Unique roles of gold nanoparticles in drug delivery, targeting and imaging applications. Molecules. 22: 1445.

Koul, B., Poonia, A.K., Yadav, D., and Jin, J.-O. 2021. Microbe-mediated biosynthesis of nanoparticles: applications and future prospects. Biomolecules. 11: 886.

Kumari, B., and Singh, D.P. 2016. A review on multifaceted application of nanoparticles in the field of bioremediation of petroleum hydrocarbons. Ecol. Eng. 97: 98–105.

Kumar, V., and Dwivedi, S.K. 2021. Mycoremediation of heavy metals: processes, mechanisms, and affecting factors. Environ. Sci. Pollut. Res. Int. 28(9): 10375–10412.

Kuzyk, S.B., Hughes, E., and Yurkov, V. 2021. Discovery of siderophore and metallophore production in the aerobic anoxygenic phototrophs. Microorganisms 9(5): 959.

Lang, C., Schüler, D., and Faivre, D. 2007. Synthesis of magnetite nanoparticles for bio- and nanotechnology: genetic engineering and biomimetics of bacterial magnetosomes. Macromol. Biosci. 7: 144–151.

Lemire, J.A., Harrison, J.J., and Turner, R.J. 2013. Antimicrobial activity of metals: mechanisms, molecular targets and applications. Nat. Rev. Microbiol. 11: 371–384.

Li, Q., Liu, F., Li, M., Chen, C., and Gadd, G.M. 2021. Nanoparticle and nanomineral production by fungi. Fungal Biol Rev. In press. Doi: 10.1016/j.fbr.2021.07.003.

Liao, W., Yu, Z., Lin, Z., Lei, Z., Ning, Z., Regenstein, J.M., Yang, J., and Ren, J. 2015. Biofunctionalization of Selenium nanoparticle with *Dictyophora Indusiata* polysaccharide and its antiproliferative activity through death-receptor and mitochondria-mediated apoptotic pathways. Sci. Rep. 5. 18629.

Maiti, D., Tong, X., Mou, X., and Yang, K. 2019. Carbon-based nanomaterials for biomedical applications: A recent study. Front Pharmacol. 9: 1401.

Maiyo, F. and Singh, M. 2017. Selenium nanoparticles: potential in cancer gene and drug delivery. Nanomed. 12: 1075–1089.

'Mandeep, S.P., and Shukla, P. 2020. Microbial nanotechnology for bioremediation of industrial wastewater. Front Microbiol. 11: 2411.

Medina Cruz, D., Mi, G., and Webster, T.J. 2018. Synthesis and characterization of biogenic selenium nanoparticles with antimicrobial properties made by *Staphylococcus aureus*, methicillin-resistant *Staphylococcus aureus* (MRSA), *Escherichia coli*, and *Pseudomonas aeruginosa*. J. Biomed. Mater Res A. 106: 1400–1412.

Mekky, A.E., Farrag, A.A., Hmed, A.A., and Sofy, A.R. 2021. Preparation of zinc oxide nanoparticles using *Aspergillus niger* as antimicrobial and anticancer agents. J. Pure Appl. Microbiol. 15(3): 1547–1566.

Mo, D., Hu, L., Zeng, G., Chen, G., Wan, J., Yu, Z., Huang, Z., He, K., Zhang, C., and Cheng, M. 2017. Cadmium-containing quantum dots: properties, applications, and toxicity. Appl. Microbiol. Biotechnol. 101: 2713–2733.

Moghaddam, A.B., Namvar, F., Moniri, M., Tahir, P.M., Azizi, S., and Mohamad, R. 2015. Nanoparticles biosynthesized by fungi and yeast: A review of their preparation, properties, and medical applications. Molecules. 20: 16540.

Moloi, M.S., Lehutso, R.F., Erasmus, M., Oberholster, P.J., and Thwala, M. 2021. Aquatic environment exposure and toxicity of engineered nanomaterials released from nano-enabled products: Current status and data needs. Nanomater. 11: 2868.

Mukherjee, A, Sarkar, D., and Sasmal, S. 2021. A review of green synthesis of metal nanoparticles using algae. Front Microbiol. 12: 693899.

Mukherjee, S., Ray, S., and Thakur, R.S. 2009. Solid lipid nanoparticles: A modern formulation approach in drug delivery system. Ind. J. Pharma. Sci. 71: 349.

Nagajyothi, P.C., Sreekanth, T.V.M., Lee, J. il, and Lee, K.D. 2014. Mycosynthesis: Antibacterial, antioxidant and antiproliferative activities of silver nanoparticles synthesized from *Inonotus obliquus* (Chaga mushroom) extract. J. Photochem. Photobiol. B: Biol. 130: 299–304.

Nande, A., Raut, S. Michalska-Domanska, M., and Dhoble, S.J. 2021. Green synthesis of nanomaterials using plant extract: A review. Curr. Pharm. Biotechnol. 22(13): 1794–1811.

Oladipo, O.G., Awotoye, O.O., Olayinka, A., Bezuidenhout, C.C., and Maboeta, M.S. 2018. Heavy metal tolerance traits of filamentous fungi isolated from gold and gemstone mining sites. Braz. J. Microbiol. 49: 29.

Oliveira, P.F.M. de, Torresi, R.M., Emmerling, F., and Camargo, P.H.C. 2020. Challenges and opportunities in the bottom-up mechanochemical synthesis of noble metal nanoparticles. J. Mater Chem. A. 8: 16114–16141.

Owaid, M.N. 2020. Biomedical applications of nanoparticles synthesized from mushrooms. pp. 289–303. *In*: Patra, J.K., Fraceto, L.F., Das, G., and Campos, E.V.R. (Eds.). Green nanoparticles, nanotechnology in the life sciences. Springer International Publishing, Cham.

Ozkan, M. 2004. Quantum dots and other nanoparticles: what can they offer to drug discovery? Drug Discov. Today. 9: 1065–1071.

Pamungkas, D.P.W., Amaliyah, S., Sabarudin, A., and Safitri, A. 2021. Biosynthesis of Cu_2O/CuO-NP and AgNP Using *Rhizopus oligosporus* as reductor agent. J. Pure Appl. Chem. Res. 10: 165–174.

Pavani, K.V., Kumar, N.S., and Sangameswaran, B.B. 2012. Synthesis of lead nanoparticles by *Aspergillus* species. Pol. J. Microbiol. 61(1): 61–63.

Pereira, L., Dias, N., Carvalho, J., Fernandes, S., Santos, C., and Lima, N. 2014. Synthesis, characterization and antifungal activity of chemically and fungal-produced silver nanoparticles against *Trichophyton rubrum*. J. Appl. Microbiol. 117: 1601–1613.

Piacenza, E., Presentato, A., and Turner, R.J. 2018a. Stability of biogenic metal(loid) nanomaterials related to the colloidal stabilization theory of chemical nanostructures. Crit. Rev. Biotechnol. 38: 1137–1156.

Piacenza, E., Presentato, A., Zonaro, E., Lampis, S., Vallini, G., and Turner, R.J. 2018b. Selenium and tellurium nanomaterials. Phys. Sci. Rev. 3(5): 20170100.

Piacenza, E., Presentato, A., Bardelli, M., Lampis, S., Vallini, G., and Turner, R.J. 2019. Influence of bacterial physiology on processing of selenite, biogenesis of nanomaterials and their thermodynamic stability. Molecules. 24(14): 2532.

Prabhu, S., and Poulose, E.K. 2012. Silver nanoparticles: mechanism of antimicrobial action, synthesis, medical applications, and toxicity effects. Intern. Nano. Lett. 2: 1–10.

Priyadarshini, E., Priyadarshini, S.S., Cousins, B.G., and Pradhan, N. 2021. Metal-fungus interaction: Review on cellular processes underlying heavy metal detoxification and synthesis of metal nanoparticles. Chemosphere. 274: 129976.

Qian, Y., Yu, H., He, D., Yang, H., Wang, W., Wan, X. et al. 2013. Biosynthesis of silver nanoparticles by the endophytic fungus *Epicoccum nigrum* and their activity against pathogenic fungi. Bioproc. Biosyst. Eng. 36: 1613–1619.

Raheman, F., Deshmukh, S., Ingle, A., Gade, A., and Rai, M. 2011. Silver nanoparticles: novel antimicrobial agent synthesized from an endophytic fungus *Pestalotia* sp. isolated from leaves of *Syzygium cumini* (L). Nano Biomed Eng., 3(3): 174–178.

24 *Mycosynthesis of Nanomaterials: Perspectives and Challenges*

Rahimi, H.-R., and Doostmohammadi, M. 2019. Nanoparticle synthesis, applications, and toxicity. In Applications of Nanobiotechnology. publisher IntechOpen.

Rahman, A., Lin, J., Jaramillo, F.E., Bazylinski, D.A., Jeffryes, C., and Dahoumane, S.A. 2020. *In vivo* biosynthesis of inorganic nanomaterials using eukaryotes-A review. Molecules. 20(14): 3246.

Rai, M., Alka, Y., Bridge, P., and Aniket, G. 2009. Myconanotechnology: a new and emerging science. Appl. Mycol. 258–267.

Rai, M., Bonde, S., Golinska, P., Trzcińska-Wencel, J., Gade, A., Abd-Elsalam, K.A., Shende, S., Gaikwad, S., and Ingle, A.P. 2021. *Fusarium* as a novel fungus for the synthesis of nanoparticles: mechanism and applications. J. Fungi. 7(2): 139.

Rattan, R., Shukla, S., Sharma, B., and Bhat, M. 2021. A mini-review on lichen-based nanoparticles and their applications as antimicrobial agents. Front Microbiol. 12: 633090.

Riddin, T.L., Gericke, M., and Whiteley, C.G. 2006. Analysis of the inter- and extracellular formation of platinum nanoparticles by *Fusarium oxysporum* f. sp. *lycopersici* using response surface methodology. Nanotechnology 17: 3482.

Robinson, J.R., Isikhuemhen, O.S., and Anike, F.N. 2021. Fungal-metal interactions: A review of toxicity and homeostasis. J. Fungi. 7(3): 225.

Rodrigues, A.G., Ping, L.Y., Marcato, P.D., Alves, O.L., Silva, M.C.P., Ruiz, R.C., Melo, I.S., Tasic, L., and Souza, A.O. de. 2013. Biogenic antimicrobial silver nanoparticles produced by fungi. Appl. Microbiol. Biotechnol. 97: 775–782.

Rose, G.K., Soni, R., Rishi, P., and Soni, S.K. 2019. Optimization of the biologicalsynthesis of silver nanoparticles using *Penicillium oxalicum* GRS-1 and their antimicrobial effects against common food-borne pathogens. Green Process Synth. 8: 144–156.

Roy, S., Mukherjee, T., Chakraborty, S., and Das, T.K. 2013. Biosynthesis, characterisation & antifungal activity of silver nanoparticles synthesized by the fungus *Aspergillus foetidus* MTCC8876. Dig. J. Nanomater. Biostruct. 8(1): 197–205.

Sanghi, R., and Verma, P. 2009. Biomimetic synthesis and characterisation of protein capped silver nanoparticles. Biores. Technol. 100: 501–504.

Shahverdi, A.R., Fakhimi, A., Mosavat, G., Jafari-Fesharaki, P., Rezaie, S., and Rezayat, S.M. 2010. Antifungal activity of biogenic selenium nanoparticles. World Appl. Sci. J. 10: 918–922.

Shahzad, A., Saeed, H., Iqtedar, M., Hussain, S.Z., Kaleem, A., Abdullah, R., Sharif, S., Naz, S., Saleem, F., Aihetasham, A., and Chaudhary, A. 2019. Size-controlled production of silver nanoparticles by *Aspergillus fumigatus* BTCB10: Likely antibacterial and cytotoxic effects. J. Nanomater. 7: 5168698.

Shakibaie, M., Forootanfar, H., Golkari, Y., Mohammadi-Khorsand, T., and Shakibaie, M.R. 2015. Anti-biofilm activity of biogenic selenium nanoparticles and selenium dioxide against clinical isolates of *Staphylococcus aureus*, *Pseudomonas aeruginosa*, and *Proteus mirabilis*. J. Trace Elem. Med. Biol. 29: 235–241.

Sastry, M., Ahmad, A., Khan, M.I. and Kumar, R. 2003. Biosynthesis of metal nanoparticles using fungi and actinomycete. Curr. Sci. 162–170.

Saxena, J., Sharma, M.M., Gubta, S., and Singh, A. 2014. Emerging role of fungi in nanoparticle synthesis and their applications. World J. Pharm. Pharmaceut. Sci. 3(9): 1586–1613.

Sharma, D., Kanchi, S., and Bisetty, K. 2015. Biogenic synthesis of nanoparticles: A review. Arab. J. Chem. 12: 3576–3600.

Sheikhloo, Z., and Salouti, M. 2011. Intracellular biosynthesis of gold nanoparticles by the fungus *Penicillium chrysogenum*. Int. J. Nanosci. Nanotechnol. 7(2): 102–105.

Shelar, G.B., and Chavan, A.M. 2014. Fungus-mediated biosynthesis of silver nanoparticles and its antibacterial activity. Arch. Appl. Sci. Res. 6: 111–114.

Si, P., Razmi, N., Nur, O., Solanki, S., Pandey, C.M., Gupta, R.K., Malhotra, B.D., Willander, M., and La Zerda, A. de. 2021. Gold nanomaterials for optical biosensing and bioimaging. Nanoscale Adv. 3: 2679–2698.

Siddiqi, K.S., and Husen, A. 2016. Fabrication of metal nanoparticles from fungi and metal salts: Scope and application. Nanoscale Rese Lett. 11: 1–15.

Sim, W., Barnard, R.T., Blaskovich, M.A.T., and Ziora, Z.M. 2018. Antimicrobial silver in medicinal and consumer applications: A patent review of the past decade (2007–2017). Antibiotics. 7(4): 93.

Sneha, K., Sathishkumar, M., Mao, J., Kwak, I.S., and Yun, Y.S. 2010. Corynebacterium glutamicum-mediated crystallization of silver ions through sorption and reduction processes. Chem. Eng. J. 162: 989–996.

Soni, N., and Prakash, S. 2012. Synthesis of gold nanoparticles by the fungus *Aspergillus niger* and its efficacy against mosquito larvae. Rep. Parasitol. 2: 1–7.

Sousa, L.M., Vilarinho, L.M., Ribeiro, G.H., Bogado, A.L., and Dinelli, L.R. 2017. An electronic device based on gold nanoparticles and tetraruthenated porphyrin as an electrochemical sensor for catechol. Roy. Soc. Open Sci. 4: 170675.

Sreenivasa Kumar G., Venkataramana, B., and Adinarayana Reddy, S. 2017. Effect of different physicochemical conditions on the synthesis of silver nanoparticles using fungal cell filtrate of *Aspergillus oryzae* (MTCC No. 1846) and their antibacterial effect. Adv. Natural Sci: Nanosci. Nanotechnol. 8: 045016.

Sudheer, S., Bai, R.G., Muthoosamy, K., Tuvikene, R., Gupta, V.K., and Manickam, S. 2022. Biosustainable production of nanoparticles via mycogenesis for biotechnological applications: A critical review. Environ. Res. 204: 111963.

Sundaravadivelan, C., and Padmanabhan, M.N. 2013. Effect of mycosynthesized silver nanoparticles from filtrate of *Trichoderma harzianum* against larvae and pupa of dengue vector *Aedes aegypti* L. Environ. Sci. Pollut. Res. 21: 4624–4633.

Suresh, A.K., Pelletier, D.A., Wang, W., Moon, J.W., Gu, B., Mortensen, N.P., Allison, D.P., Joy, D.C., Phelps, T.J., and Doktycz, M.J. 2010. Silver nanocrystallites: Biofabrication using *Shewanella oneidensis*, and an evaluation of their comparative toxicity on gram-negative and gram-positive bacteria. Environ. Sci. Technol. 44: 5210–5215.

Syed, A., and Ahmad, A. 2013. Extracellular biosynthesis of CdTe quantum dots by the fungus *Fusarium oxysporum* and their anti-bacterial activity. Spectrochim Acta A. Mol. Biomol Spectrosc. 106: 41–7.

Tamuly, C., Hazarika, M., Bordoloi, M., and Das, M.R. 2013. Photocatalytic activity of Ag nanoparticles synthesized by using *Piper pedicellatum* C.DC fruits. Mater Lett. 102–103: 1–4.

Tan, P., Li, H.S., Wang, J., and Gopinath, S.C.B. 2021. Silver nanoparticle in biosensor and bioimaging: Clinical perspectives. Biotechnol. Appl. Biochem. 68: 1236–1242.

Tran, Q.H., Nguyen, V.Q., and Le, A.T. 2013. Silver nanoparticles: synthesis, properties, toxicology, applications and perspectives. Adv. Nat. Sci: Nanosci. Nanotechnol. 4: 033001.

Tyagi, S., Tyagi, P.K., Gola, D., Chauhan, N., and Bharti, R.K. 2019. Extracellular synthesis of silver nanoparticles using entomopathogenic fungus: characterization and antibacterial potential. SN Appl. Sci. 1: 1–9.

Uddin, I., Adhynthaya, S., Syed, A., Selvaraj, K., Ahmad, A., and Poddar, P. 2008. Structure and microbial synthesis of sub-10 nm Bi2O3 nanocrystals. J. Nanosci. Nanotechnol. 8: 3909–391.

Vahabi, K., Mansoori, G.A., and Karimi, S. 2011. Biosynthesis of silver nanoparticles by the fungus *Trichoderma reesei*. Insciences J. 1: 65–79.

Varghese Alex, K., Tamil Pavai, P., Rugmini, R., Shiva Prasad, M., Kamakshi, K., and Sekhar, K.C. 2020. Green synthesized Ag nanoparticles for bio-sensing and photocatalytic applications. ACS Omega. 5: 13123–13129.

Verma, V.C., Kharwar, R.N., and Gange, A.C. 2009. Biosynthesis of antimicrobial silver nanoparticles by the endophytic fungus *Aspergillus clavatus*. Future Med. 5: 33–40.

Wisecaver, J.H., Slot, J.C., and Rokas, A. 2014. The evolution of fungal metabolic pathways. *PLoS Genetics.* 10. e1004816.

Wu, Q., Miao, W.S., Zhang, Y. du, Gao, H.J., and Hui, D. 2020. Mechanical properties of nanomaterials: A review. Nanotechnol. Rev. 9: 259–273.

Yadav, A., Kon, K., Kratosova, G., Duran, N., Ingle, A.P., and Rai, M. 2015. Fungi as an efficient mycosystem for the synthesis of metal nanoparticles: progress and key aspects of research. Biotechnol. Lett. 37(11): 2099–120.

Yamaguchi, T., Tsuruda, Y., Furukawa, T., Negishi, L., Imura, Y., Sakuda, S., Yoshimura, E., and Suzuki, M. 2016. Synthesis of CdSe quantum dots using *Fusarium oxysporum*. Materials. 9(10): 855.

Yang, Y., Mathieu, J.M., Chattopadhyay, S., Miller, J.T., Wu, T., Shibata, T., Guo, W., and Alvarez, P.J.J. 2012. Defense mechanisms of *Pseudomonas aeruginosa* pao1 against quantum dots and their released heavy metals. ACS Nano. 6: 6091–6098.

Yang, J., and Zhong, X. 2016. CdTe based quantum dot sensitized solar cells with efficiency exceeding 7% fabricated from quantum dots prepared in aqueous media. J. Mater Chem. A. 4: 16553–16561.

Yaqoob, A.A., Ahmad, H., Parveen, T., Ahmad, A., Oves, M., Ismail, I.M.I., Qari, H.A., Umar, K., and Mohamad Ibrahim, M.N. 2020. Recent advances in metal decorated nanomaterials and their various biological applications: A review. Front Chem. 8: 341.

Young, S.L., Kellon, J.E., and Hutchison, J.E. 2016. Small gold nanoparticles interfaced to electrodes through molecular linkers: a platform to enhance electron transfer and increase electrochemically active surface area. JACS 138: 13975–13984.

Xue, B., He, D., Gao, S., Wang, D., Yokoyama, K., and Wang, L. 2016. Biosynthesis of silver nanoparticles by the fungus *Arthroderma fulvum* and its antifungal activity against genera of *Candida, Aspergillus* and *Fusarium*. Int. J. Nanomed. 11: 1899–1906.

Yu, B., Zhang, Y., Zheng, W., Fan, C., and Chen, T. 2012. Positive surface charge enhances selective cellular uptake and anticancer efficacy of selenium nanoparticles. Inorg. Chem. 51: 8956–8963.

Zhai, Y., Hunting, E.R., Wouters, M., Peijnenburg, W.J.G.M., and Vijver, M.G. 2016. Silver nanoparticles, ions, and shape governing soil microbial functional diversity: Nano shapes micro. Front Microbiol. 7: 1123.

Zheng, J., Zhou, C., Yu, M., and Liu, J. 2012. Different sized luminescent gold nanoparticles. Nanoscale. 4: 4073.

Zhong, W. 2009. Nanomaterials in fluorescence-based biosensing. Anal. Bioanal. Chem. 394: 47–59.

CHAPTER 2

Yeast as a Cell Factory
Biological Synthesis of Selenium Nanoparticles

Farnoush Asghari-Paskiabi,[1] *Mehdi Razzaghi-Abyaneh*[1]
and *Mohammad Imani*[2,3,*]

Introduction

Nanoparticles (NPs) can be produced in large quantities at a given time through chemical methods ending to the particles with a specific size and morphology. However, the chemical methods are complex and costly and produce hazardous toxic wastes that are seriously harmful to the environment and human health. The use of chemicals can be avoided by adopting the enzymatic processes for the production of the NPs. The "green" methods do not use much energy compared to the chemical methods so they can be regarded as environmentally friendly methods (Li et al. 2011). The biological methods use biological organisms such as bacteria, actinomycetes, fungi and plant extracts for synthesis of nanoparticles. The reduction of metal ions using biological agents at ambient pressure and temperature is much faster.

Biological methods for synthesis of inorganic NPs have attracted much attention during the recent decade mostly due to the method flexibility, mild reaction conditions, higher biocompatibility profile of the NPs, and the use of aqueous environment. Biological methods developed for the production of non-toxic, inorganic NPs are cost-effective and environmental-friendly. Of course, these methods suffer from several shortcomings like inherent limitations in the level of morphological

---

[1] Department of Mycology, Pasteur Institute of Iran, Tehran, 1316943551, Iran.
[2] Novel Drug Delivery Systems Dept., Iran Polymer and Petrochemical Institute, P.O. Box 14975-112, Tehran, Iran.
[3] Institute for Nanoscience and Nanotechnology, Sharif University of Technology, Tehran, 14588-89694, Iran.
* Corresponding author: M.Imani@ippi.ac.ir; mohammad.imani@sharif.edu

control over the NPs, need for long-time cultures of some microorganisms, and the size and crystallinity of the NPs produced. To this end, the growth conditions of microorganisms must be optimized through adjustment of factors like incubation time, temperature, pH, metal ion concentration and size of the bio-inoculation along with various physical factors such as microwave, light and ultrasonic irradiation. Microorganisms secrete large amounts of enzymes that can reduce metal ions. The biomass used in the synthesis of NPs is easily recycled to nature, the synthesis is done at ambient pressure and temperature, and fewer chemicals are used. The synthesis process is also cost-effective, non-toxic and green hence, it is preferable to chemical and physical synthesis methods (Shankar et al. 2004). Synthesis of NPs by filamentous fungi has its own benefits. Mycelia provide a large surface area in biomass, are easy to produce, and are easy to work with in addition to being cost-effective (Salvadori et al. 2014). The nutrients they need are not complex, the capacity of the metal to attach to their walls is high, so the metals are easily absorbed into the cell (Hemath Naveen et al. 2010). Cell walls of fungi benefit from the presence of a diverse range of biomolecules and functional groups, i.e., lipids, melanins, amines, sulfates, phosphates, carboxyl, and hydroxides, which enable them to absorb metal ions (Salvadori et al. 2014). Yeasts can also produce enzymes that reduce metal ions and convert them into elemental NPs. If metal ions are present, enzymes may be secreted both inside and outside the cells. For example, selenium nanocomposites with a particle size of 30 to 100 nm were biosynthesized under aerobic conditions by the yeast *Saccharomyces cerevisiae* (Hariharan et al. 2012). Lead sulfide (Seshadri et al. 2011), cadmium (Dameron et al. 1989), and silver (Selvaraj 2013) NPs were also produced using yeasts as bio-factories.

Selenium exists at least in 25 different human selenoproteins and enzymes in the form of selenocysteine (Wadhwani et al. 2016) which are critical for the proper functioning of the immune system and thyroid glands and are necessary to protect the cells from oxidative damage (Sharma et al. 2014). Se NPs (NPs) exhibit excellent anticancer and antimicrobial activities but show lower cytotoxicity compared to selenium bulk, (Sharma et al. 2014; Wadhwani et al. 2016). The significance of Se NPs is increasing due to their catalytic, semiconductor, and photoelectric properties too (Wadhwani et al. 2016). Here, we review the literature reporting on the biosynthesis of selenium NPs, especially by using yeasts.

Geochemistry of selenium

Selenium distribution on the earth's surface follows a heterogeneous pattern. While there are surface waters and contaminated soils that are high in selenium, most areas have small amounts of selenium. The role of methylation in the biogeochemical cycle of selenium has been investigated. Methylated selenium compounds are commonly produced in nature. Methylation of selenium species is involved in their redistribution in nature. The high mobility of methylated selenium compared to aquatic and solid-state species allows them to be emitted into the atmosphere more quickly and efficiently. In fact, methylation can make the biogeochemical cycle much more mobile (Vriens 2015).

Plants take selenium from soil through the roots, methylate it and prevent it from entering the atmosphere. Due to the heterogeneity of selenium distribution in the environment and the narrow boundary existing between essential and toxic doses of selenium for the human body health, the selenium cycle must be considered and closely observed in natural and agricultural systems. Improving selenium bioavailability via increasing its content in plant foods is a way to reduce the health problems associated with selenium deficiency in some areas. For example, soil can be improved by adding selenium, or cultivars of plants that absorb selenium can be used, or genetically engineered plants can be considered as a solution just to name a few. Selenium-containing fertilizers have been added to soil in Finland and the United States to obtain selenium-fortified food products. This method may be suitable in short term, but in the long term can be toxic to ecosystems. There are reports on soil bacteria that can reduce selenium to zero valences (selenium NPs) starting from selenate and selenite. But the level of significance of this capability in natural environments and the fate of these biologically-produced NPs in soil systems is not completely clear (Winkel et al. 2015).

Using *Cyperus laevigatus* biomass, biologically-synthesized silver NPs were used to remove selenium from wastewater. Optimized levels of factors involved in this process including pH, initial selenium concentration, adsorption dose concentration, and contact time with the biomass were obtained accurately by adopting statistical experimental design through a response-surface methodology. The selenium adsorption capacity was increased as a function of the granular texture of silver NPs, the resulting porosity and enhanced surface area according to TEM micrographs (Badr et al. 2020).

Chemical methods for synthesis of selenium NPs

The chemical synthesis of selenium NPs was initially considered by several research groups around the globe. Some polymer stabilizers (PSt)-based, biogenic selenium-coating nanocomplexes were investigated in terms of morphology, density and kinetic parameters of self-organization. To obtain red selenium NPs; ascorbic acid was used to reduce selenious acid:

$$H_2SeO_3 + 2C_6H_8O_6 \rightarrow Se + 3H_2O + 2C_6H_6O_6 \tag{1}$$

Polymer-nanoparticle complexes are involved in the stages of nuclei formation and growth, NPs adsorption of polymer and rearrangement and ordering of polymer molecules and increase of density of the formed nanostructures. The nature of PSts was critical in polymer-nanoparticle structure. For example, when polyvinylpyrrolidone (PVP) or poly(trimethylammonioethylmethacrylate)chloride (PTMAEM) were used as PSt, the size of NPs was 18 nm. The size of the NPs varied in the range of 18–30 nm when PMAA or PAMS were used as PSt. However, the particle size of the basic core of selenium NPs was 6 nm in all samples. The sizes were measured through small-angle light scattering (SAXS) (Valueva et al. 2013).

Selenium (Se) NPs and silver phosphate (SP) NPs were compared in terms of their antibacterial properties against *Staphylococcus aureus*. Selenium NPs were obtained by dissolving $Na_2SeO_3 \cdot 5H_2O$ in water and adding 3-mercaptopropionic acid

30 *Mycosynthesis of Nanomaterials: Perspectives and Challenges*

consequently, adjusting the pH to eight and stirring for two hours. The particle size of small spherical SeNPs was in the range of 50–100 nm. During bacterial cell culture, *S. aureus* growth inhibition of SeNPs at 300 µM (7.0 ± 0.5 mm) concentration was more than twice that of the growth inhibition by SPNPs (3.0 ± 0.5 mm) at the same concentration. At 300 µM Se or SPNPs concentration in the microplate, SeNPs completely inhibited *S. aureus* growth while SPNPs inhibited the growth by 37.5% at the same concentration (Chudobova et al. 2014).

Metallic selenium (Se0) in the 500–600 nm size range was obtained upon exposure of aqueous selenium ions to bovine serum albumin at pH 7 and 121°C for 20 min. Also, SeNPs in 100–200 nm size range were obtained through keratin-mediated synthesis at pH 11 at 65°C for 20 min. Moreover, shrinkage, rounding structures, and decreased adhesion to the plate were observed for H9C2 cells plus oxidative damage by ROS generation. Programmed cell death arising from ethanol was inhibited by SeNPs. The NPs also reduced ethanol-induced pericardial edema in embryos of zebrafish model (Kalishwaralal et al. 2015).

Different forms of selenium including an inorganic form (sodium selenite), organic forms (selenium methionine and Zn-Se-Meth) and selenium NPs as solid particles or aqueous dispersion were used in the diet of broilers at two levels of 0.15 and 0.30 ppm. Examination of bird carcasses supported increased concentration of this element in liver and thigh tissues and improved growth as a function of the presence of organic and nano-forms of selenium, as well as increasing the concentration of selenium in the diet to 0.30 ppm. It is worth noting that selenium accumulation in the liver was more significant than in the thigh. However, selenium content in the abdominal fat of carcasses, giblets and malondialdehyde in the thigh muscle was not affected by the levels and sources of the selenium fed. Adding selenium or Zn-Se-Meth yeast (an organic form of selenium) at a concentration of 0.30 ppm as well as nano-forms of this element to the diet or drinking water of broilers may affect their growth rate and meat quality (Selim et al. 2015).

Antibacterial effect of selenium NPs against methicillin-resistant *Staphylococcus aureus* (MRSA) was investigated compared to ampicillin. Selenium NPs were synthesized by the reduction of sodium selenite using L-cysteine amino acid as a reducing agent. To load antibiotics on the surface of selenium NPs, ampicillin was gradually added to the reaction flask where the nanoparticle dispersion was being shaken. The disk diffusion method was employed to investigate the response exhibited by the MRSA strain. Minimum inhibitory concentration (MIC) was obtained and was also determined by the broth microdilution method. Adsorption of the antibiotic on the surface of selenium NPs improved the antimicrobial properties of ampicillin. The diameter of the non-growth zone for selenium NPs was between 20 and 25 mm but it was increased to 22 to 28 mm for samples containing ampicillin antibiotic adsorbed on the selenium NPs, while for ampicillin alone this value was between 8 to 17 mm. MIC of selenium NPs and antibiotic-carrying NPs were both between 7.8 ppm and 62.5 ppm, while the MIC for the free form of ampicillin was between 31.2 to 250 ppm for *S. aureus* species (Mehrbakhsh Bandari et al. 2018).

Selenium NPs with spherical morphology and 15 to 18 nm particle size were obtained from sodium selenite via a simple precipitation method using ascorbic acid as a chemical reducing agent. These NPs were able to stop *Staphylococcus aureus*,

Pseudomonas aeruginosa and *Escherichia coli* growth successfully (Ananth et al. 2019).

Selenium-containing glutathione peroxidase 4 (GPx 4) and selenoprotein W (SelW) protect cells against damage by reactive oxygen species (ROS). SelP is a selenoprotein containing selenocysteine, which is responsible for transporting selenium in the blood. Selenium is also involved in spermatogenesis according to a study performed on the effects of selenium NPs and sodium selenite on the expression of *SelP*, *SelW* and *GPx 4* genes in broiler breeder roosters. The results showed a higher expression rate for these genes by adding selenium NPs as a supplement to the animal's diet. The observed increase was significant compared to the control group and the group to which sodium selenite was added to the diet (Jafarzadeh et al. 2020).

Silver NPs were deposited on the surface of electrospun polycaprolactone fibers containing carbonated hydroxyapatite, which was doped with selenite ions by pulsed laser deposition technique. The effect of this coating on the microstructure and attachment of these structures to the human osteoblast cell line (HFB4) as well as its antibacterial activity against *S. aureus* and *E. coli* were investigated. As the deposition time increased, its roughness increased. This increase led to improved HFB4 cell growth and increased antibacterial properties (Menazea et al. 2020).

Selenium antioxidant properties and its effect on the ram semen were investigated during the freezing-thawing process of the semen. Selenium NPs at 1 or 2 $\mu g.mL^{-1}$ concentrations was added to frozen sperm on days 0, 15 and 30 after storage. According to the results, selenium NPs significantly reduced acrosome membrane damage at 1 $\mu g.mL^{-1}$ concentration as well as the percentage of abnormal sperms. Plasma membrane cohesion, overall and progressive sperm motility, and survival rate were also increased. Malondialdehyde concentration, as an indicator of oxidative stress, was significantly reduced compared to the control group. The same decrement was observed in terms of lipid peroxidation extent (Nateq et al. 2020).

Using an IR dyeing machine, selenium NPs were deposited on polypropylene (PP) fabrics during a one-step process under hydrothermal conditions. Color strength and UV protection factor (UPF) were increased significantly in selenium NPs-deposited PP fabrics. These fabrics also demonstrated antibacterial activity against gram-positive bacteria (*S. aureus* and *Bacillus cereus*) and gram-negative bacteria, i.e., *E. coli* and *Pseudomonas aeruginosa*. No toxic effect was observed after incubation of the fabrics with WI-38 cell line. The combination of these features in addition to their conductivity makes the resulting fabric a suitable choice for use in the new generation of medical textiles, curtains and automotive interior parts made of PP (AbouElmaaty et al. 2021).

Antifungal and antibacterial effects of chemically produced selenium NPs were investigated after adding to quail diets. Effects of these NPs on growth, hematology profile, blood biochemical composition, immunity, and intestinal microbiota of carcass were also investigated. The results showed that with increasing selenium NPs concentration, its antibacterial activity was also increased. Among gram-positive and gram-negative bacteria, the highest growth inhibition zone and the lowest inhibitory concentration belonged to the three species of gram-positive bacteria: *S. aureus* (MTCC 1809), *Listeria monocytogenes* (ATCC 15313) and *B. cereus* (ATCC 11778).

32 *Mycosynthesis of Nanomaterials: Perspectives and Challenges*

Antimicrobial properties of the said selenium NPs were more significant than the precursor for NPs synthesis, i.e., sodium selenite.

Antifungal properties of selenium NPs against *Candida albicans* (ATCC 4862), *Candida glabrata* (ATCC 64677), *Candida parapsilosis* (ATCC 22019) and *Candida guilliermondii* (ATCC 6260) were investigated and proved, of which *C. albicans* was more sensitive to selenium NPs. Adding selenium NPs to the quail diet improved their growth in terms of body weight and weight gain. But it did not have a significant effect on the carcasses and their organs, except that it increased the weight of the liver compared to the control group. Platelet count and hemoglobin concentration were increased in quail blood, while other blood factors remained unchanged. Alanine aminotransferase (ALT) and lactate dehydrogenase (LDH), as well as creatinine levels, were increased compared to the control group. Aspartate aminotransferase (AST) and urea did not change much. Triglycerides, total cholesterol, and very-low-density lipoprotein (VLDL) levels were significantly reduced. On the other hand, the amount of high-density lipoproteins (LDH) was increased. Superoxide dismutase (SOD) activity, glutathione peroxidase activity and reduced-glutathione (GSH) level was increased significantly. Malondialdehyde levels decreased compared to the control group but immunoglobulin G (IgG), IgM and IgA concentrations were increased. Overall, the health and physiological status of quail-fed selenium appear to be improved in the presence of selenium NPs (Alagawany et al. 2021).

Selenium NPs effect on the antioxidant system was investigated when Tibetan gazelles (*Procapra picticaudata*) were deficient in terms of selenium micronutrient. The concentration of selenium in the blood and hair of the animals was very low compared to healthy animals due to severe selenium deficiency in soil and pasture grass. Selenium deficiency had led to a decrease in red blood cell count, hemoglobin and higher uric acid levels. Serum AST, ALT, LDH, creatine kinase (CK), SOD, glutathione peroxidase (GSH-Px) and catalase (CAT) activity were also decreased along with the total antioxidant capacity (T-AOC). Malondialdehyde was significantly higher than the healthy animal control group. Oral treatment with selenium NPs as a nutritional supplement increased selenium concentration significantly in the blood during 5 days. Blood selenium level was stabilized after 20 days of continuous feeding with nano-selenium at 1.5 mg.kg^{-1} dose per body weight. Antioxidant capacity was also greatly improved the same as GSH-Px and SOD activity (Shen et al. 2021).

Biosynthesis of selenium NPs with bacteria

The bacterial synthesis of selenium NPs has received much attention during the recent decade. Selenium NPs were made using *Proteus mirabilis* YC 801, which were highly resistant to selenite and isolated from the gut of an adult herbivorous insect *Monochamus alternatus*. This strain was able to successfully convert selenite into red selenium NPs. These NPs were 178.3 ± 11.5 nm in diameter and were produced extracellularly. In this process, NADH and NADPH were electron donors, and the cytoplasmic enzymes such as thioredoxin reductase and some similar proteins were responsible for reducing the ions to selenium NPs. In addition to the insect stomach, *P. mirabilis* is also found in abundance in natural waters and soil. Therefore, these bacteria are suitable options for selenium biotransformation, bioremediation and nature protection (Wang et al. 2018).

In order to study the possibility of optimizing antibacterial activity, selenium NPs were synthesized with the bacterial isolate of *Stenotrophomonas maltophilia* SeITEO2 and their activity was measured against several references or isolated bacteria. In a study by Cremonini et al., the effect of NPs coating on their performance was investigated. To this end, the protein coating of the bio-NPs was denatured using 10% SDS solution and boiling for 10 and 30 minutes. The greater the power of denaturing, the more organic carbohydrates and proteins of the nanoparticle surface lost, so that the NPs size increased from 181 nm in the untreated samples to 270 nm in the samples with the highest treatment. MIC values increased with increasing denaturation of the organic coating layer. Therefore, the antibacterial properties of selenium NPs depend on the presence/absence of coating and size, the smaller the size, the greater the inhibition of growth. Basically, the inherent nature of metals, their size and surface structure are the main factors determining the antibacterial properties of metal and metalloid NPs. The presence of a bio-cap surrounding the bio-NPs prevents them from aggregating and also changes the zeta potential to more negative values; while near-neutral values tend to coalesce (Cremonini et al. 2018).

Many studies have shown that enriching animal feed with organic selenium is preferable to bioavailability and efficiency compared to its inorganic source of sodium selenite. The possibility of using bacterial organic selenium as a selenium dietary supplement for humans and animals looks very promising, so the possibility of using bacterial organic selenium in animal nutrition and studying the antioxidant status of selenium was studied by Abd Alla. *Enterobacter cloacae* (three isolates) and *Klebsiella pneumoniae* (one isolate) of rumen origin, as well as *Stenotrophomonas maltophilia* of hot spring origin, were able to absorb approximately 50% of inorganic selenium and convert it into selenium-containing proteins. All of these proteins showed antioxidant properties *in vitro*. Selenium (organic and inorganic) increased animal life, antioxidant properties and accumulation of selenium in tissues. It also improved water storage capacity in meat and caused mRNA up-regulation of some selenoproteins. The antioxidant capacity of birds fed with organic selenium was better than that of inorganic selenium. These properties became even more effective with the consumption of vitamin E (Abd Alla 2018). Biogenic amorphous spherical Se nanoparticles were obtained using anoxygenic phototrophic bacteria, *Rhodobacter capsulatus* B10 and *Rhodobacter sphaeroides* 2R in sizes of 280 nm and 200 nm respectively. These amorphous NPs under ambient conditions of ultrasonic dispersion (500 W, 15 min) and autoclaving (121°C, 15 min) gradually started to transform into trigonal Se crystals (Komova et al. 2018).

Selenium salt reducing bacteria have been isolated from contaminated water of industrial pollution origin and domestic wastewater. In this regard, *Bacillus cereus* strain AJK3, which was able to reduce selenite to selenium, was isolated from contaminated water from industrial effluents and domestic wastewater. The resulting selenium NPs were spherical but polydisperse in terms of particle size distribution and their average size was 93 nm. Extracellular proteins released from bacterial cells and membrane and cell wall proteins were responsible for the metalloid reduction. Due to the binding of bacterial proteins on selenium NPs, their zeta potential had a

34 *Mycosynthesis of Nanomaterials: Perspectives and Challenges*

high negative charge (−31.1 ± 4.9 mV.). The diffused pattern of selected area electron diffraction (SAED) and XRD demonstrated the amorphous character of the Se NPs. These bacteria appear to be suitable for decontaminating selenite from contaminated natural spaces (Kora 2018).

Selenium NPs with an average diameter of 50 to 156 nm were synthesized using methicillin-resistant *E. coli*, *P. aeruginosa* and *S. aureus*. The NPs produced by these bacteria showed different antibacterial effects against the tested strains. Among them, NPs made by *S. aureus* had the lowest IC50 (246.48 ± 23.32 $\mu g.mL^{-1}$). The diameter of these NPs was 180 nm (Medina Cruz et al. 2018). Selenium NPs synthesized using *Bacillus* sp. Msh-1 was able to successfully inhibit fluconazole-resistant *C. albicans* at a concentration of 100 $\mu g.mL^{-1}$ and fluconazole-sensitive type at a concentration of 70 $\mu g.mL^{-1}$. Quantitative PCR assay showed that these NPs reduced the expression of fluconazole resistance genes CDR1 and ERG11 (Parsamehr 2017).

Stenotrophomonas maltophilia SeITE02 was isolated from the soil of *Astragalus bisulcatus* rhizosphere from legumes that could accumulate selenium. *Ochrobactrum* sp. MPV1 was isolated from the waste material of the roasted pyrite landfills. These two bacteria were used as cell factories to make selenium nanostructures and to determine the biological stages of selenium synthesis and release during this process. Both isolates were able to use glucose and pyruvate as carbon and energy sources. MPV1 cells grown in the presence of glucose and to a lesser extent those grown in the presence of pyruvate entered the death phase after 24 and 72 hours, respectively. The addition of SeO_3^{2-} reduced biomass over time. In the early stages, neither bacterium reduced oxyanion, which was probably due to SeO_3^{2-} toxicity. Toxicity due to SeO_3^{2-} caused the intracellular production of ROS and resulted in cell death. But in later stages, SeITE02 synthesized extracellular NPs and MPV1 synthesized intracellular selenium nanomaterials (Piacenza et al. 2018).

Selenium nanowires with a diameter of 10 to 30 nm were fabricated using a probiotic bacterium, *Bacillus licheniformis*, microbial exopolymer (MEP). These one-dimensional nanomaterials, at a concentration of 100 $\mu g.mL^{-1}$, were able to inhibit Gram-positive and Gram-negative bacteria and at a concentration of 75 $\mu g.mL^{-1}$ also inhibited biofilm formation. Antioxidant properties were investigated using DPPH [2,2-diphenyl-1-picryl hydrazyl], i.e., a free radical probe with hydrogen accepting characteristics toward antioxidant. The level of antioxidant activity of scavenging activity was directly related to the concentration of nanomaterials and its maximum was observed at a concentration of 100 $\mu g.mL^{-1}$. But this property was significantly lower than the positive control, ascorbic acid. These nanomaterials were relatively toxic in hemolytic acidosis. Values of LC50 less than 10 $\mu g.mL^{-1}$ were also tested on *Culex quinquefasiatus* 3rd instar larvae and *Aedes aegypti* larvae, which are Zika virus vectors. With histological techniques, the damage done to the insect's midgut was examined. It seems that these nanomaterials can be used to control arbovirus-carrying insects (Abinaya et al. 2019). The antioxidant properties of selenium NPs synthesized by *Lactobacillus casei* ATCC 393 were evaluated on intestinal epithelial barrier dysfunction. To do this, a model was made of human mucosal epithelial cells in the colon that had oxidative damage. According to the results, pretreatment with selenium NPs reduces the destructive effects of H_2O_2 on mitochondria. To evaluate the effect of selenium NPs on intestinal epithelial permeability, the effect of selenium NPs on the passage of fluorescein isothiocyanate-dextran (FITC-dextran) through the NCM460 membrane, and also their effect on transendothelial electrical resistance (TER) were investigated. FITC-dextran secretion increased in the presence of selenium NPs. It also reduced the electrical resistance caused by H_2O_2, indicating a reduction in oxidative stress due to H_2O_2. Thus, the selenium NPs produced by *L. casei* ATCC 393 can protect cells from oxidative stress, promote intestinal epithelial cell survival, improve mitochondrial function, and maintain intestinal epithelial barrier cohesion (Cremonini et al. 2018). Diclofenac, a non-steroidal anti-inflammatory drug, is resistant to biodegradation processes. Selenium NPs obtained from actinomycetes in the presence of UV light was used to remove diclofenac. A concentration of 32 µg.mL^{-1} of spherical selenium NPs (particle size between 48.2 and 135.7 nm), the presence of H_2O_2 0.05 mM at pH 5.5 and UV light intensity of 30 W.M^2 were the optimal conditions for diclofenac decomposition (Ameri et al. 2020). Amorphous selenium NPs in size of approximately 130 nm were produced by *Lysinobacillus* sp. NOSK via selenite (SeO$_3{}^{2-}$) reduction. These NPs inhibited the growth of *E. coli* and *S. aureus* and inhibited the production of *P. aeruginosa* biofilms while did not show toxicity to keratinocytes (HaCaT) even at the highest concentrations (San Keskin et al. 2020). Selenium crystalline spherical NPs with a diameter of 106.1 nm were synthesized using *Lactobacillus pentosus* ADET MW861694. The activity of these NPs was tested against a number of food pathogens, with the highest activity observed against *Salmonella arizonae* bacteria (Christianah et al. 2021).

Spherical selenium NPs (46-nanometer diameter in average) were fabricated using *Lactobacillus acidophilus* ML14 to be used to control crown/root rot diseases (CRDs) of wheat caused by *Fusarium*. These NPs at concentrations of 20 to 40 µg.mL^{-1} completely inhibited the growth of *Fusarium*. The radical scavenging activity of selenium NPs was 92% against 2,2′-azinobis-(3-ethylbenzothiazoline-6-sulfonic acid) ABTS radicals and 88% against DPPH radicals. Under greenhouse conditions, when wheat was seeded with selenium NPs at a concentration of 100 µg.mL^{-1}, its growth and yield increased significantly and the amount of CRDs decreased by up to 75% (El-Saadony et al. 2021). Interestingly, *Fusarium* has previously been used as a manufacturer of selenium nanoparticles (Asghari-Paskiabi et al. 2018).

Biosynthesis of selenium NPs through fungi

Fungi play an important role in the biosynthesis of nanoparticles and have also been used in the synthesis of selenium nanoparticles. Selenium is important in many ways in the physiology of the human body. It is absorbed in the human body through the amino acid selenocystein and is involved in mechanisms involved in antioxidant defense, immunity, redox signaling and thyroid metabolism. At the same time, if selenium concentration in the body passes a threshold, it can cause the formation of reactive oxygen species. Therefore, its excess is toxic to the body and damages the liver, spleen, pancreas, kidneys, and heart, and may cause cancer. The body's selenium is usually supplied from plant selenomethionine. The predominant oxidation state in selenium is Se IV and VI, which are found in the form of selenite (SeO$_3{}^{2-}$) and selenate (SeO_4^{2-}) in solution. These oxyanions can be reduced to elemental selenium with the help of reducing agents. Elemental Se is the most thermodynamically stable allotrope of selenium which appears as a gray material. This material is called gray, black, metallic, gamma, or trigonal selenium. This insoluble selenide as Se (–II) is also stable under acidic and reducing conditions. Biogenic selenium is initially found in the amorphous red phase. In the case of organic selenium compounds, the most common oxidation state is with Se (–II), which is usually synthesized by microorganisms and accumulated in them or in biological tissues (Ruocco 2011).

Statins are designed to competitively inhibit 3-hydroxy-3-methyl glutaryl coenzyme A (HMG-CoA) reductase, a cholesterol biosynthesis enzyme, which consequently lowers the blood cholesterol level. The mixture of selenium and lovastatin NPs, one of the first statins approved to lower blood cholesterol, has antioxidant properties and scavenging free radicals activity. The oral administration of this compound was tested in strains that suffered from oxidative damage due to gamma radiation. According to the results, the oxidative status of protein carbonyl group, thiobabituric acid reactive substances (TBARS), catalase (CAT), SOD, and xanthine oxidoreductase (XOR) in the heart improved. GSH, nitric oxide (NO) and blood selenium levels also improved dramatically after administration of the compound. Therefore, lovastatin-Se mixture can reduce lipid and protein oxidization (El-Batal et al. 2012).

The effect of probiotics *Aspergillus awamori* with or without selenium NPs was examined on the growth and other vital characteristics of broilers. The results showed that body weight and chest muscle weight increased compared to the control group, while food absorption decreased. It also showed an increase in protein digestibility and energy. Glucose, triglycerides and plasma total cholesterol were reduced. Serum total protein and HDL were increased, and α-tocopherol and muscle total fat were increased (Saleh 2014).

Phanerochaete chrysosporium was evaluated for its ability to reduce oxidized forms of selenium, i.e., selenite and selenate. Both salts affected the fungus, but the effect of selenite was stronger at a concentration of 10 mg.L^{-1} where it inhibited the growth of the fungus. The fungus absorbed 40% of the selenite in the environment, while this value was less than 10% for selenate. In these processes, *P. chrysosporium* was able to reduce selenite to elemental selenium, but this did not happen with selenate. Selenium NPs ranging in size from 30 to 400 nm were produced inside the fungal cells and accumulated in mycelia. Selenite at the lowest concentration, i.e., 2 mg.L^{-1}, also reduced the growth and decreased the dry weight of the fungal biomass. This decreasing trend increased with increasing in selenite concentration. Selenite also affected the morphology of *P. chrysosporium* pellets and made them smaller. Pellets are usually formed due to the interaction of hyphae, spores and (nano)particles when the cell is under stress. The presence of selenite, causing oxidative stress in the fungus, also affected the pellets, making them more compact and softer than in the absence of selenite. The ultimate goal of using these fungi to decontaminate the selenium oxyanions of contaminated effluents seems to be achievable (Espinosa-Ortiz et al. 2015). But can the combination of selenium NPs and the pellets remove selenium and heavy metals from wastewater at the same time? Espinosa-Ortiz et al. (2016) performed this experiment on zinc. To this end, the adsorption capacity of selenium NPs-pellet to remove zinc from aqueous media was investigated. The capacity of selenium NPs to absorb zinc increased in the presence of pellets. In this process, the initial concentrations of zinc, biomass, and the environment pH were decisive (Espinosa-Ortiz et al. 2016).

Antifungal effect of silver and selenium NPs against *Alternaria solani*, a fungal pathogen for potatoes, showed that the fungus stopped growing at low concentrations of silver and selenium NPs. While silver NPs appear to be a viable alternative to conventional fungicides, selenium NPs, with their antioxidant properties, enhance plant immunity (Ismail et al. 2016). The ability of marine *Aspergillus terreus* to absorb selenium and recover in a bioreactor was proven. It was found that the fungus was able to absorb selenium during 120 hours of contact, at a pH of 6–7 with an efficiency of 86%. Therefore, this fungus possesses the ability for bioremediation and selenium recovery from the marine environment (Raja et al. 2016).

Selenium NPs, in addition to anti-cancer, antioxidant and antibacterial properties, also have antifungal properties that can be used to protect foods from contamination by fungi and their toxins. The effect of selenium NPs with a diameter of 32 nm on the growth and toxin production of three toxic fungal species was investigated. At concentrations of 3000, 7000 and 9000 $\mu g.mL^{-1}$, the growth of *Aspergillus nidulans*, *Aspergillus parasiticus* and *Aspergillus ochraceus* was completely stopped. Production of sterigmatocystin, aflatoxins, and ochratoxin A was completely stopped at concentrations of 800, 200, and 200 $\mu g.mL^{-1}$, respectively. These findings suggest that selenium NPs can play an important role in the safety of food and agricultural products (Abdel-Kareem and Ahmed Zohri 2017).

The fungicidal properties of chitosan and selenium NPs were considered to prevent future biofilm-related infections against *C. albicans*. Selenium NPs were fabricated by irradiation method with the help of pulsed laser-ablation in chitosan-containing liquid medium to make chitosan capping agent of selenium NPs. Chitosan and selenium NPs (IC50 21.7 ppm) alone showed antifungal properties against *C. albicans* adult biofilms and when combined (IC50 3.5 ppm), their inhibition, as well as synergistic interaction, increased with the dose used. Doses lower than IC50 were antagonistic (Lara et al. 2018).

Selenium nanocomposites were fabricated with the help of *Lentinula edodes*, *Grifola umbellata*, *Ganoderma lucidum*, *Pleurotus ostreatus* and *Laetiporus sulphureus*. Antibacterial effect of these nanocomposites was tested against a gram-positive bacterium called *Clavibacter michiganensis* subsp. *sepedonicus*. The most antibacterial properties were related to selenium nanocomposites produced extracellularly by *L. edodes* and *G. lucidum*. Selenium nanocomposites, depending on the type, had different and significant effects on reducing the ability of *Clavibacter michiganensis* to produce biofilms (Perfileva et al. 2018).

Selenium NPs were also synthesized using *Trichoderma harzianum*. This fungus itself has antifungal activity. In fact, *Trichoderma* fungi play an important role in the food cycle and improve soil health by controlling environmental microbes biologically. In some cases, they can even alter mycotoxins into their sulfated and glycosylated derivatives or cause down-regulation of mycotoxin synthesis genes. At the same time, they produce peptides, enzymes and organic compounds that are able to synthesize NPs and increase their antifungal properties. In this way, they

38 *Mycosynthesis of Nanomaterials: Perspectives and Challenges*

can control mycotoxin contamination more efficiently. The interaction of selenium NPs produced by the fungus *T. harzianum* and its metabolites increased the control feature of the function of deoxynivalenol, *Alternaria* toxins and fumonisin B1 toxins and the expression of *TR15, TR16, FUM1* and *PA* genes, which are key genes for the synthesis and production of mycotoxins, was significantly reduced (Hu et al. 2019). Hexagonal crystalline spherical and quasi-spherical selenium NPs with an average size of 87.82 ± 2.71 nm were synthesized using a cell-free extract of a novel yeast called *Magnusiomyces ingens*. These NPs were able to stop the gram-positive bacterium *Arthrobacter* sp. W1, but showed no effect on the gram-negative bacterium *E. coli* BL21 (Lian et al. 2019).

The feasibility of reducing the oxyanions of selenium and tellurium metalloids to NPs was investigated by a number of fungal strains. After 10 days of incubation, *Mortierella humilis* and *Aureobasidium pullulans* were able to synthesize selenium NPs with an approximate diameter of 48 nm (at 1463 µg.L^{-1}) and 60 nm (at 1079 µg.L^{-1}) from selenate, respectively. Nanoparticle diameters were measured with single particle inductively coupled plasma mass spectrometry (SP-ICPMS). The metalloids were first placed in the hyphal matrix and then precipitated as NPs by interacting with proteins and other extracellular polymeric materials. Therefore, it is possible to absorb tellurium and selenium from the environment with these fungi and convert them into NPs to recover them from the environment (Liang et al. 2019). Selenium NPs with an average diameter of 126 nm were synthesized extracellularly through *Bacillus cereus* isolates isolated from soil. The best production performance at a concentration of 6.4 mM of SeO$_2$ was obtained in 72 hours at 33°C and pH 9. The resulting selenium NPs showed strong antioxidant activity. The antioxidant activity of selenium NPs at a concentration of 200 µg.L^{-1} was about 37.6% and with increasing the concentration to 400 µg.L^{-1}, its activity reached 56.5% (Akçay and Avcı 2020). *Phoma glomerata*, a pathogenic fungus was used as a model organism to reduce selenite or tellurite into NPs internally and extracellularly. The fungus also interacted directly with selenium and tellurium in sulfide and volcanogenic sediments. Therefore, *P. glomerata* can accumulate these metabolites in environment for the purpose of bioremediation, biorecovery, and immobilization (Liang et al. 2020).

Biological synthesis of selenium NPs by yeasts

Yeasts and microfungi play an important role in the selenium distribution cycle. A few scientific reports are summarized in Table 1 to provide a base to compare different approaches utilized in the biosynthesis of selenium nanoparticles by yeasts. A marine yeast, *Rhodotorula mucilaginosa* strain 13B was isolated from the sludge sediments of the salt marshes of the mid-Atlantic coast of the United States. This microorganism was able to reduce selenium oxyanion to elemental selenium. Elemental selenium accumulated during the exponential and stationary phases of the yeast. More than 20% accumulated in cell pellets in aerated cultures and 1% dispersed in volatile form, indicating the role of this microorganism in the selenium cycle, i.e., the reduction of selenium oxyanion to elemental selenium (Ruocco 2011).

The effect of dissolved oxygen on selenium/protein production in *Saccharomyces cerevisiae* was investigated. According to the results comparing

Table 1. A Comparison of biosynthesis of selenium nanoparticles by yeasts.

Yeast	Se NP Size (nm)	Se NP morphology	Precursor	Temp. (°C)	Duration Time (h.)	Application	Intra/Extra Cellular	References (Year)
Saccharomyces cerevisiae	100	Spherical	selenite	30	-	-	extra	Zhang et al. (2012)
Yarrowia lipolytica	30–60	Spherical	Sodium selenite	30	-	Growth and survival of *Artemia salina*	extra	Hamza et al. (2017)
Pichia pastoris	70–180	Spherical	Selenium dioxide	30	96	-	intra	Elahian et al. (2017)
Saccharomyces cerevisiae	20–30	Spherical	Sodium selenite	30	48	-	intra	Pereira et al. (2018)
Nematospora coryli	50–250	spherical	Sodium selenite	28	48	Anti-candida/anti-oxidant activities	intra	Rasouli (2019)
Magnusiomyces ingens	70–90	Spherical/ quasi-spherical	Selenium dioxide	30	24	Antimicrobial	extra	Lian et al. (2019)
Saccharomyces cerevisiae var. boulardii	235	-	Selenium dioxide	35	12	-	intra	Bartosiak et al. (2019)
Saccharomyces cerevisiae	2.3–6.1	Spherical	Sodium selenite & Sodium selenate	28	168	-	intra	Garcia et al. (2020)
Saccharomyces cerevisiae var. boulardii	235	Spherical	Selenium dioxide	35	12	-	intra	Borowska et al. (2020)
Candida utilis	20–30	Spherical	Sodium selenite	-	24	Study of Se metabolism in yeast cells	-	Kieliszek et al. (2020)
Saccharomyces cerevisiae	75–709	Spherical	Sodium selenite	32	96	-	intra	Faramarzi et al. (2020)
Kluyveromyces lactis	80–150	Spherical	Sodium selenite	30	24	Anti-inflammatory	intra	Song et al. (2021)
Baker's yeast	71	Spherical	Sodium selenite	30	36	Antioxidant activity	intra	Wu et al. (2021)
Baker's yeast	4 & 51	Spherical	Sodium selenite	30	24	Anti-foodborne pathogens activity	extra	Salem (2022)

aerobic and anaerobic fermentation conditions, when the yeast was exposed to a limited amount of oxygen, it converted more selenite to extracellular selenium NPs. It seems that a certain amount of oxygen is required for the enzymatic reduction of selenite to elemental selenium, and amounts higher or lower than the optimal value are considered a disadvantage. These NPs, along with proteins, were removed from *Saccharomyces* cells as vesicle-like structures. Apparently, selenium accumulated initially in the vacuolar portions of the *Saccharomyces* cell where it was reduced to elemental selenium as an internal detoxification mechanism. For selenium NPs to be released from the vesicles, energy was required, which obligates the presence of oxygen at that specific stage and caused respiratory activity to take place. After the vesicles left the cell, the biological molecules acted to break down the structure of the vesicles (Zhang et al. 2012). Figure 1A shows these vesicles lined up below the cell membrane. Yeast cell walls and membranes are completely darkened compared to control (1B) due to selenium uptake (Asghari-Paskiabi et al. unpublished data).

In order to improve the health of aquaculture species, *Yarrowia lipolytica* strain was used to synthesize selenium NPs and evaluate its effect on *Artemia salina* shrimp. When *A. salina* was fed the biomass of *Y. lipolytica*, rich in selenium NPs, life expectancy increased to 96.66% compared with the control group of *S. cerevisiae* (60%). The size of larvae fed in this way was also larger compared to the control group. The larvae survived up to 70% against Vibrio infection. The combination of selenium NPs and *Y. lipolytica* can protect farmed aquatic animals through an eco-friendly method (Hamza et al. 2017). The amount of selenium (S^0) in the yeast was quantitatively measured and the presence of selenium NPs in selenium-rich yeasts was confirmed, examined, and identified by SP-ICPMS. This technique provides information about the mass of the element in each nanoparticle, which, in addition to the information given by TEM and EDX about composition, shape, and density, can be used to obtain particle size. The instrument showed that in selenium-rich yeasts, selenium NPs were present in sizes between 40 and 200 nanometers (Jiménez-Lamana et al. 2018).

The researchers looked at selenium in the yeast *S. cerevisiae*. This yeast which contains selenomethionine, is used as dietary supplements for animals and humans. To prepare them, the cells were incubated with sodium selenite and then dried. Large amounts of selenomethionine were produced inside them, and

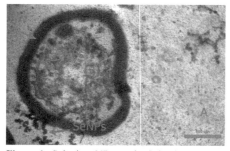 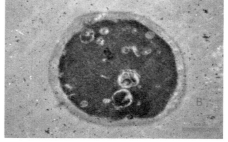

Figure 1. Selenium NPs synthesis by *S. cerevisiae*. The yeast cell wall and membrane were completely darkened when exposed to selenium salts due to selenium uptake (A). Cells in the control group that were not exposed to the selenium salt (B). (Asghari-Paskiabi et al. unpublished data).

approximately 48 hours after exposure to sodium selenite, spherical amorphous selenium NPs were also detectable in approximately 20 to 30 nm in size. After these NPs were synthesized inside the cell, they left the cells over time and gradually resembled germination. These yeasts can easily introduce selenium NPs into nature, and organisms can take advantage of the non-toxicity of this element over equal doses of its bulk (Pereira et al. 2018). The yeast *S. cerevisiae* was also used to produce selenium sulfide NPs. To this end, a combination of selenious acid substrates (as a source of selenium) and ammonium sulfate (as a source of sulfur), as well as an independent substrate of sodium selenosulfate (as a source of sulfur and selenium) were used, of which, selenium sulfide NPs with an average size of 153 and 6 nm, measured by SEM, were synthesized, respectively. These NPs were able to stop the growth of *Aspergillus fumigatus, Aspergillus flavus, Alternaria alternata, Microsporum canis, Trichophyton rubrum,* and *Candida krusei* (Asghari-Paskiabi et al. 2019).

Functionalization of the nanoparticle surface and the ability to form a corona through the functionalization process plays an important role to decide on the type of nanoparticle applicable in specific targeted therapy, heavy metal toxicity, and drug delivery. Using microwave plasma optical emission spectrometry technique, the chemical composition of the selenium NPs surface and their particle distribution alone and in conjugated state with human albumin serum, in the form of nano-powder, were investigated. It was found that selenium NPs, when synthesized with the help of microwave and in the presence of yeast extract, are four times more efficiently conjugated to human serum albumin compared to their synthesis using live yeast cells. The binding rate was 16% when yeast extract was used and 4% when live yeast cells were used (Borowska et al. 2020).

The effect of selenium on the growth of the yeast *Candida utilis* grown in a selenium-rich environment was investigated. Selenium exerted an inhibitory effect on fungal growth and at the same time, the activity of antioxidant enzymes including thioredoxin reductase, glutathione reductase, glutathione S-transferase, and glutathione peroxidase increased significantly compared to the control group that grew without selenium addition. During this process, selenium NPs were also fabricated, possibly due to stress conditions and detoxification operations (Kieliszek et al. 2020).

A genetic engineering approach was designed for a yeast, *Pichia pastoris* in order to overexpress the enzyme cytochrome b5 reductase (Cyb 5 R) of the fungus *Mucor racemosus* metal resistant. This yeast is used to synthesize silver and selenium nanoparticles. *M. racemosus* was isolated from polluted industrial effluents and wastewaters. Cloning of Cyb 5 R enzyme caused *P. pastoris* to grow in the presence of metal ions and was able to produce metal nanoparticles by presenting cofactors and coenzymes in the form of eco-friendly. Wild parental yeast strains were resistant to the oxyanion only up to 1 mM. Through this experiment, silver nitrate and selenium oxide were used at 0 to 40 mM concentrations to study the growth kinetics of these yeasts. The transformants were able to grow maximally and reduce ions to nanoparticles in the presence of silver nitrate (6 mM) and selenium oxide (4 mM). This efficient biotransformation and high-efficiency bioabsorption led to the production of 70 to 180 nm silver and selenium nanoparticles with the

42 *Mycosynthesis of Nanomaterials: Perspectives and Challenges*

help of *P. pastoris*. These nanoparticles showed at least 10 times less toxicity against human epithelial breast cancer (T47D), human gastric carcinoma (EPG 85–257), and primary human dermal fibroblasts (HDF) cell lines compared to selenium dioxide and silver nitrate (Elahian et al. 2017).

Biosynthesis of selenium nanoparticles by *S. cerevisiae* was investigated from a point of view to the type of initial substrate required and effect of the initial concentration of the substrate also the properties of the obtained selenium nanoparticles. Selenium nanoparticles were fabricated intracellularly. The mean size of the nanoparticles was in their minimal (75 µm) and their antioxidant activity was 48.5%, at the lowest amount of sodium selenite (5 µg) added. More uniform nanoparticles were obtained at the highest (25 µg) amount of selenite used but average diameter of the nanoparticles increased to 709 nm and the antioxidant activity decreased to 20.8 (Faramarzi et al. 2020).

In another study, glucose, ascorbic acid, and the yeast *Saccharomyces boulardii* were used separately as green chemical reductants for the synthesis of selenium nanoparticles. Using the continuous photochemical vapor generation (PCVG) technique paired with microwave-induced plasma (MIP) and UV spectrophotometry, a method for direct monitoring of selenium nanoparticle biosynthesis was developed. When glucose was used to synthesize selenium nanoparticles, selenium ions were reduced to spherical nanoparticles with 120 nm and glucose was oxidized to gluconic acid. This study showed that the reaction time and glucose concentration had a significant role in the efficiency of selenium nanoparticle biosynthesis. When ascorbic acid was used as a reducing agent, ascorbic acid was oxidized to dihydroascorbic acid during the production process of selenium nanoparticles. With the increasing concentration of this powerful reductant, the nanoparticle size increased from 18 nm to 118 nm. When *S. boulardii* was used to produce selenium nanoparticles, there was no need to add a stabilizing agent due to the presence of natural organic molecules in the yeast culture medium. After synthesis, the nanoparticles were separated from the yeasts. The average nanoparticle size was 235 nm, and the particles were uniform with a dispersion index of less than 0.1 (Bartosiak et al. 2019).

The yeast *Kluyvermyces lactis* 66799 was able to successfully synthesize selenium nanoparticles intracellularly with a diameter of 80 and 150 nm using sodium selenite as a substrate. The isolated nanoparticles had protein and carbohydrate corona covering. When these bioparticles were added to the diet of mice at a dose of 0.6 mg selenium per kg, they were able to reduce oxidative stress as well as intestinal inflammation usually in ulcerative colitis (UC) made by dextran sodium sulfate in mice animal model (fan Song et al. 2021). Spherical selenium nanoparticles (4–51 nm in diameter) made by the baker's yeast were able to successfully work against foodborne pathogens, *A. niger*, *A. fumigatus*, *E. coli,* and *S. aureus* (Salem 2022). Spherical selenium nanoparticles with a diameter of 50–250 nm were synthesized using a yeast called *Nematospora coryli* and were able to show antifungal activity against *C. albicans*. They were also able to neutralize free radicals and DPPH by donating protons, indicating their potency as an antioxidant (Rasouli 2019).

A genetic engineering approach was designed for a yeast, *Pichia pastoris* in order to overexpress the enzyme cytochrome b5 reductase (Cyb 5 R) of the fungus *Mucor racemosus* metal resistant. This yeast is used to synthesize silver and selenium nanoparticles. *M. racemosus* was isolated from polluted industrial effluents and wastewaters. Cloning of Cyb 5 R enzyme caused *P. pastoris* to grow in the presence of metal ions and was able to produce metal nanoparticles by presenting cofactors and coenzymes in the form of eco-friendly. Wild parental yeast strains were resistant to the oxyanion only up to 1 mM. Through this experiment, silver nitrate and selenium oxide were used at 0 to 40 mM concentrations to study the growth kinetics of these yeasts. The transformants were able to grow maximally and reduce ions to nanoparticles in the presence of silver nitrate (6 mM) and selenium oxide (4 mM). This efficient biotransformation and high-efficiency bioabsorption led to the production of 70 to 180 nm silver and selenium nanoparticles with the help of *P. pastoris*. These nanoparticles showed at least 10 times less toxicity against human epithelial breast cancer (T47D), human gastric carcinoma (EPG 85–257), and primary human dermal fibroblasts (HDF) cell lines compared to selenium dioxide and silver nitrate (Elahian et al. 2017).

Biosynthesis of selenium nanoparticles by *S. cerevisiae* was investigated from a point of view to the type of initial substrate required and effect of the initial concentration of the substrate also the properties of the obtained selenium nanoparticles. Selenium nanoparticles were fabricated intracellularly. Mean particle size of the nanoparticles were in their minimal (75 μm) and their antioxidant activity was 48.5%, at the lowest amount of sodium selenite (5 μg) added. More uniform nanoparticles were obtained at the highest (25 μg) amount of selenite used but average diameter of the nanoparticles increased to 709 nm and the antioxidant activity decreased to 20.8 (Faramarzi et al. 2020).

Mechanism of biosynthesis of selenium nanoparticles by yeasts

Usually, the concentration of the element must be adjusted to about 1–1000 ppm in the yeast culture medium to obtain the desired nanoparticles via biosynthesis by the yeasts. This roughly estimated concentration range can be recommended provided that the element is not toxic to the yeast at that concentration. The reduction of metal salts to metal nanoparticles may be governed by metabolites like proteins, phenols, alkaloids, terpenoids or other reducing molecules containing appropriate functional groups like amine, amide, or carbonyl functional groups. Due to the rapid growth of yeasts and easy access to the nutrients required for their cultivation, these microorganisms are good candidates for the mass production of nanoparticles in general. Quinones and membrane-bound oxidoreductases appear to be the major reductants. Also, elevated pH levels within the yeast's environment contribute to the activity of oxidoreductase enzymes. The cascade of enzymatic activity can be attributed to the response of the organism to metal-induced stress, during which glutathione and phytochelatin synthase, which are both responsible for relieving internal stress, are produced. Both of these compounds show nucleophilic and redox properties, and ultimately the reduction works in favor of nanoparticle formation (Skalickova et al. 2017).

In previous research, a bakery yeast that was used to synthesize selenium nanoparticles produced spherical and amorphous nanoparticles with a diameter of about 71 nm. By adding 0.3–0.8 mg.kg^{-1} of selenium nanoparticles to the diet of mice that had previously received cyclophosphamide (CTX), an immunosuppressive drug. According to the findings, the health status of the mice was significantly improved which can be attributed to their ability to respond to stress and also increased level of immunity. The nanoparticles protected mice from damage by ROS and free radicals. As a cofactor, selenium nanoparticles improved the activity of antioxidant enzymes SOD, GSP-X, and catalase. These enzymes scavenge free radicals and reduce MDE, indicating that they delay fat peroxidation. Biochemical indexes of the spleen, liver and kidney health of the rats were also improved, indicating that selenium NPs can reduce the immunosuppression effect induced by CTX in rats. Serum immunoglobulins also increased and humoral immunity improved. Due to the antibiotic characteristics of selenium nanoparticles, it seems that their application in animal diet can play an effective role in animal health (Figure 2) (Wu et al. 2021).

The growth kinetics of *S. cerevisiae* was investigated by the same authors of this chapter by counting the colony-forming units (CFU). CFU counts were correlated to the biosynthesis of selenium sulfide NPs at 5-time points after inoculation at 100-minute intervals. The precursor added to the culture medium was a combination of selenious acid and sodium sulfite. According to the results, NPs produced during the process affect their own cellular factories in turn and reduce the CFU. In addition, it was found that sulfite reductase activity, the main enzyme-producing selenium sulfide NPs, increased with decreasing cell population. The expression of the main genes regulating sulfite reductase enzyme, namely Met 5 and Met 10, changed according to the cell population. NPs cause toxicity in the yeast cells by altering the redox state and creating oxidative stress through ROS which can damage proteins, DNA, lipids and cellular carbohydrates when exceeded a certain concentration level. Genes responsible for managing oxidative stress response including GST and SOD2 genes were also activated at times the yeasts confront ROS. The said genes are responsive to stress and superoxide dismutase activity. One of the main mechanisms of metal toxicity is GSH deficiency, which is at the forefront of cell antioxidant activity. The same is true for NPs (Figure 3).

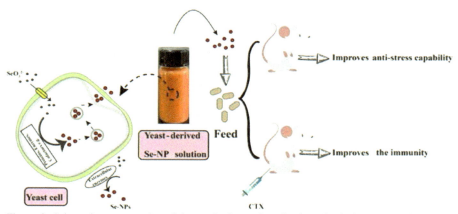

Figure 2. Schematic representation of the production and application of selenium nanoparticles with yeast. (Wu et al. 2021, reproduced from ACS Omega, an open access journal under Creative Commons public use license).

Selenium is absorbed through binding to carbohydrates positioned in the hydrophobic cell wall of the yeasts in a direct proportion to its concentration in the environment up to a limit. Selenium is naturally present in the structure of GSH peroxidase enzyme and increases its enzymatic activity and thus plays a significant role in cellular defense against oxidative stress. GSH peroxidase activity depends on GSH, the main constituent of which is sulfite reductase. Therefore, sulfite reductase activity is increased due to GSH deficiency, which is in favor of reducing selenium ions to selenium sulfide NPs. For this reason, the cell population decreased over time and with an increment of the NPs concentration. Selenium, therefore, played a dual role in this process. On the one hand, it helped the antioxidant system, and on the other hand, it caused increased production of the NPs and their ROS properties, and consequently the consumption of glutathione and the need for more selenium (Asghari-Paskiabi et al. 2020).

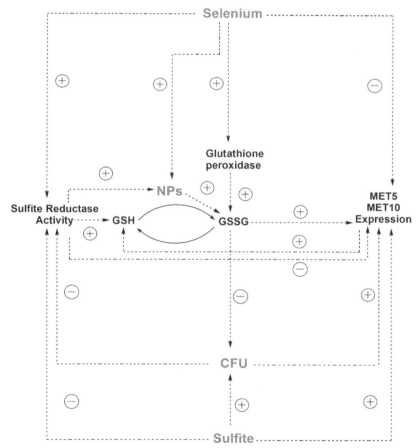

Figure 3. Correlation between nanoparticle synthesis and enzymes involved in yeast cell antioxidant activity (Asghari-Paskiabi et al. 2020, reproduced from Frontiers in Microbiology, an open access journal under e Creative Commons Attribution License (CC BY)).

Conclusions

Selenium NPs are superior to bulk selenium in meeting the health needs of various species. It can be produced by chemical or biological methods. Its bio-production has the certain advantage of serving the environment by bioremediation as an example. In fact, in addition to being a less risky method, it can decontaminate the environment. There are numerous reports of biosynthesis of selenium NPs using plant extracts, bacteria, and fungi as cell factories. Physicochemical properties of these NPs are characterized and the optimal growth conditions of their productive microorganisms are also obtained. But, we still do not know how they can be produced and put into operation in an industrial scale. It is hoped that this will happen in the near future.

References

Abd Alla, D.A.A.M. 2018. Selenium-enriched bacterial protein as a source of organic selenium in broiler chickens. Ph.D. dissertation, Universiti Putra Malaysia.

Abdel-Kareem, M.M., and Ahmed Zohri, A.A. 2017. Inhibition of three toxigenic fungal strains and their toxins production using selenium nanoparticles. Czech Mycol. 69(2): 193–204.

Abinaya, M., Vaseeharan, B., Rekha, R., Shanthini, S., Govindarajan, M., Alharbi, N.S., Kadaikunnan, S., Khaled, J.M., and Al-Anbr, M.N. 2019. Microbial exopolymer-capped selenium nanowires– Towards new antibacterial, antibiofilm and arbovirus vector larvicides? J. Photochem. Photobiol. B: Biol. 192(5): 55–67.

AbouElmaaty, T., Abdeldayem, S.A., Ramadan, S.M., Sayed-Ahmed, K., and Plutino, M.R. 2021. Coloration and multi-functionalization of polypropylene fabrics with selenium nanoparticles. Polymers 13(15): 24–45.

Akçay, F.A., and Avcı, A. 2020. Effects of process conditions and yeast extract on the synthesis of selenium nanoparticles by a novel indigenous isolate *Bacillus* sp. EKT1 and characterization of nanoparticles. Arch. Microbiol. 202(8): 2233–2243. https://doi.org/10.1007/s00203-020-01942-8.

Alagawany, M., El-Saadony, M., Elnesr, S., Farahat, M., Attia, G., Mahmoud, M., Madkour, M., and Reda, F. 2021. Use of chemical nano selenium as antibacterial and antifungal agent in quail diets and its effect on growth, carcasses, antioxidant, immunity and caecal microbe. Res. Square 23(1): 1–21.

Ameri, A., Shakibaie, M., Pournamdari, M., Ameri, A., Foroutanfar, A., Doostmohammadi, M., and Forootanfar, H. 2020. Degradation of diclofenac sodium using UV/biogenic selenium nanoparticles/ H_2O_2: Optimization of process parameters. J. Photochem. Photobiol. A: Chem. 392(4): 1123–1182.

Ananth, A., Keerthika, V., and Rajan, M. R. 2019. Synthesis and characterization of nano-selenium and its antibacterial response on some important human pathogens. Curr. Sci. 116(2): 28–41.

Asghari-Paskiabi, F., Imani, M., Razzaghi-Abyaneh, M., and Rafii-Tabar, H. 2018. *Fusarium oxysporum*, a bio-factory for nano selenium compounds: synthesis and characterization. Sci. Iran. 25(3): 1857–1863.

Asghari-Paskiabi, F., Imani, M., Rafii-Tabar, H., and Razzaghi-Abyaneh, M. 2019. Physicochemical properties, antifungal activity and cytotoxicity of selenium sulfide nanoparticles green synthesized by *Saccharomyces cerevisiae*. Biochem. Biophys. Res. Commun. 516(4): 1078–1084.

Asghari-Paskiabi, F., Imani, M., Eybpoosh, S., Rafii-Tabar, H., and Razzaghi-Abyaneh, M. 2020. Population kinetics and mechanistic aspects of *Saccharomyces cerevisiae* growth in relation to selenium sulfide nanoparticle synthesis. Front. Microbiol. 11(2): 1–11.

Badr, N.B., Al-Qahtani, K.M., and Mahmoud, A. 2020. Factorial experimental design for optimizing selenium sorption on *Cyperus laevigatus* biomass and green-synthesized nano-silver. Alex. Eng. J. 59(6): 5219–5229.

Bartosiak, M., Giersz, J., and Jankowski, K. 2019. Analytical monitoring of selenium nanoparticles green synthesis using photochemical vapor generation coupled with MIP-OES and UV–Vis spectrophotometry. Microchem. J. 145(9): 1169–1175.

Borowska, M., Pawlik, E., and Jankowski, K. 2020. Investigation of interaction between biogenic selenium nanoparticles and human serum albumin using microwave plasma optical emission spectrometry operating in a single-particle mode. Monatsh. Chem. 151(8): 1283–1290.

Christianah, B., Olawunmi, B., and Omoniyi, S. 2021. Antibacterial activity of intracellular greenly fabricated selenium. Int. J. Biotechnol. 10(1): 39–51.

Chudobova, D., Cihalova, K., Dostalova, S., Ruttkay-Nedecky, B., Merlos Rodrigo, M.A., Tmejova, K., Kopel, P., Nejdl, L., Kudr, J., and Gumulec, J. 2014. Comparison of the effects of silver phosphate and selenium nanoparticles on *Staphylococcus aureus* growth reveals potential for selenium particles to prevent infection. FEMS Microbiol. Lett. 351(2): 195–201.

Cremonini, E., Boaretti, M., Vandecandelaere, I., Zonaro, E., Coenye, T., Lleo, M.M., Lampis, S., and Vallini, G. 2018. Biogenic selenium nanoparticles synthesized by *Stenotrophomonas maltophilia* Se ITE 02 loose antibacterial and antibiofilm efficacy as a result of the progressive alteration of their organic coating layer. Microb. Biotechnol. 11(6): 1037–1047.

Dameron, C.T., Reese, R.N., Mehra, R.K., Kortan, A.R., Carroll, P.J., Steigerwald, M.L., Brus, L.E., and Winge, D.R. 1989. Biosynthesis of cadmium sulphide quantum semiconductor crystallites. Nature 338: 596–597.

El-Batal, A.I., Thabet, N.M., Osman, A., Ghaffar, A., and Azab, K.S. 2012. Amelioration of oxidative damage induced in gamma irradiated rats by nano selenium and lovastatin mixture. World Appl. Sci. J. 19(7): 962–971.

El-Saadony, M.T., Saad, A.M., Najjar, A.A., Alzahrani, S.O., Alkhatib, F.M., Shafi, M.E., Selem, E., Desoky, E.-S.M., Fouda, S.E., and El-Tahan, A.M. 2021. The use of biological selenium nanoparticles to suppress *Triticum aestivum* L. crown and root rot diseases induced by *Fusarium* species and improve yield under drought and heat stress. Saudi J. Biol. Sci. 10(3): 1–15.

Elahian, F., Reiisi, S., Shahidi, A., and Mirzaei, S.A. 2017. High-throughput bioaccumulation, biotransformation, and production of silver and selenium nanoparticles using genetically engineered *Pichia pastoris*. Nanomed.: Nanotechnol. Biol. Med. 13(3): 853–861.

Espinosa-Ortiz, E.J., Gonzalez-Gil, G., Saikaly, P.E., van Hullebusch, E.D., and Lens, P.N. 2015. Effects of selenium oxyanions on the white-rot fungus *Phanerochaete chrysosporium*. Appl. Microbiol. Biotechnol. 99(5): 2405–2418.

Espinosa-Ortiz, E.J., Shakya, M., Jain, R., Rene, E.R., van Hullebusch, E.D., and Lens, P.N. 2016. Sorption of zinc onto elemental selenium nanoparticles immobilized in *Phanerochaete chrysosporium* pellets. Environ. Sci. Pollut. Res. 23(21): 21619–21630.

Song, X., Qiao, L., Yan, S., Chen, Y., Dou, X., and Xu, C. 2021. Preparation, characterization, and *in vivo* evaluation of anti-inflammatory activities of selenium nanoparticles synthesized by *Kluyveromyces lactis* GG799. Food Funct. 12(14): 6403–6415. https://doi.org/10.1039/D1FO01019K.

Faramarzi, S., Anzabi, Y., and Jafarizadeh-Malmiri, H. 2020. Nanobiotechnology approach in intracellular selenium nanoparticle synthesis using *Saccharomyces cerevisiae*—fabrication and characterization. Arch. Microbiol. 202(5): 1203–1209.

García, R.Á.-F., Corte-Rodríguez, M., Macke, M., LeBlanc, K., Mester, Z., Montes-Bayón, M., and Bettmer, J. 2020. Addressing the presence of biogenic selenium nanoparticles in yeast cells: analytical strategies based on ICP-TQ-MS. Analyst 145(4): 1457–1465.

Hamza, F., Vaidya, A., Apte, M., Kumar, A.R., and Zinjarde, S. 2017. Selenium nanoparticle-enriched biomass of Yarrowia lipolytica enhances growth and survival of *Artemia salina*. Enzyme Microb. Technol. 106(5): 48–54.

Hariharan, H., Al-Dhabi, N.A., Karuppiah, P., and Rajaram, S.K. 2012. Microbial synthesis of selenium nanocomposite using *Saccharomyces cerevisiae* and its antimicrobial activity against pathogens causing nosocomial infection. Chalcogenide Lett. 9: 509–515.

Hemath Naveen, K.S., Kumar, G., Karthik, L., and Bhaskara Rao, K.V. 2010. Extracellular biosynthesis of silver nanoparticles using the filamentous fungus *Penicillium* sp. Arch. Appl. Sci. Res 2: 161–167.

Hu, D., Yu, S., Yu, D., Liu, N., Tang, Y., Fan, Y., Wang, C., and Wu, A. 2019. Biogenic *Trichoderma harzianum*-derived selenium nanoparticles with control functionalities originating from diverse recognition metabolites against phytopathogens and mycotoxins. Food Control 106(10): 1048–1067.

Ismail, A.-W.A., Sidkey, N.M., Arafa, R.A., Fathy, R.M., and El-Batal, A.I. 2016. Evaluation of in vitro antifungal activity of silver and selenium nanoparticles against *Alternaria solani* caused early blight disease on potato. Biotechnol. J. Int. 6(2): 1–11.

Jafarzadeh, H., Allymehr, M., Talebi,, A. ASRI REZAEI, S., and Soleimanzadeh, A. 2020. Effects of nano-selenium and sodium selenite on SelP, GPx4 and SelW genes expression in testes of broiler breeder roosters. Bulg. J. Vet. Med. 23(2): 1–14.

Jiménez-Lamana, J., Abad-Álvaro, I., Bierla, K., Laborda, F., Szpunar J., and Lobinski, R. 2018. Detection and characterization of biogenic selenium nanoparticles in selenium-rich yeast by single particle ICPMS. J. Anal. At. Spectrom. 33(3): 452–460.

Kalishwaralal, K., Jeyabharathi, S., Sundar, K., and Muthukumaran, A. 2015. Sodium selenite/selenium nanoparticles (SeNPs) protect cardiomyoblasts and zebrafish embryos against ethanol induced oxidative stress. J. Trace. Elem. Med. Biol. 32: 135–144.

Kieliszek, M., Bierla, K., Jiménez-Lamana, J., Kot, A.M., Alcántara-Durán, J., Piwowarek, K., Błażejak, S., and Szpunar, J. 2020. Metabolic response of the yeast *Candida utilis* during enrichment in selenium. Int. J. Mol. Sci. 21(15): 52–87.

Komova, A., Aliev, R., Mel'nikova, A., Kamyshinskii, R., Presnyakov, M.Y., Kal'sin, A., and Namsaraev, Z. 2018. Fabrication and characterization of biogenic selenium nanoparticles. Crystallogr. Rep. 63(2): 12–24.

Kora, A.J. 2018. *Bacillus cereus*, selenite-reducing bacterium from contaminated lake of an industrial area: A renewable nanofactory for the synthesis of selenium nanoparticles. Bioresour. Bioprocess 5(1): 1–12.

Lara, H.H., Guisbiers, G., Mendoza, J., Mimun, L.C., Vincent, B.A., Lopez-Ribot, J.L., and Nash, K.L. 2018. Synergistic antifungal effect of chitosan-stabilized selenium nanoparticles synthesized by pulsed laser ablation in liquids against *Candida albicans* biofilms. Int. J. Nanomedicine. 13(4): 26–97.

Li, X., Xu, H., Chen, Z.-S., and Chen, G. 2011. Biosynthesis of nanoparticles by microorganisms and their applications. J. Nanomater. 2011(3): 1–14.

Lian, S., Diko, C.S., Yan, Y., Li, Z., Zhang, H., Ma, Q., and Qu, Y. 2019. Characterization of biogenic selenium nanoparticles derived from cell-free extracts of a novel yeast *Magnusiomyces ingens*. Biotech 9(6): 1–8.

Liang, X., Perez, M.A.M.-J., Nwoko, K.C., Egbers, P., Feldmann, J., Csetenyi, L., and Gadd, G.M. 2019. Fungal formation of selenium and tellurium nanoparticles. Appl. Microbiol. Biotechnol. 103(17): 7241–7259.

Liang, X., Perez, M.A.M.J., Zhang, S., Song, W., Armstrong, J.G., Bullock, L.A., Feldmann, J., Parnell, J., Csetenyi, L., and Gadd, G.M. 2020. Fungal transformation of selenium and tellurium located in a volcanogenic sulfide deposit. Environ. Microbiol. 22(6): 2346–2364.

Medina Cruz, D., Mi, G., and Webster, T. 2018. Synthesis and characterization of biogenic selenium nanoparticles with antimicrobial properties made by *Staphylococcus aureus*, methicillin-resistant Staphylococcus aureus (MRSA), *Escherichia coli,* and *Pseudomonas aeruginosa*. J. Biomed. Mater. Res. A. 106(5): 1400–1412.

Mehrbakhsh Bandari, M.A., Asadpour, L., and Pourahmad, A. 2018. Antibacterial effect of synthetized selenium nanoparticles and Ampicillin-selenium nanoparticles against clinical isolates of methicillin resistant *Staphylococcus aureus*. Iran. J. Microbiol. 11(6): 184–191.

Menazea, A., Abdelbadie, S.A., and Ahmed, M. 2020. Manipulation of AgNPs coated on selenium/carbonated hydroxyapatite/ε-polycaprolactone nano-fibrous via pulsed laser deposition for wound healing applications. Appl. Surf. Sci. 508(2): 145–199.

Nateq, S., Moghaddam, G., Alijani, S., and Behnam, M. 2020. The effects of different levels of Nano selenium on the quality of frozen-thawed sperm in ram. J. Appl. Anim. Res. 48(1): 434–439.

Parsameher, N., Rezaei, S., Khodavasiy, S., Salari, S., Hadizade, S., Kord, M., and Mousavi, S.A.A. 2017. Effect of biogenic selenium nanoparticles on *ERG11* and *CDR1* gene expression in both fluconazole-resistant and-susceptible *Candida albicans* isolates. Curr. Med. Mycol. 3(3): 16–20. doi: 10.29252/cmm.3.3.16.

Pereira, A.G., Gerolis, L.G.L., Gonçalves, L.S., Pedrosa, T.A., and Neves, M.J. 2018. Selenized *Saccharomyces cerevisiae* cells are a green dispenser of nanoparticles. Biomed. Phys. Eng. Express. 4(3): 1–18.

Perfileva, A., Tsivileva, O., Koftin, O., Anis'Kov, A., and Ibragimova, D. 2018. Selenium-containing nanobiocomposites of fungal origin reduce the viability and biofilm formation of the bacterial phytopathogen *Clavibacter michiganensis* subsp. *Sepedonicus*. Nanotechnologies Russ. 13(5): 268–276.

Yeast as a Cell Factory: Biological Synthesis of Selenium Nanoparticles 49

Piacenza, E., Presentato, A., Ambrosi, E., Speghini, A., Turner, R.J., Vallini, G., and Lampis, S. 2018. Physical–chemical properties of biogenic selenium nanostructures produced by *Stenotrophomonas maltophilia* SeITE02 and *Ochrobactrum* sp. MPV1. Front. Microbiol. 9(3): 31–48.

Raja, C.P., Jacob, J.M., and Balakrishnan, R.M. 2016. Selenium biosorption and recovery by marine *Aspergillus terreus* in an upflow bioreactor. J. Environ. Eng. 142(9): 10–21.

Rasouli, M. 2019. Biosynthesis of selenium nanoparticles using yeast *Nematospora coryli* and examination of their anti-candida and anti-oxidant activities. IET Nanobiotechnol. 13(2): 214–218.

Ruocco, M.H.W. 2011. Biotransformation and distribution of selenium in cultures of a salt-marsh yeast. Master's Thesis, pp. 1–79, University of Delaware, http://udspace.udel.edu/handle/19716/10677.

Saleh, A.A. 2014. Effect of dietary mixture of *Aspergillus* probiotic and selenium nano-particles on growth, nutrient digestibilities, selected blood parameters and muscle fatty acid profile in broiler chickens. Anim. Sci. Pap. Rep. 32(3): 65–79.

Salem, S.S. 2022. Bio-fabrication of selenium nanoparticles using baker's yeast extract and its antimicrobial efficacy on food borne pathogens. Appl. Biochem. Biotechnol. 19(2): 1–13.

Salvadori, M.R., Ando, R.A., Oller do Nascimento, C.A., and Corrêa, B. 2014. Intracellular biosynthesis and removal of copper nanoparticles by dead biomass of yeast isolated from the wastewater of a mine in the Brazilian Amazonia. PLoS One 9(1): e87968. https://doi.org/10.1371/journal.pone.0087968.

San Keskin, N.O., Akbal Vural, O., and Abaci, S. 2020. Biosynthesis of noble selenium nanoparticles from *Lysinibacillus* sp. NOSK for antimicrobial, antibiofilm activity, and biocompatibility. Geomicrobiol. J. 37(10): 919–928.

Selim, N., Radwan, N., Youssef, S., Eldin, T.S., and Elwafa, S.A. 2015. Effect of inclusion inorganic, organic or nano selenium forms in broiler diets on: 1-growth performance, carcass and meat characteristics. Int. J. Poult. Sci. 14(3): 135–145.

Selvaraj, A. 2013. Extracellular biosynthesis and biomedical application of silver nanoparticles synthesized from Baker's Yeast. Int. J. Res. Pharm. Biomed. Sci. 3(4): 822–828.

Seshadri, S., Saranya, K., and Kowshik, M. 2011. Green synthesis of lead sulfide nanoparticles by the lead resistant marine yeast, *Rhodosporidium diobovatum*. Biotechnol. Prog. 27(5): 1464–1469.

Shankar, S.S., Rai, A., Ankamwar, B., Singh, A., Ahmad, A., and Sastry, M. 2004. Biological synthesis of triangular gold nanoprisms. Nat. Mater. 3(7): 482–488.

Sharma, G., Sharma, A.R., Bhavesh, R., Park, J., Ganbold, B., Nam, J.-S., and Lee, S.-S. 2014. Biomolecule-mediated synthesis of selenium nanoparticles using dried Vitis vinifera (raisin) extract. Molecules 19(3): 2761–2770.

Shen, X., Huo, B., and Gan, S. 2021. Effects of nano-selenium on antioxidant capacity in Se-deprived Tibetan gazelle (*Procapra picticaudata*) in the Qinghai–Tibet Plateau. Biol. Trace Elem. Res. 199(2): 981–988.

Skalickova, S., Baron, M., and Sochor, J. 2017. Nanoparticles biosynthesized by yeast: a review of their application. Kvas. prům. 63(6): 290–292.

Valueva, S., Vylegzhanina, M., Lavrent'ev, V., Borovikova, L., and Sukhanova, T. 2013. Biogenic nanosized systems based on selenium nanoparticles: Self-organization, structure, and morphology. Russ. J. Phys. Chem. 87(3): 484–489.

Vriens, B.P. 2015. The role of methylation in the biogeochemical selenium cycle, ETH Zurich.

Wadhwani, S.A., Shedbalkar, U.U., Singh, R., and Chopade, B.A. 2016. Biogenic selenium nanoparticles: current status and future prospects. Appl. Microbiol. Biotechnol. 100(1): 2556–2566.

Wang, Y., Shu, X., Hou, J., Lu, W., Zhao, W., Huang, S., and Wu, L. 2018. Selenium nanoparticle synthesized by *Proteus mirabilis* YC801: an efficacious pathway for selenite biotransformation and detoxification. Int. J. Mol. Sci. 19(12): 38–45.

Winkel, L.H., Vriens, B., Jones, G.D., Schneider, L.S., Pilon-Smits, E., and Bañuelos, G.S. 2015. Selenium cycling across soil-plant-atmosphere interfaces: a critical review. Nutrients 7(6): 4199–4239.

Wu, Z., Ren, Y., Liang, Y., Huang, L., Yang, Y., Zafar, A., Hasan, M., Yang, F., and Shu, X. 2021. Synthesis, characterization, immune regulation, and antioxidative assessment of yeast-derived selenium nanoparticles in cyclophosphamide-induced rats. ACS Omega 6(38): 24585–24594.

Zhang, L., Li, D., and Gao, P. 2012. Expulsion of selenium/protein nanoparticles through vesicle-like structures by *Saccharomyces cerevisiae* under microaerophilic environment. World J. Microbiol. Biotechnol. 28(12): 3381–3386.

Chapter 3

Penicillium species as an Innovative Microbial Platform for Bioengineering of Biologically Active Nanomaterials

Hamed Barabadi,[1,*] *Kamyar Jounaki,*[1] *Elahe Pishgahzadeh,*[1] *Hamed Morad*[2,3] *and Hossein Vahidi*[1]

Introduction

Various organisms in the fungal kingdom have been used for different purposes for many years and their functions in pharmaceutical sciences are well established (Crawford et al. 1995). Novel advances in mycology have provided for special potential in bioengineering bioactive materials (Hesham et al. 2020). Nanobiotechnology has played a unique role in combination with most of the sciences, and mycology has demonstrated that it could be one of the successful areas for this integration. Myconanotechnology (myco=fungi; nanotechnology=evaluation of the properties of materials in the size range of 1–100 nm) has been selected as a new term for the combination of these two major fields (Rai et al. 2009). Briefly, myconanotechnology is determined as the synthesis of nanostructures by utilizing fungi and evaluating their functional properties (Abd-Elsalam 2021). Generally, the properties of materials are determined by the organization of building blocks in three-dimensional structures. The properties of nanomaterials are entirely different from bulk materials. These differences are related to the quantum confinement of

---

[1] Department of Pharmaceutical Biotechnology, School of Pharmacy, Shahid Beheshti University of Medical Sciences, Tehran, Iran.

[2] Department of Pharmaceutics, Faculty of Pharmacy, Mazandaran University of Medical Sciences, Sari, Iran.

[3] Ramsar campus, Mazandaran University of Medical Sciences, Ramsar, Iran.

* Corresponding author: barabadi@sbmu.ac.ir

structure in the nanoscale (Zehetbauer and Zhu 2009; Nabi et al. 2018; Tahir et al. 2020). Nanomaterials possess at least one dimension on the scale of nanometer and the major characteristic of each nanostructure is based on dimensionality. Hence, based on the number of dimensions, they are categorized as zero-dimensional (0D) like quantum dots (QDs), hollow spheres, nano lenses, core-shell QDs, one-dimensional (1D) like nanotubes, nanowires, nanorods, two-dimensional (2D) like nanofilms, nanocoating, and nanolayers, and three-dimensional (3D) like dispersion of NPs, bundles of nanowires and multi nano-layers (Pokropivny and Skorokhod 2008; Rafique et al. 2018). Nanoparticles (NPs) as one of the major parts of the nanostructures family and are categorized into two types, namely, organic and inorganic NPs. Inorganic NPs group, consists of magnetic, semiconductor, and noble metal NPs. Inorganic NPs comprise advantages like fine biocompatibility, availability, and rich functionality. These NPs have had an obvious impact on various industrial fields like pharmaceutical sciences, and electronics (Singh et al. 2011). In addition, their optical features are their most interesting properties. For instance, gold NPs (AuNPs), silver NPs (AgNPs), platinum NPs (PtNPs), and palladium NPs (PdNPs) have various colors like vine red, yellowish-grey, and black, respectively (Mittal et al. 2013; Heera and Shanmugam 2015; Ovais et al. 2016; Barabadi 2017).

The two main approaches for the fabrication of NPs include: top-down and bottom-up. Briefly, during the top-down approach, the physical or chemical shears would be applied to break down the bulk material to particles in nanoscale size. In the bottom-up approach, the NPs would be fabricated from smaller components by manipulating the molecules with chemical or biological procedures. Wet and dry grinding are two main routes of the top-down method. Shock, hammer mill, jet mill, and roller mill are some methods of grinding. Examples of the bottom-up approach are chemical vapor deposition, physical vapor deposition, and gas-phase technique (Wang and Xia 2004; Mittal et al. 2013; Heera and Shanmugam 2015; Ovais et al. 2016; Barabadi 2017).

Since physical methods providing high pressure and temperature consumes large amounts of energy, this would put a financial burden on the manufacturers (Iravani et al. 2014; Kulkarni and Muddapur 2014; Fawcett et al. 2017). On the other hand, despite the chemical methods being more cost-benefitial for industrial scales, the risk of toxic solvents residuals in finished products and also hazardous by-products during the procedures are the main obstacles against the selection of this method for pharmaceuticals and biopharmaceuticals productions (Iravani et al. 2014; Barabadi et al. 2017). In contrast, nanobiotechnology has initiated a new era of studies that has provided green and economical fabrication methods with a high yield and biocompatibility. The trend of replacing nanobiotechnology-based technologies with the chemical synthesis of NPs is under the spotlight. There is a crucial need for more penetration of nanobiotechnology in the pharmaceutical and biopharmaceutical industries (Dahoumane et al. 2017a; Dahoumane et al. 2017b).

Utilizing biological systems for the biosynthesis of metal-based NPs (MNPs) could be conducted by many biological resources, like fungi (Raveendran et al. 2003; Shankar et al. 2003), algae (Xie et al. 2007; Mata et al. 2009), bacteria (Lengke et al. 2006; He et al. 2008), and plants (Philip 2009; 2010; Rajasekharreddy et al. 2010). Meanwhile, the mycosynthesis of MNPs is a type of bottom-up approach and could lead to the biosynthesis of MNPs by intracellular or extracellular harvesting. Hence, it could be a reasonable and safe alternative to previous methods (Zhang et al. 2011; Hulkoti and Taranath 2014). Briefly, the intracellular pathway includes procedures of transporting the metal ions through the biological cell and the production of MNPs in the presence of enzymes. In the extracellular pathway, the cells would secrete the synthetic enzymes which synthesize the NPs in the outer medium of cells (Hulkoti and Taranath 2014). It seems that the extracellular pathway is more reasonable compared to the intracellular procedure, which needs a further chemical and physical process to drain out the fabricated MNPs from cells (Zhang et al. 2011; Dahoumane et al. 2017a). Polydispersity is one of the main challenges of the mycosynthesis method. Controlling the size of MNPs and reaching monodispersity depends on the type of microorganism, growth medium, and synthesis conditions. Moreover, the shape of MNPs also depends on the type of microorganism (Zhang et al. 2011). The microorganism-based biosynthesis of MNPs has remained on the laboratory scale. However, the application of fungi for scaling-up proceedings has been more noticeable. More attention to the fungi is related to their advantages like efficient extracellular secretion of enzymes, economic viability, facile biomass application, easy biomass harvesting, and downstream processing (Boroumand Moghaddam et al. 2015). The fungal-based biosynthesis of MNPs in the extracellular pathway has been conducted by various fungi. For example, *Aspergillus niger* strains have been studied and showed that a mycotoxin named ochratoxin A would be produced during the process of MNPs production (Schuster et al. 2002). Investigation of the related studies on *Aspergillus* genus demonstrated that the *Aspergillus flavus* species produces aflatoxin as a by-product which is a potent toxin and causes hepatocellular carcinoma. Moreover, *Fusarium solani* strains showed that may induce endocarditis and lung disease (Duran et al. 2010). Hence, there is a need to select the safe genus of fungi to eliminate any probable risk in MNPs production. The *Penicillium* species have just shown a great candidate for biosynthesis of the MNPs and recently were highlighted for this application. *Penicillium* species have great distribution in the whole world and are widely reachable around in soils, foods, drinks, and even indoor air (Leitão 2009). Some *Penicillium* species are utilized in the food industry like cheese production or fermented sausages. (Visagie et al. 2014). Up to now, many species of *Penicillium* have been recognized for bioengineering of MNPs, and these NPs exhibited different biological activities. This chapter provides recent advances in the *Penicillium* species-mediated bioengineering of MNPs and their biological applications.

Penicillium species: An overview

Penicillium is a genus of anamorphic ascomycetous fungi which consist of more than 300 species with worldwide distribution and various habitats such as soil, air, food product, and different environment (water-deficient, alkaline, acidic, low/high-temperature) (Yadav et al. 2018). *Penicillium*, *Furcatum*, *Biverticillium*, and *Aspergilloides* are four subgenera of the genus *Penicillium*. The application of subgenus *Penicillium* has been much more investigated and diverse, from the natural environment to food and valuable bioactive secondary metabolites (Kozlovskiĭ et al. 2013).

For centuries, different *Penicillium* species have been used for food production, such as *P. camemberti* utilized to produce white mold cheese, and *P. roqueforti* made blue fermented cheese (Frisvad 2014). Moreover, the bioactive compounds produced by *Penicillium* contain ergot alkaloids, quinolines, quinazolines, diketopiperazines, and polyketides. The specific structure of ergot alkaloids consists of a tetracyclic ergoline core. Ergot alkaloid is considered an essential component in which the therapeutic activities include decreased vasomotor activities, uterine contraction, hyperthermia, vomiting, antibiotic and cytotoxic activities. Besides, diketopiperazines is a biologically active compound with antibiotic and antitumor effects. In addition, quinolones and quinazolines compounds have an effect against Gram-positive and Gram-negative bacteria, yeast, and fungi. Mycotoxin activity was seen in polyketides as the numerous group of the active compound (Kozlovskiĭ et al. 2013; Kumar et al. 2018).

Mycotoxins cause food spoilage which are the toxic secondary metabolites produced by some *Penicillium* species (Houbraken et al. 2014). Among these mycotoxins, ochratoxins A is a critical nephrotoxic agent whose level is controlled in food precisely. Citrinin is another mycotoxin isolated from *P. citrinum, P. miczynskii, P. hirsutum, P. verrucosum, P. westlingi, P. expansum,* and *P. steckii* (Kumar et al. 2018).

Conversely, *P. caseifulvum, P. digitatum, P. nalgiovense, P. olsonii, P. solitum, P. thymicola,* and *P. ulaiense* species do not produce mycotoxins. Furthermore, some of the *Penicillium* species produce mycotoxin only in the laboratory environment (Frisvad et al. 2004).

Green nanotechnology: An innovative approach for bioengineering of nanomaterials

Green nanotechnology is an approach that integrates green chemistry and green engineering. In this technology, fewer materials and renewable inputs would be used that could reduce energy and fuel consumption. Moreover, any processes, products, or applications based on nanotechnology would save raw materials and natural resources like water or energy which enters fewer greenhouse gases or hazardous by-products to the environment. Accordingly, green nanotechnology provides climate and environmental protection. Hence, the main advantages of green nanotechnology are enhancement in energy efficiency, decrease the number of wastes and greenhouse pollution gas and reduction in utilizing non-renewable resources. This technology provides an opportunity for prevention proceedings (Hullmann and Meyer 2003; Zou et al. 2008). Over the last decade, nanotechnology research has been boosted due to its various application in medicine, agriculture, food, environmental science, etc. The valuable properties of nanoparticles, such as size (between 1 to 100 nanometers), morphology, high surface to volume ratio, etc., led to creating innovative ways for nanomaterial synthesis (Dikshit et al. 2021).

Green synthesis became highlighted during recent years. This fact is related to the beneficial advantages of this method compared with conventional physical and

54   *Mycosynthesis of Nanomaterials: Perspectives and Challenges*

chemical methods. Easy and cost-effectiveness, safeness and eco-friendly procedures are some main reasons (Pal et al. 2019). The biological method utilized microbial systems such as bacteria, yeast, fungi, viruses, and plant systems to synthesize MNPs. This method is safe, rapid, and eco-friendly, and due to these exceptional factors, green synthesis is preferred over traditional methods. During the green synthesis of MNPs, oxidation/reduction that occurs by biological molecules causes the formation of NPs. Furthermore, different types of microorganisms and environmental conditions such as pH and temperature determine the final properties of nanoparticles (Hashem et al. 2021).

Utilizing plants to produce NPs is one of the most environmentally friendly and practical methods of green synthesis. In this method, the bioactive molecules such as phenols, terpenoids, and flavonoids of plants are extracted by drying, floating in hot water, and then filtration. These bioactive molecules provide a situation for reducing and stabilizing for MNPs precursors. Furthermore, live plants have also been utilized for nanomaterial synthesis; for example, AgNPs were formed in the living alfalfa sprouts by reducing inside the plant.

Moreover, bacterial-mediated MNPs synthesis is one of the important approaches due to their different advantages and controllability. Different bacterial species such as *Escherichia coli, Bacillus subtilis,* and *Morganella psychrotolerans*, etc., have been considered to produce different types of MNPs (Huston et al. 2021). There are some biochemical pathways like enzymatic reduction, either intracellularly or extracellularly synthesized MNPs. Further investigations revealed that NADH-dependent reductase enzyme and nitrate dependent—reductase were involved in the bioreduction of MNPs (Grasso et al. 2020).

Besides, yeasts in both forms, the live cells, and cell extracts were utilized to synthesize silver, gold, cadmium sulfide, lead sulfide, ferrous oxide, selenium NPs. Like bacteria, enzymatic oxidation and reduction, trapping metal ions in the membrane and 1,3-glucan synthase-mediated formation resulted in MNPs formation and stabilized the complex. For instance: *Saccharomyces cerevisiae* was utilized to produce different types of MNPs (Huston et al. 2021).

In addition, viruses are other attractive microorganisms for the green synthesis of MNPs. Viruses consist of genetic material and shell of a protein termed as capsid. Due to the dependency of viruses on the host body and lack of contagiousness outside the host, microorganisms make viruses the safe target for MNPs biosynthesis. The virus cell wall has appropriate features such as different amino acids and functional groups which can adsorb the metal ions (Khan and Lee 2020). Genetic engineering can manipulate the capsid protein to obtain more desirable features for synthesizing different nano-compositions (Koul et al. 2021). For example, M13 bacteriophage has been utilized to produce cadmium sulfide (CdS) and zinc sulfide (ZnS) NPs (Mao et al. 2003).

Recently, fungi-based MNPs production has been well noticed as an effective method. The great stability and fine dispersion are the desirable characteristics of the MNPs which would be achieved from the fungi. Generally, myconanotechnology became an eco-friendly and cost-benefit technology due to special features like providing a significant volume of related enzymes by fungi, easy-working, particle size manipulation, and economic scale-up (Gade et al. 2010; Youssef et al. 2017). A large range of MNPs could be fabricated by filamentous fungi including silver, gold, magnetic and bimetallic NPs (Sastry et al. 2003; Molnár et al. 2018). Mycosynthesis can be conducted by two synthesis locations, including extracellular and intracellular pathways, via reduction enzymes or biomolecules (Chhipa 2019). The intracellular mechanism uses different steps to synthesize MNPs. The first step is electrostatic interaction and trapping of metal ions at the cell wall. In the next step, the metal ions were reduced by reducing enzymes, and then MNPs were synthesized. Another mechanism is the extracellular pathway in which metal ions reduction was conducted by NADPH-dependent nitrate reductase enzyme and cysteine containing proteins in the solution. Eventually, the extracellular mechanism due to higher yield compared to the intracellular pathway is more desirable (Youssef et al. 2017). Mycosynthesis approaches make more advantageous than other biosynthesis processes due to high metal uptake capacity and tolerability and fewer downstream steps. Moreover, it has mycelial growth, which provides a high surface area and makes it economically desirable with a higher yield than other microorganisms (Chhipa 2019). Further investigations revealed that the size and morphology of MNPs are affected by pH, temperature, ionic concentration, incubation time, synthesis mechanism, and reduction agents. For instance: Increasing pH value causes synthesis process enhancement, and different shapes of MNPs such as triangles, hexagons, spheres, and rods were produced in a different range of pH values (Rana et al. 2020; Koul et al. 2021).

Bioengineering of nanomaterials using *Penicillium* species

Different studies revealed the potential of *Penicillium* species for producing a diverse range of MNPs. Table 1 represents the different *Penicillium* species for the biomanufacturing of MNPs. Majeed et al. reported biofabrication of spherical AgNPs by using *P. decumbens* (MTCC-2494). The particle size of these NPs was reported to be between 30–60 nm, which were produced via an extracellular mechanism (Majeed et al. 2016). In a separate study conducted by Solanki et al., *P. brevicompactum* was utilized to produce spherical AgNPs extracellularly with an average particle size of 6.28–15.12 ± 0.872 (Solanki et al. 2016). Likewise, Tyupa et al. reported biosynthesis of spherical AgNPs by using *P. glabrum* in an extracellular approach in the range of 20–30 nm (Tyupa et al. 2016). Besides, in a study reported by Chandrappa et al., extracellular cubic AgNPs were synthesized by *Penicillium* sp. in the range of 33.71–65.92 nm (Chandrappa et al. 2016). Moreover, Ammar et al. reported the fungus-mediated synthesis of spherical AgNPs in the range of 14–25 nm by using *P. expansum* HA2 N in an extracellular approach (Ammar and El-Desouky 2016). Alternatively, Khan and Jameel biosynthesized AgNPs extracellularly by utilizing *P. fellutanum* in the range of 10–100 nm (Khan and Jameel 2016). Furthermore, Sheet et al. biosynthesized AgNPs extracellularly by using *P. chrysogenum*. These NPs were mostly spherical within the size distribution of 55–56 nm (Sheet et al. 2017). Likewise, Yassin et al. showed *P. citrinum* produced AgNPs via extracellular pathways, and their morphology reported mostly spherical in the range of 3–13 nm (Yassin et al. 2017). Likewise, in a study reported by Verma et al., spherical AgNPs were biosynthesized using *Penicillium* spp. in the range of 149–397 nm via

Table 1. Biological synthesis of metal nanoparticles by using *Penicillium* species.

Scientific name	Metal nanoparticles	Characterization techniques	Extracellular or intracellular pathways	Size distribution (nm)	Morphology	References
Penicillium decumbens (MTCC-2494)	Silver	FTIR, AFM, FE-SEM, UV-vis	Extracellular	30–60	Spherical	(Majeed et al. 2016)
Penicillium brevicompactum	Silver	UV-vis, TEM, SEM-EDX, FTIR, XRD	Extracellular	6.28–15.12 ± 0.8	Spherical	(Solanki et al. 2016)
Penicillium glabrum	Silver	UV-vis, TEM, EDS, SEM, FTIR, TEM, DLS, XRD	Extracellular	20–30	Spherical	(Tyupa et al. 2016)
Penicillium sp	Silver	UV-vis, XRD, SEM	Extracellular	33.71–65.92	Cubic	(Chandrappa et al. 2016)
Penicillium expansum HA2 N	Silver	UV-vis, DLS, FTIR, TEM	Extracellular	14–25	Spherical	(Ammar and El-Desouky 2016)
Penicillium fellutanum	Silver	UV-vis	Extracellular	10–100	No data	(Khan and Jameel 2016)
Penicillium chrysogenum	Silver	UV-vis, FTIR, XRD, TEM	Extracellular	55–65	Mostly spherical	(Sheet et al. 2017)
Penicillium citrinum	Silver	XRD, EDS, SEM, TEM	Extracellular	3–13	Mostly spherical	(Yassin et al. 2017)
Penicillium spp	Silver	UV-vis, SEM	Extracellular	149–397	Spherical	(Verma et al. 2013)
Penicillium aculeatum	Gold	UV-vis, SEM, AFM, DLS, FTIR	Extracellular	60	Spherical	(Barabadi et al. 2017)
Penicillium aculeatum	Silver	TEM, HR-TEM, XRD, FTIR	Extracellular	4–55	Spherical	(Barabadi et al. 2017)
Penicillium funiculosum BL1	Gold	UV-vis, TEM, DLS, SEM, EDS	Extracellular	18–28	Spherical	(Maliszewska et al. 2017)
Penicillium citrinum	Gold	UV-vis, FE-SEM, XRD, FTIR, DLS	Extracellular	60–80	Spherical	(Manjunath et al. 2017)

Penicillium chrysogenum strain FGCC/BLS1	Silver	UV–vis, FTIR, TEM, DLS	Extracellular	96.8	Ellipsoidal	(Saxena et al. 2017)
Penicillium chrysogenum PTCC 5031	Tellurium	AFM, SEM, DLS, EDX, FTIR	Extracellular	50.16	Spherical	(Barabadi et al. 2018)
Penicillium cyclopium	Silver	UV–vis, SEM, TEM, FTIR	Intracellular	16 ± 6	Irregular shape	(Wanarska and Maliszewska 2019)
Penicillium chrysogenum NG85	Silver	TEM, DLS, FTIR	Extracellular	9–17.5	Spherical	(Khalil et al. 2019)
Penicillium aculeatum (PTCC 5167), Penicillium notatum (PTCC 5074) and Penicillium purpurogenome (PTCC 5212)	Zirconium	SEM, AFM, DLS, EDX, FTIR	Extracellular	< 100 nm	Spherical	(Ghomi et al. 2019)
Penicillium polonicum ARA 10	Silver	UV–vis, FTIR, Raman spectroscopy, HR-TEM, EDX	Extracellular	10–15	Spherical or near to spherical	(Neethu et al. 2018)
Penicillium oxalicum	Silver	XRD, SEM, UV–vis,	Extracellular	67	Cubical	(Feroze et al. 2020)
Penicillium crustosum	Silver	UV-vis, DLS, TEM	Extracellular	30–40	Spherical to oval shape	(El-Sayed and Ali 2018)
Penicillium duclauxii	Silver	XRD, TEM, SEM, UV-vis	Extracellular	32	Mostly visible in spherical	(Almaary et al. 2020)
Penicillium italicum	Silver	UV–vis, XRD, FTIR, SEM, EDX, TEM	Extracellular	39.5	Spherical	(Taha et al. 2019)
Penicillium corylophilum	Selenium	UV–vis, FTIR, TEM, EDX, XRD, DLS	Extracellular	29.1–48.9	Spherical	(Salem et al. 2021)
Penicillium citrinum CGJ-C2	Silver	UV–vis, HR-TEM, XRD, FTIR	Extracellular	2–20	Spherical	(Danagoudar et al. 2020)
Penicillium chrysogenum (PTCC 5031)	Selenium	PCS, SEM, AFM, EDX, XRD, FTIR	Extracellular	24.65	Spherical	(Vahidi et al. 2020)

Table 1 contd. ...

...*Table 1 contd.*

Scientific name	Metal nanoparticles	Characterization techniques	Extracellular or intracellular pathways	Size distribution (nm)	Morphology	References
Penicillium expansum RCMB 001001(2)	Silver	UV–vis, EDX, Zeta Potentials, TEM	Extracellular	9–18	Irregular, spherical, hexagonal	(Ismail et al. 2021)
Penicillium oxalicum	Cadmium oxide	XRD, UV–vis, FTIR, SEM, EDS	Intracellular	22.94	Spherical	(Asghar et al. 2020)
Penicillium radiatolobatum strain AN003	Silver	XRD, FTIR, TEM, EDS, UV–vis, Zeta potential	Extracellular	5.09–24.85	Spherical and few in triangle and hexagonal	(Naveen et al. 2021)
Penicillium chrysogenum	Magnesium Oxide	UV–vis, XRD, TEM, DLS, EDX, FTIR, XPS	Extracellular	7–40	Spherical	(Fouda et al. 2021)
Penicillium verrucosum	Silver	UV–vis, TEM, XRD, SEM, EDS	Extracellular	10–12	Irregular	(Yassin et al. 2021)
Penicillium chrysogenum (PTCC 5031)	Silver	UV–vis, TEM, AFM, XRD, DLS, Zeta potential, FTIR	Extracellular	48.2	Spherical shape	(Barabadi et al. 2021)
Penicillium toxicarium KJ173540.1	Silver	UV–vis, FTIR, TEM, EDX, SEM	Extracellular	59.22 ± 8.38	Spherical	(Korcan et al. 2021)
Penicillium expansum (ATTC 36200)	Selenium	UV–vis, FTIR, XRD, SEM, TEM	Extracellular	4–12.7	Spherical	(Hashem et al. 2021)
Penicillium chrysogenum F9	Silver	UV–vis, FTIR, XRD, TEM, DLS, zeta potential	Extracellular	18–60	Spherical	(Soliman et al. 2021)
Penicillium chrysogenum (PTCC 5031)	Tellurium	SEM, DLS, XRD	Extracellular	33.8 ± 6.833	Spherical	(Vahidi et al. 2021)

Penicillium species-Mediated Biosynthesis of Nanomaterials 59

an extracellular pathway (Verma et al. 2013). Additionally, Barabadi et al. reported mycosynthesis of AuNPs using *P. aculeatum* with spherical shapes and an average particle size of 60 nm (Barabadi et al. 2017). Moreover, biosynthesis of AuNPs was reported by Maliszewska et al. These NPs were produced extracellularly by using *P. funiculosum* BL1 with spherical morphology and a particle size between 18–28 nm (Maliszewska et al. 2017). Besides, Manjunath et al. reported the induction of spherical AuNPs in the range of 60–80 nm extracellularly by using *P. citrinum* (Manjunath et al. 2017). Alternatively, Saxena et al. biosynthesized AgNPs by using *P. chrysogenum* strain FGCC/BLS1 via an extracellular mechanism with ellipsoidal morphology and an average particle size of 96.8 nm (Saxena et al. 2017). In addition, Barabadi et al. reported biosynthesis of tellurium NPs (TeNPs) produced by *P. chrysogenum* (PTCC 5031) with a spherical shape and an average hydrodynamic size of 50.16 nm (Barabadi et al. 2018). Furthermore, a study reported the biosynthesis of AgNPs in the range of 16 ± 6 nm by utilizing *P. cyclopium*. These NPs were synthesized on the mycelia surface with various shapes (Wanarska and Maliszewska 2019). Moreover, Khalil et al. reported extracellular AgNPs biosynthesis by using *P. chrysogenum* NG85 in the range of 9 to 17.5 nm (Khalil et al. 2019). Besides, in a study, Ghomi et al. showed that zirconium NPs (ZrNPs) with spherical shape below 100 nm were synthesized extracellularly by three *Penicillium* species including *P. aculeatum* (PTCC 5167), *P. notatum* (PTCC 5074), and *P. purpurogenome* (PTCC 5212) (Ghomi et al. 2019). Remarkably, Neethu et al. reported extracellular biosynthesis of AgNPs with particle size between 10–15 nm by employing *P. polonicum* ARA 10 isolated from *Chetomorpha antennina* (Neethu et al. 2018). Similarly, Feroze and colleagues investigated *P. oxalicum* to synthesize AgNPs with cubical morphology and the particle size of 67 nm by extracellular pathway (Feroze et al. 2020). Furthermore, a study reported that *P. crustosum* can biosynthesize AgNPs in the range of 30–40 nm extracellularly with spherical to oval shapes (El-Sayed and Ali 2018). Besides, in a study conducted by Almaary et al., the extracellular biosynthesis pathway of AgNPs was seen using *P. duclauxii* with an average particle size of 32 nm (Almaary et al. 2020). Furthermore, Taha et al. reported the extracellular biosynthesis of AgNPs by using *P. italicum* with an average particle size of 39.5 nm (Taha et al. 2019). Besides, in a study, selenium NPs (SeNPs) were fabricated by using *P. corylophilum*. The SeNPs were in spherical morphology and size distribution range between 29.1–48.9 nm (Salem et al. 2021). Remarkably, Danagoudar et al. reported the fabrication of AgNPs by using *P. citrinum* CGJ-C2. These NPs had spherical morphology in the range of 2-20 nm (Danagoudar et al. 2020). Alternatively, in a separate study, Vahidi et al. reported the extracellular formation of SeNPs by utilizing *P. chrysogenum* (PTCC 5031). They revealed that the average hydrodynamic size of 24.65 nm in pH value of 7. The SEM and AFM confirmed that the SeNPs were fabricated with spherical morphology (Vahidi et al. 2020). In addition, Ismail et al. biosynthesized AgNPs in an extracellular approach by exploited *P. expansum* RCMB 001001(2) in the range of 9–18 nm within an irregular, spherical and hexagonal shape (Ismail et al. 2021). Moreover, the fabrication of cadmium oxide NPs (CdONPs) was studied by Asghar et al. In this study, CdONPs biosynthesis induced by *P. oxalicum* through intracellular mechanism with cubic crystalline shape and particle size of 22.94 nm (Asghar et al. 2020). In a study, Naveen et al. biosynthesized

60 *Mycosynthesis of Nanomaterials: Perspectives and Challenges*

mycogenic AgNPs by using *P. radiatolobatum* strain AN003. These NPs were spherical and few in a triangle and hexagonal shape in the range of 5.09 to 24.85 nm (Naveen et al. 2021). Furthermore, Fouda et al. reported biosynthesis of magnesium oxide NPs (MgONPs) induced by *P. chrysogenum* in the range of 7–40 nm with a spherical shape (Fouda et al. 2021). In a study, *P. verrucosum* was used to synthesize AgNPs. These mycogenic NPs with a range of 10–12 nm and irregular morphology were fabricated in a cell-free extract (Yassin et al. 2021). Besides, Barabadi et al. reported mycosynthesis of AgNPs by using *P. chrysogenum* (PTCC 5031) with an average hydrodynamic diameter of 48.2 nm with a polydispersity index (PDI) of 0.3 and a spherical morphology (Barabadi et al. 2021). Likewise, Korcan et al. biosynthesized (AgNPs) extracellularly by using *P. toxicarium* KJ173540.1 within a spherical shape and an average particle size of 59.22 ± 8.38 nm (Korcan et al. 2021). Moreover, Hashem et al. reported biosynthesis of (SeNPs) induced by the cell-free filtrate of *P. expansum* (ATTC 36200) in the range of 4 to 12.7 nm (Hashem et al. 2021). Additionally, Soliman et al. reported extracellular biosynthesis of (AgNPs) by exploited *P. chrysogenum* F9 with spherical morphology and a size range between 18 to 60 nm (Soliman et al. 2021). Besides, Vahidi et al. reported the biogenic fabrication of (TeNPs) using *P. chrysogenum* (PTCC 5031) in the range of 33.8 ± 6.833 nm and spherical shape in an extracellular approach (Vahidi et al. 2021).

Antimicrobial activity of bioengineered nanomaterials using *Penicillium* species

Many studies stated the *in vitro* antibacterial, antibiofilm, antifungal, and antiparasitic activity of biogenic nanosized metal particles synthesized from *Penicillium* species. In a study, endophytic fungus *P. radiatolobatum* was used to biosynthesize spherical AgNPs with an average particle size of 72.04 nm. They also tested the antimicrobial potential of the mycogenic AgNPs in five targeted pathogenic bacteria by well diffusion method. Their results exhibited the zones of inhibition (ZOI) in bacterial species and concentration reliant. The AgNPs at the concentrations of 0.1 to 1 mg/mL exhibited antibacterial activity with ZOI ranged from 8.2 ± 0.25 to 14.7 ± 0.36 mm for *Bacillus cereus*, 18.9 ± 0.25 to 25.1 ± 0.32 mm for *Staphylococcus aureus*, 12 ± 0.25 to 19 ± 0.30 mm for *E. coli*, 20 ± 0.25 to 25 ± 0.15 mm for *Salmonella enterica*, and 10.3 ± 0.81 mm to 17.8 ± 0.41 mm for *Listeria monocytogenes*. Based on their outcomes, the AgNPs inhibited the growth of *S. aureus* and *S. enterica* higher than the *B. cereus*, *E. coli*, and *L. monocytogenes*. Based on the TEM analysis of the AgNPs effect on damaging the *S. enterica* and *S. aureus* cells (Figure 1A), the authors attributed the bacterial killing activity of the AgNPs to their higher cellular penetration in bacterial cells due to the nanosized and high surface area of particles and alteration in the cell wall and plasma membrane as a result of the inactivation of proteins and peroxidation of lipids present in the membrane. The authors believed that the damage to the membrane structural integrity is causing bacterial cell death through metabolic alternation and potassium leakage (Naveen et al. 2021). Besides, in a recently published study, the authors reported an investigation on the multifunctional properties of spherical and well-dispersed MgONPs with a particle size range of 7 to 40 nm fabricated through

Penicillium species-Mediated Biosynthesis of Nanomaterials   61

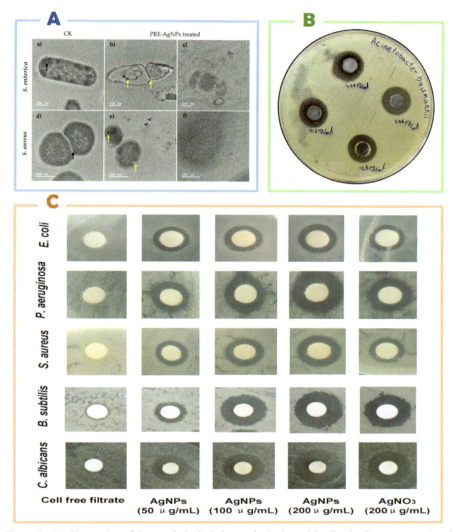

Figure 1. (A) Observation of the morphological changes in the bacterial cells viz. *S. enterica* (a–c) and *S. aureus* (d–f) in different magnification by transmission electron microscopy, (a & d) control cells, (b–c & e–f) cells after the treatment of AgNPs. Black arrows indicate the electron-dense cytoplasmic material (a) and membrane intact cell division septa (d) in control cells. Yellow arrows indicate the nicks in the plasma membrane and the presence of AgNPs inside the interior of the bacterial cells (b & e). In higher magnification, the destruction of the plasma membrane and bacterial cell debris of AgNPs treated cells can be seen (c & f) [Reprinted with permission from (Naveen et al. 2021)]; **(B)** Preliminary antibacterial activity of AgNPs against *A. baumannii* by using well diffusion assay [Reprinted with permission from (Barabadi et al. 2021)]; **(C)** The antimicrobial effects of biosynthesized AgNPs evaluated by the standard Kirby-Bauer disc diffusion method on MHA plates [Reprinted with permission from (Barabadi et al. 2017)].

harnessing metabolites secreted by *P. chrysogenum*. In this study, the potentiality of biosynthesized MgONPs as an antimicrobial agent was investigated to inhibit the pathogenic *S. aureus*, *B. subtilis*, *Pseudomonas aeruginosa*, *E. coli*, and *Candida*

62 *Mycosynthesis of Nanomaterials: Perspectives and Challenges*

albicans, and exhibited ZOIs of 12.0 ± 0.0, 12.7 ± 0.9, 23.3 ± 0.8, 17.7 ± 1.6, and 14.7 ± 0.6 mm, respectively at the highest MgONPs concentration (200 µg/mL). They discovered that when MgONPs concentrations grew, so did the inhibitory zone in a dose-dependent manner. Based on their MIC data, defined as the minimum MgO concentration that can inhibit bacterial growth, the Gram-negative bacteria were more sensitive to biogenic MgONPs than Gram-positive bacteria. The authors attributed the mentioned antimicrobial activity to the attraction between positively charged NPs and the negative charge on the lipopolysaccharide layer of Gram-negative bacteria. In addition to the previous mechanism, the authors regard the antimicrobial activity of MgONPs to inhibition of cellular functions due to blocking of the quorum sensing between cells, dissociation of Mg^{2+} ions inside the microbial cells, and alkaline effects of the formed water zone on the MgONPs surface, which due to greater pH leads to destroying the cell wall and cell membranes. The authors also checked the efficacy of biosynthesized MgONPs as larvicidal against *Anopheles stephensi,* and their results illustrated that the biogenic MgONPs exhibited high efficacy against different larvae instar and pupa of *Anopheles stephensi* with LC_{50} (50% lethal concentration) values of 12.5–15.5 ppm for I–IV larvae instar and 16.5 ppm for the pupa. Additionally, 5 mg/cm^2 of MgONPs showed the highest protection percentages as a repellent agent against adults of *A. stephensi* with values of 100% for 150 min and 67.6% ± 1.4% for 210 min. The authors summarized the mechanisms of MgONPs as mosquitocidal into two major points; first is their efficacy to produce reactive oxygen species (ROS), and second is their efficacy to destabilize the cellular equilibrium and discharge of cellular components due to an increase in the concentration of Mg^{2+} which eventually leads to mosquito cell death (Fouda et al. 2021).

Additionally, in a study, the cell-free filtrate of *P. verrucosum* was utilized to bioreduce the silver nitrate, and as a result, monodisperse fungus-based colloidal AgNPs with a size range of 10 to 12 nm were fabricated. The fungal growth inhibition assay of biogenic AgNPs against two phytopathogenic fungi via nanomaterial seeding media revealed that the AgNPs inhibited the linear growth of *Fusarium chlamydosporum* and *A. flavus* by about 50% at the concentration of 150 ppm and by more than 50% at 200 ppm. The authors proposed that the antifungal activity of NPs could be attributed to the size, shape, and capping proteins attached to the AgNPs. The authors also related the antifungal potential of AgNPs to their high concentration in the solution at which they can saturate and adhere to hyphae and lead to the damaged expression of ribosomal subunit proteins due to the loss of DNA replication ability in fungal cells (Yassin et al. 2021). Moreover, a study evaluated the antimicrobial properties and biofilm inhibitory potential of *P. chrysogenum*-derived nanosized Ag particles with spherical morphology and a mean hydrodynamic diameter of 48.2 nm against the standard and infectious *A. baumannii* via a 96-well microtiter plate-based technique. The results of their well diffusion assay, as a preliminary antibacterial activity of mycogenic AgNPs with ZOIs of 17, 18, 20, and 25 mm demonstrated dose-dependent antibacterial activity against *Acinetobacter baumannii* at concentrations of 128, 256, 512, and 1024 g/mL, respectively, as illustrated in Figure 1B. The authors attributed the mechanisms of the antimicrobial ability of AgNPs to disruption of the structural integrity of cell membrane, which raises the permeability of the bacterial membrane with the indication of potassium ions leakage, induction of apoptosis, and necrosis due to ROS overgeneration, and the proclivity of AgNPs to bind to the phosphorus and sulfur groups of phospholipid cell membranes, membrane proteins and nucleic acids, which can result in intracellular protein and DNA degradation/inactivation and eventually cell death. The MIC and MBC of AgNPs were determined to be 4 and 32 µg/mL, respectively, whilst the MIC and MBC of tetracycline were discovered to be 1024 and 8192 µg/mL, respectively. The authors suggested that fungus-based AgNPs and silver ions have the potential to interfere with transporter proteins, intracellular enzymes, and DNA, resulting in DNA fragmentation and cell destruction. In addition, their mycogenic AgNPs had significant biofilm inhibitory impact against both standard and pathogenic *Acinetobacter baumannii* at the investigated concentrations, whereas tetracycline showed dose-dependent and inferior biofilm inhibition efficacy against some of the bacterial species. The AgNPs inhibited *A. baumannii* biofilms at a concentration of 0.5*MIC (2 µg/mL), while tetracycline inhibited *A. baumannii* biofilms at a concentration of 4*MIC (4096 µg/mL). The authors proposed that the biofilm inhibitory effect of AgNPs was due to their capability to inhibit or block bacterial cell secretion of exopolysaccharides (EPSs). It was also proposed that biosorption may lead to the interaction of AgNPs with free thiol groups found in glutathione and L-cysteine of intracellular proteins, which could be the major mechanism accountable for the inactivation of biofilm formation by AgNPs (Barabadi et al. 2021).

Furthermore, in a recently published study, the extracellular synthesis of AgNPs was carried out using the cell-free filtrate of *P. toxicarium*. TEM images of fungus-mediated synthesized silver particles displayed spherical and tiny shapes of AgNPs ranging from 48.03 to 73.04 nm with an average size of 59.22 ± 8.38 nm and clear morphology. The AgNPs were tested for antibacterial activity against model organisms *S. aureus* (Gram-positive), *E. coli* (Gram-negative), and *B. subtilis* (spore-forming bacteria) using the agar well diffusion method. Their results showed that the synthesized AgNPs had antimicrobial activity on the tested bacteria except *E. coli* in a dose-dependent manner in which 25% dilution of biosynthesized AgNPs did not show any antibacterial effects. In comparison, 100 and 50% concentration of AgNPs showed antibacterial activity on *S. aureus* with ZOI of 18.67 ± 1.15 mm for 100% concentration and 12.67 ± 1.15 mm for 50% concentration and on *B. subtilis* with ZOI of 16.67 ± 2.08 mm and 11.33 ± 1.53 mm, respectively. They also tested the biofilm inhibitory effect of biomimetic silver nanoparticles against *P. aeruginosa*. The results of the biofilm formation assay showed 80.69, 48.32, 28.41, and 25.49% decrease in biofilm formation of *P. aeruginosa* at 100, 50, and 25% of mycosynthesized AgNPs and 2.5 mM $AgNO_3$, respectively. They mentioned that the biofabricated AgNPs with their large surface areas could interact with sulfuric compounds in the bacterial cell membrane, phosphorus-containing compounds in protein, and DNA, leading to the substantial inhibition effect of AgNPs on biofilm formation and bacterial cell growth (Korcan et al. 2021). Besides, a study stated a green and eco-friendly approach to the biofabrication of SeNPs by exploiting *P. expansum* with an average size ranging from 4 to 12.7 nm. In the mentioned study, the researchers assessed the antibacterial and antifungal potential and the MIC of mycosynthesized SeNPs against human pathogenic microbial strains. The inhibitory action of SeNPs represented more effective on Gram-positive than Gram-negative and

64 *Mycosynthesis of Nanomaterials: Perspectives and Challenges*

unicellular and multicellular fungi with ZOI of 36.3 ± 0.88, 30.3 ± 1.09, 28.3 ± 0.33, 26 ± 0.55, 25.6 ± 0.66, 23.7 ± 0.14, and 22.9 ± 0.49 mm for *S. aureus*, *B. subtilis*, *E. coli*, *P. aeruginosa*, *C. albicans*, *A. fumigatus*, and *A. niger*, respectively. The MIC for *S. aureus* and *B. subtilis* were found to be 62.5 µg/mL, while it was 125 µg/mL for *E. coli*, *P. aeruginosa*, and *C. albicans*. The MIC for *A. fumigatus* and *A. niger* was 250 µg/mL. The authors proposed that the mechanism of SeNPs may be attributed to NPs surface, which is electrostatically attracted to the microbial cell membrane and infiltrate into it, causing physical damage, and consequently, leakage of cellular constituents, which inhibits respiratory enzymes and leads to death of the microbial cell. Furthermore, the inhibitory effects of nanomaterials may be associated with the destruction of the DNA structure or disruption of enzyme functions by producing free radicals of hydroxyl (Hashem et al. 2021).

Interestingly, in a study, a green approach was developed to overcome the resistance pattern of *Candida* spp. by using biosynthesized crystallographic spherical AgNPs fabricated via *P. chrysogenum* with an average size of 18 to 60 nm. They assessed the susceptibility pattern of *Candida* spp. to AgNPs as compared with fluconazole and amphotericin B by using the broth micro-dilution method. Their findings revealed that sixty *Candida* isolates (100%) were susceptible to biosynthesized AgNPs, while 25 isolates (41.6%) and 30 isolates (50%) were susceptible to fluconazole and amphotericin B, respectively. On the other hand, resistant isolates towards amphotericin B (50%) were more than those recorded for fluconazole (28.3%), while 30% of Candida isolates were susceptible in a dose-dependent manner to fluconazole. The lowest MIC_{50} due to AgNPs treatment was recorded for four isolates of *C. parapsilosis* (31.5 µg/mL) as compared to other *Candida* spp., which was 62.5 µg/mL for *C. albicans* and 125 µg/mL for other species, while MIC_{90} for AgNPs was 250 µg/mL for *C. tropicalis* and *C. krusei*, while it was 125 µg/mL for *C. albicans*, *C. parapsilosis*, and *C. glabrata*. The recorded minimum inhibitory concentration 50/90 ($MIC_{50/90}$) due to other treatments was 16/64 µg/mL for fluconazole and 1/4 µg/mL for amphotericin B against *C. parapsilosis* and *C. albicans,* respectively (Soliman et al. 2021). Furthermore, a study reported not only antibacterial activity of extracellularly synthesized TeNPs with a mean diameter of 33.8 ± 6.83 nm by employing the supernatant of *P. chrysogenum* against *E. coli*, *K. pneumoniae*, *S. aureus*, and *S. epidermidis*, but also examined the anti-yeast potential of TeNPs against *C. albicans*, and *S. cerevisiae*. According to the broth micro-dilution assay, no MIC for mycofabricated TeNPs up to 250 µg/mL of elemental Te was detected in the current study against all investigated bacteria and yeasts. Higher concentrations of potassium tellurite, on the other hand, demonstrated antibacterial activity with a MIC of 5 mM (equal to 638 µg/mL of elemental Te) against *E. coli*, *K. pneumoniae*, *S. aureus* and *S. epidermidis*. Furthermore, at a MIC of 10 mM (equal to 1276 µg/mL of elemental Te), potassium tellurite demonstrated lower anti-yeast potency against *C. albicans* and *S. cerevisiae*. Their findings indicated that lower concentrations of TeNPs and potassium tellurite lacked good antimicrobial properties. They attributed this conflict of results rather than similar studies to different particle sizes and morphologies, the presence of a variety of biological-based compounds on the surface of NPs, and strain-specific antimicrobial properties of TeNPs (Vahidi et al. 2021).

Alternatively, a separate study reported a biogenic synthesis of AgNPs as prospective antibiotics from fungal metabolites of *P. oxalicum* with a nearly spherical shape with an average size of the NPs ranging from 60 to 80 nm. Using a well diffusion technique and optical density measurements in liquid broth, the antibacterial effect of biosynthesized AgNPs was assessed against *S. aureus*, *S. dysenteriae*, and *S. typhi*. The maximum ZOI recorded against *S. aureus* and *S. dysenteriae* was found to be 17.5 ± 0.5 mm for both species and 18.3 ± 0.60 mm for *S. typhi*. The lowest optical density (OD) recorded for *S. aureus*, *S. dysenteriae*, and *S. typhi* at 5 mM of AgNPs was found to be 0.031, 0.021, and 0.013, respectively after 8 h. Therefore, their results indicated that biosynthesized AgNPs showed less antibacterial activity against gram-positive strains that may be due to the thick cell wall present in these bacteria. The authors proposed that microbes cell wall contains a negative charge, whereas MNPs contain a positive charge. Due to this, an electrostatic attraction developed between microbes and targets the cell surface. This electrostatic attraction caused death immediately by oxidizing the microbes. In addition, the authors also attributed the AgNPs antibiotic mechanism to the potential of NPs to discharge ions that reacted with the thiol group (-SH) of the proteins found on the surface of bacteria. These proteins allowed the transport of nutrients by protruding out through the bacterial cell membrane. Thereby the AgNPs decreased membrane permeability, inactivated the protein, and ultimately caused cell death (Feroze et al. 2020). On the other hand, a study reported an ecofriendly, effective protocol for the biogenic synthesis of AgNPs using fungi *P. expansum* with a well dispersed, more or less spherical particles and size range of 9 to 18 nm. The results of the antibacterial assay of biosynthesized AgNPs (800 µg/mL) against four different pathogenic bacteria including methicillin-resistant *S. aureus* (MRSA), *B. subtilis*, *Klebsiella pneumoniae*, and *E. coli* showed the ZOI of 20 ± 0.09 mm, 18 ± 0.07 mm, 13 ± 0.03 mm and 15 ± 0.02 mm, respectively. The authors proposed that due to the size and crystallographic structure of the AgNPs, particles can disrupt the bacteria cell wall and cause distortion and dysfunction in the cell membrane followed by the leakage of intracellular contents and cell death (Ismail et al. 2021).

In one publication, the antimicrobial effect of biosynthesized AgNPs from *P. aculeatum* with sizes ranging from 4 to 55 nm was assessed against *E. coli*, *P. aeruginosa*, *S. aureus*, *B. subtilis*, and *Candida albicans* using the standard Kirby-Bauer disc diffusion technique. In this investigation, 200 µg/mL Ag nanoparticles exhibited the highest antimicrobial effect against all of the tested strains, with a significant difference ($P < 0.05$) when compared to 100 µg/mL AgNPs and an exceedingly significant difference ($P < 0.01$) when opposed to 200 µg/mL AgNO$_3$. The results of the mentioned antimicrobial assay were illustrated in Figure 1C (Barabadi et al. 2017). Furthermore, a study reported the biosynthesis of copper oxide NPs (CuONPs) via *P. chrysogenum* filtrate utilizing copper sulfate as a simple reduction method of copper ions and by the aid of various doses of gamma rays to evaluate the antimicrobial potential of CuONPs against selected crop pathogenic microbes. Their prepared CuONPs possessed a mean particle size diameter at 9.70 nm and exhibited maximum antifungal activity against *Fusarium oxysporum* (ZOI = 37.0 mm) followed by *Alternaria solani* (ZOI = 28.0 mm), *Aspergillus niger* (ZOI = 26.5 mm), and *P. citrinum* (ZOI = 20.7 mm). On the other hand, the mycogenic nanosized CuO particles were active as an antibacterial agent against *Ralstonia solanacearum* (ZOI = 22.0 mm) and *Erwinia amylovora* (ZOI = 19.0 mm). The antifungal activity of CuONPs was probably due to the reaction with the media components which promoted the liberation of Cu^{+2} ions. The significantly larger discharge of Cu^{+2} ions in the media may be linked to the behavior of the oxide layer on the CuONPs and the reaction with chloride ions in the media. The formed Cu^{+2} ions may produce free radicals and induce intracellular oxidative stress to the fungal cells. In addition, the authors stated that the purpose for the higher sensitivity of microbes to the CuONPs might be assigned to a higher excess of the carboxyl groups and amines on the microbial cell surfaces and elevated susceptibility of copper to these groups. The released Cu^{+2} ions may connect with the DNA and direct to the dysfunction of the helical formation by cross-linking within the nucleic acid strands. Copper ions also can interrupt the biochemical cycles inside the fungal cells. Overall, CuONPs altered bacterial cell wall and membrane configuration changed the layer permeability, created the expression of oxidative stress response genes in the bacterial cell (due to H_2O_2 generation) (El-Batal et al. 2020).

In a study, researchers stated not only antibacterial property of extra/intracellular biosynthesis of Ag nanoparticles by *P. citreonigrum* using microdilution method against *B. subtilis, S. aureus, Salmonella typhimorium, E. coli, P. aeruginosa*, but also demonstrated significant antifungal activity against *A. niger, P. expansum, P. notatum, Synthphalastrum racemosum*, and *C. albicans*. The mycosynthesized NPs were found to have a spherical morphology with particle size between 10 and 50 nm. In addition, this study showed cytoprotective efficacy and antiviral effect against the *Herpes simplex* type 2 virus. In detail, extracellularly generated AgNPs had a significant antiviral impact at concentrations of 50 μg/mL, moderate antiviral activity at concentrations of 25 μg/mL, and a weak outcome at 12.5 μg/mL, but intracellular AgNPs had considerably less antibacterial effect at concentrations of 50 and 25 μg/mL. The authors proclaimed that in the viral envelope, the disulfide bond sections of a binding domain through the glycoprotein subunit might bind to nanoparticles smaller than 10 nm because of their surface plasmon resonance and relatively high effective scattering cross-section of individual silver nanoparticles (Ali et al. 2014).

Moreover, the authors of a separate study dealt with mycosynthesis and characterization of SeNPs using *P. chrysogenum* to evaluate their antibacterial activity. Their obtained MNPs were fairly uniform with an average hydrodynamic size of 24.65 nm. The *P. chrysogenum*-mediated fabricated SeNPs revealed antibacterial activity against gram-positive bacteria, including *S. aureus* and *L. monocytogenes*, through well diffusion assay with ZOI of 10 and 13 mm, respectively. However, no ZOI was found against gram-negative bacteria, including *E. coli, S. typhimurium*, and *P. aeruginosa*. The authors proposed that owing to the significantly lesser membrane negative surface charge in Gram-positive bacteria than Gram-negative bacteria, the deposition of SeNPs on the surface of Gram-positive bacteria and inducing bacterial damage was more probable. Thereby, the Gram-negative bacteria tend to resist to SeNPs due to electrostatic repulsion between SeNPs and bacterial membrane charge. In addition, due to the absence of lipopolysaccharide membrane in Gram-positive bacteria, SeNPs could penetrate easily into that type of microbes by chemisorption

Penicillium species-Mediated Biosynthesis of Nanomaterials 67

(Vahidi et al. 2020). With that being said, a study focused on using the fungal strain *P. chrysogenum* for biogenesis of statically regulated different sizes of nano-silver to accomplish all conceivable combinations of input factors that can impact the form and size of NPs. Herein, the results of their best combination of variables to achieve the highest antibacterial effect showed AgNPs with a spherical uniform, and small-sized ranging from 3.59 to 9.25 nm, with a mean of 6.30 ± 2.19 nm had the best antimicrobial potential with ZOI of 22 and 23 mm for *S. aureus* and *E. coli*, respectively; and shown substantial antifungal activity against the unicellular yeast *C. tropicalis* (15 mm) as well as the filamentous fungus *F. solani* and *A. niger* (22 and 21 mm, respectively). The authors ascribed the antifungal efficacy of AgNPs to their effects on fungal mycelia along with targeting yeast cell membranes and altering membrane potential. Additionally, they concluded that the biological activity of microbial-mediated nanomaterials is mostly determined by the particle's nature, morphology, and size, rather than by the source. As a result, they discovered that smaller particles have a larger surface area in contact with bacterial and fungal cells and can easily diffuse or penetrate a bacterial cell membrane, inhibiting growth by interfering with natural metabolism processes or interacting with biomolecules such as DNA, lipids, and proteins, resulting in microbial death (Abd El Aty et al. 2020).

Furthermore, a study described mycofabrication of SeNPs as anti-vector malaria by employing *P. corylophilum* in the presence of ascorbic acid as a reducing agent. The SeNPs characterization findings confirmed the ability of *P. corylophilum* to build up SeNPs in a spherical shape with an average size of 29.1 to 48.9 nm. Results demonstrated the high potency of SeNPs against larvae, pupa, and adults of the third instar of *Anopheles stephensi* mosquitoes even at low concentration in which by increasing the concentration of SeNPs, the larvicidal activities reached 100% at 100 ppm of SeNPs. Also, the LC_{50} of the biosynthesized SeNPs against mosquito larvae was established at 25 ppm. The activity of SeNPs as insecticidal may be attributed to the reaction of NPs and the domain engaged with them with –SH group-containing amino acids or phosphorus-containing compounds such as nucleic acid, which denatured the larvae organelles. On the other hand, the authors mentioned that NPs causing inactivation of enzymes after penetrating through midget membranes, and then peroxidase released caused cell death (Salem et al. 2021). Besides, a study prepared cadmium oxide NPs (CdONPs) with an average particle size ranging between 0.5 and 3.0 μm utilizing *P. oxalicum* and cadmium acetate solution via coprecipitate method for antibacterial applications. The results of their bactericidal assay indicated the inhibitory effect of CdONPs on bacterial growth at 100 μg/mL used against *E. coli*, *S. aureus*, *B. cereus*, and *P. aeruginosa* was 17.6 ± 0.4, 19.6 ± 0.6, 17.5 ± 0.2, and 17.9 ± 0.4 mm, respectively. In addition, to assess bacterial susceptibility of all selected strains against prepared CdONPs, OD was measured, and among all strains, 100 μg/mL CdONPs were found to exhibit the maximum inhibitory effect in liquid broth. The authors related the result of the aforementioned test to hyperosmotic stress-induced by NPs against bacteria that disrupts cell permeability leading to cell death. Besides, according to the authors, the mechanism involving the interaction of NPs to the bacterial membrane can generate oxidative stress, impair phagocytosis, inhibit cell proliferation, reduce cell viability, and finally cell death (Asghar et al. 2020).

Notably, a study investigated the larvicidal potential of biosynthesized endophytic fungal AgNPs using *P. citrinum*. The fabricated NPs were roughly spherical, and uniform in sizes ranging from 2 to 20 nm. The results of their test on larvicidal activity biogenic AgNPs against third instar larvae of *Culex quinquefasciatus* exhibited larval mortality in a dose-dependent manner. A total of 100% mortality was achieved at 500 ppm concentrations of AgNPs at 48 h. The LC_{50} values after 24, 48, and 72 h of treatment with AgNPs were found to be 7.24, 2.38, and 1.68 ppm, respectively (Danagoudar et al. 2020). Moreover, a study reported a green biosynthetic approach to generate an antibacterial surface with mycosynthesized AgNPs, which were fabricated by the enzymatic reduction of silver nitrate using the marine algicolous endophytic fungus *P. polonicum*. Biomimetic Ag nanostructures were spherical, and the most observed sizes were between 10 and 15 nm. The results

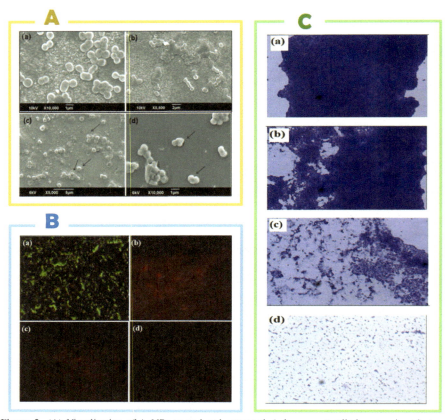

Figure 2. (A) Visualization of AgNPs treated and untreated *A. baumannii* cells by scanning electron microscope, (a) Untreated cells (b) treated cells at 1 h (c) treated cells at 2 h (d) treated cells at 5 h with AgNPs (black arrows indicate the damaged cells); **(B)** Fluorescence microscopic images of AgNPs treated *A. baumannii*, viable cells show green fluorescence and the dead cells show red fluorescence, (a) Bacterial cells at 0 h (b) at 1 h (c) at 2 h and (d) 5 h after treatment; **(C)** Imaging of antibiofilm activity by crystal violet staining, (a) Biofilm formation (before treatment) (b) Biofilm at 2 h (after treatment with MBCs of AgNPs) (c) Biofilm at 4 h of treatment (d) Planktonic cells at 24 h after treatment. [Reprinted with permission from (Neethu et al. 2020)].

confirmed that the mycogenerated AgNPs had very specific and potent bactericidal activity against *A. baumannii* as visualized under microscopic images (Figure 2A, and 2B) with the MIC and MBC of 15.6 and 31.2 µg/mL, respectively. However, they can also eradicate an established biofilm of *A. baumannii* in a rapid manner (Figure 2C). Also, as mentioned earlier, the authors formulated a bio-nanocomposite coating for the central venous catheter (CVC) using the mycogenerated AgNPs and polydopamine. The results revealed a clear inhibition zone (23.9 ± 0.8 mm) around CVC-polydopamine- AgNP nanocomposite (DCVC-AgNP), which showed significant antibacterial potency. Moreover, the result of their antibiofilm efficacy of DCVC-AgNPs proved the complete eradication of biofilms on nanoparticle-modified CVC samples. The antibiofilm activity could result from the direct interaction of AgNPs to the bacterial cell wall. The interaction and subsequent penetration resulted in the infiltration of AgNPs through the biofilm structure. The infiltration causes physical changes in the biofilms, further leading to biofilms dispersion and lysis of the bacterial cell (Neethu et al. 2020).

Anticancer activity of bioengineered nanomaterials using *Penicillium* species

Several studies stated the anticancer potential of mycosynthesized MNPs biofabricated via *Penicillium* species against different cancer cell lines. Recently, a study reported extracellular biosynthesis of AgNPs from endophytic fungus *P. radiatolobatum* to evaluate their physicochemical properties and biological activities. Their prepared AgNPs had an almost spherical shape with an average zeta potential and size of –22.2 ± 0.87 mV and 72.04 nm, respectively. Biocompatibility and cytotoxicity of biogenic AgNPs were tested in human embryonic kidney cells (HEK293) and human lung adenocarcinoma cells (A549) using WST cell viability assay. Their results showed that the higher concentration of AgNPs (120 µg/mL) did not show any cytotoxicity to HEK293 cells. However, it caused significant cytotoxicity to the A549 cells with the IC_{50} of 19.67 ± 3.09 µg/ mL through activation of ROS-induced cell death. The authors attributed the cytotoxic nature of the green synthesized AgNPs to their potential to activate the apoptotic-associated signaling cascade, thereby inducing cell death. In addition, they proposed that AgNPs from *Penicillium* sp. might be triggering the oxidative stress associated with mitochondrial-mediated apoptosis in the A549 cell line (Naveen et al. 2021). Besides, a study reported that the cell-free filtrate of *P. toxicarium* was utilized to biosynthesize AgNPs with a size distribution ranging between 232.7 and 485.9 nm with a mean hydrodynamic diameter of 340.25 ± 97.06 nm. The *Allium cepa* anaphase-telophase test was implemented to observe the cyto-genotoxicity effect of mycosynthesized AgNPs on the mitotic index (MI) and chromosome aberrations (CAs) in *A. cepa* root tips. The AgNPs exhibited cyto–genotoxicity in *Allium cepa* root meristem cells by statistically decreasing MI values from 57.17 ± 0.86 to 50.59 ± 0.56 compared to the distilled water as the negative control group (59.35 ± 0.64) in a dose-dependent manner. The reduction in MI at 25, 50, and 100% of AgNPs was statistically lower than the positive control (methyl methanesulfonate). The authors proposed that the reduction in MI could be due to the accumulation of silver compounds within the cells resulting generation of

70   *Mycosynthesis of Nanomaterials: Perspectives and Challenges*

ROS or the slower progression of cells from the S phase to the M phase of the cell cycle. In addition, a dose-dependent increase of total CAs, including chromosome laggard's stickiness, polyploidy, anaphase bridges, and disturbed ana-telophase caused by the AgNPs, was observed in ana-telophase cells of *A. cepa*. The cyto–genotoxicity mechanism of AgNPs is still unknown. However, the authors reported that AgNPs induced toxicity in *A. cepa*, leading to reduction of MI and increasing CAs, which may be attributed to the interference of AgNPs during DNA repair and the changes in protein profile (Korcan et al. 2021).

Additionally, a study reported the production of extracellular SeNPs using *P. expansum* cell filtrate. The average size of SeNPs was found in the range of 4 to 12.7 nm with good dispersion and spherical morphology. The cytotoxicity of mycosynthesized SeNPs was determined using the MTT protocol. Their findings illustrated that the biofabricated SeNPs were considered nontoxic and safe for the human normal cell line with the IC_{50} (half maximal inhibitory concentration) of 316.73 µg/mL on the normal Vero cell line (CCL-81). Moreover, the results of their investigation on the anticancer activity of SeNPs at different concentrations against prostate cancer (PC3) cell line showed the potential anticancer activity with IC_{50} of 99.25 µg/mL. The effective SeNPs concentration against PC3 cells was 125 µg/mL, because this concentration inhibited PC3 cells with 67.37% without any toxicity on human cells, where the toxicity of SeNPs against normal Vero cells was 0.74%, and cell viability was 99.26%. The authors explained the mechanism behind SeNPs cytotoxicity with the diffusion of the SeNPs into the cell membrane via the ion channels and the contact with the nitrogen bases of DNA or intracellular proteins to cause cell cycle arrest, mitochondrial dysfunction, DNA fragmentation, and eventually cell apoptosis (Hashem et al. 2021). Moreover, a study depicted biosynthesis of TeNPs with a mean hydrodynamic size of 50.16 nm and spherical shape by using *P. chrysogenum* to evaluate their physicochemical and cytotoxic characteristics on murine normal fibroblast cells (L929) and human breast cancer cells (MCF-7). Their MTT assay results (Figure 3A) showed that TeNPs have considerable anticancer activity against MCF-7 cells, with an IC_{50} of 39.83 µg/mL of elemental Te after 48 hours of treatment, whereas no IC_{50} was detected against normal L929 cells up to 50 µg/mL after 48 hours of treatment. According to the authors, the cytocompatibility of TeNPs can be contributed to phytochemicals surrounding the surface of TeNPs, which inhibit the release of Te ions and, avoid cell damage. The alteration in the morphological shape of unhealthy cells after exposure of cells to TeNPs as compared to control (untreated cell lines) was depicted in Figure 3B (Vahidi et al. 2021).

Moreover, a study depicted the production of AgNPs with an average hydrodynamic diameter of 49.99 nm via *P. expansum* to evaluate the cytotoxic and antitumor effects of AgNPs. Their findings of cytotoxicity investigations of different biosynthesized AgNPs concentrations (0–100 µg/mL) on fibroblast cells, as a normal cell line showed $IC_{50} \geq 100$ µg/mL. Also, their biological synthesized AgNPs against cancerous MCF-7 cell line showed a cytotoxic effect in a dose-dependent approach with the IC_{50} of 50.9 µg/mL after 24 h incubation. Hence, it was revealed that AgNPs were cytotoxic against cancerous MCF-7 cell line, while being nontoxic toward normal fibroblast cell line. The authors tried to attribute the antitumor effect of AgNPs to the potential ROS generation of particles by using a ROS assay and

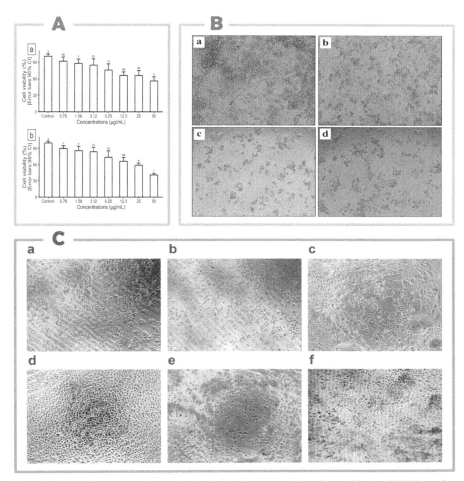

Figure 3. (A) MTT assay results confirming the in vitro cytotoxicity of mycofabricated TeNPs against MCF-7 cells after (a) 24 h and (b) 48 h of treatment [Reprinted with permission from (Vahidi et al. 2021)]; (B) Morphology of MCF-7 cells after 48 h of treatment with the maximum tested concentrations in 96-well plate (Optical Microscope; 10X): (a) Control negative (cells treated with normal saline), (b) MCF-7 cells treated with potassium tellurite at 100 μM, (c) MCF-7 cells treated with TeNPs at 50 μg/mL of elemental Te, (d) MCF-7 cells treated with doxorubicin at 50 μg/mL [Reprinted with permission from (Vahidi et al. 2021)]; (C) The morphology of HBE and A549 cells before and after exposure to AgNO$_3$ and AgNPs for 24 h (100 × magnification). (a) Untreated HBE cells, (b) HBE cells exposed to 20 μg/mL AgNO$_3$, (c) HBE cells exposed to 20 μg/mL AgNPs, (d) untreated A549 cells, (e) A549 cells exposed to 50 μg/mL AgNO$_3$, and (f) A549 cells exposed to 50 μg/mL AgNPs [Reprinted with permission from (Barabadi et al. 2017)].

morphological evaluation of AgNP-treated tumor cells. Their results showed that cells with degenerative changes lose their cell morphology and release their contents due to the ROS production of AgNPs, which can damage the genetic content and oxidize the intracellular content of cells, leading to the reduction of cell number (Ismail et al. 2021). Likewise, a study demonstrated the formation of extracellular AgNPs with particle sizes within 2–20 nm using an aqueous extract of *P. citrinum*.

72   *Mycosynthesis of Nanomaterials: Perspectives and Challenges*

The anti-proliferation potential of AgNPs was assayed using XTT against the MCF-7, A431 (skin carcinoma), HepG2 (hepatoma), and HEK-293 cell lines. The AgNPs inhibited the growth of three cancer cells in a concentration-dependent manner without affecting the non-cancer origin cells. The IC_{50} values of the AgNPs against A431, HepG2, MCF-7, and HEK-293 cells were found to be 2.09 ± 0.60, 3.16 ± 0.12, 2.40 ± 0.25, and 210.81 ± 3.12 µg/mL, respectively after 48 h incubation. This study suggested that the AgNPs had more toxicity towards cancer cells than the non-cancer cells by promoting their apoptosis pathway and inducing DNA damage in the cells (Danagoudar et al. 2020).

On the other hand, a study reported the biosynthesis of statically regulated different sizes of AgNPs with diverse biological activities using *P. chrysogenum*. The results of their cytotoxicity assay revealed that bioinspired NPs with small sizes (about 9.74 ± 5.23 nm) possessed cytotoxic action against colon cancer (HCT-116) and breast cancer (MCF-7) cell lines, with IC_{50} values of 106.9 and 90.3 µg/mL, respectively. They also tested the cytotoxicity of AgNPs on human normal cell lines to determine their safety on normal cells. Their finding showed a mildly cytotoxic impact (33.6%) on the normal melanocyte cell line (HFB4), indicating that their AgNPs sample was far more cytotoxic to tumor cell lines than the normal cell line. The authors proposed that AgNPs could be taken up by cells via diffusion or endocytosis mechanisms such as macropinocytosis, clathrin-mediated endocytosis, fluid-phase endocytosis, and PI3/Akt mediated endocytosis, causing cell damage caused by excessive generation of ROS such as superoxide radical (O^{-2}) and H_2O_2, mitochondrial dysfunction, along with activation of signaling pathways inhibiting cell proliferation (Abd El Aty et al. 2020). Besides, a study reported extracellular biogeneration of AgNPs by exploiting cell-free filtrate of *P. aculeatum* as a reducing agent. Their microscopic measurements demonstrated that AgNPs were spherical or approximately spherical, with a size between 4 and 55 nm. As illustrated in Figure 3C, This research concluded that biosynthesized AgNPs were far more biocompatible than $AgNO_3$ solution for human bronchial epithelial (HBE) cells, but it was significantly toxic in a dose-dependent manner with an IC_{50} of 48.73 µg/mL against A549 cells, indicating a potential influence on the inhibition of proliferation in human lung adenocarcinoma (A549) cells (Barabadi et al. 2017).

Moreover, a study utilizing *P. decumbens* for extracellular biological synthesis of AgNPs with spherical morphology and particle size ranging from 30 to 60 nm. Their findings revealed that biogenic AgNPs displayed cytotoxic activity against the malignant A549 cell line (IC_{50}: 80 µg/mL after 24 h), while considerably less toxicity against the normal Vero cell line at the same concentration. The authors rationalized their findings by claiming that AgNPs produce oxidative stress, which not only damages vital enzymes and causes DNA strand breaks, but also induces necrosis at higher concentrations (Majeed et al. 2016). Additionally, a study revealed extracellular creation of AgNPs utilizing *P. chrysogenum* grown under microgravity and normal conditions to assess their physicochemical and biological activity. The microgravity-fabricated AgNPs obtained showed predominantly monodisperse, spherical shape, and the size distribution ranging from 10 to 15 nm. Their finding showed that microgravity-synthesized AgNPs were significantly more cytotoxic against Hep-G2 cancer cell lines than normal gravity-synthesized AgNPs, with IC_{50}

values of 37.05 and 52.07 µg/mL, respectively. However, the IC_{50} values observed for microgravity-synthesized AgNPs were 35.10 and 43.53 µg/mL for normal gravity-synthesized AgNPs, respectively against 3T3-L1 normal fibroblasts cells. The authors attributed the cytotoxic effect of NPs against host cells to the physiochemical interaction of AgNPs with major functional groups of cellular proteins, nitrogen and phosphate groups in DNA. Moreover, regarding the size-dependent cytotoxic action, the authors hypothesized that smaller size of microgravity-synthesized AgNPs may trigger the efficiency of cytotoxicity capacity compared to bigger size normal gravity-synthesized AgNPs on cancer cells, and also they believed that the microgravity-synthesized AgNPs were likely taken up through the intracellular membrane into the internal nucleus due to their small size and showed more cytotoxic effect (Sheet et al. 2017).

Moreover, a study reported simple, green extra/intracellular biosynthesis of AgNPs using *P. citreonigrum*. Biosynthesized NPs were found to have spherical morphology and size between 10 and 50 nm. AgNPs were analyzed for cytotoxicity against MCF-7, HCT-116 colon carcinoma cell line, and HepG2 cell line. Notably, in all three malignant cell lines, the extracellular AgNPs exhibited considerably more inhibitory activity than intracellular NPs. In this context, the authors proposed that the disruption of the mitochondrial respiratory chain by AgNPs may result in the generation of ROS, which disturbed ATP production and finally led to DNA damage (Ali et al. 2014). Also, in a separate study, an eco-friendly and convenient approach for synthesizing AgNPs using *P. italicum* was reported to explore these NPs *in vitro* therapeutic potentials. The cytotoxic activities of synthesized AgNPs against MCF-7 cancer cells were assessed after 24 h using 3-(4,5-dimethylthiazol-2-yl)-2, 5-diphenyltetrazolium bromide (MTT) reduction assay. The results obtained with this colorimetric assay demonstrated a significant concentration-dependent increase in cytotoxicity against MCF-7 cells. The highest cytotoxicity (87%) was found at a concentration of 80 µg/mL of AgNPs, while at 5 µg/mL, 16% of cells were lost compared to those of control. In order to further investigate the potential of AgNPs to stimulate the death of MCF-7 cells, the authors carried out combined staining with acridine orange-ethidium bromide. The cancer cells treated with 10 µg/mL of AgNPs demonstrated a more aggressive loss of membrane integrity and chromatin condensation as compared to untreated cells. The nucleus of an apoptotic cell has a color that ranges between red and orange and chromatin with different levels of condensation or fragmentation. The authors signified that the morphological alterations in cells treated with synthesized AgNPs indicate apoptosis rather than necrosis-induced cell death. Additionally, they attributed the anticancer activities of AgNPs to the bioproducts of the fungal species that adhered to the outer surface of AgNPs (Taha et al. 2019).

Another study proposed the biogenic synthesis of SeNPs by introducing *P. corylophilum* filtrate to sodium selenite as a precursor to the fabrication of fungus-mediated nanoscaled selenium particles with a mean hydrodynamic diameter of 72 nm. The results obtained from their cell viability assay against human normal lung fibroblast (Wi 38) and human cancer colorectal adenocarcinoma epithelial (Caco-2) cell lines signified that alteration occurred in the cells was represented as loss of their typical shape, partial or complete loss of monolayer, granulation, shrinking or cell

74 *Mycosynthesis of Nanomaterials: Perspectives and Challenges*

rounding with IC_{50} value of 171.8 and 104.3 ppm for Wi 38 and Caco-2 cell lines, respectively revealing the high toxicity of SeNPs towards cancer cells compared to normal cells (Salem et al. 2021).

Other biomedical activities of bioengineered nanomaterials using *Penicillium* species

A number of studies reported the antioxidant, catalytic activity, enzyme inhibitory, and other biomedical activities of bioinspired MNPs synthesized from *Penicillium* species. Recently, a study investigated the antioxidant activity of biogenic SeNPs produced via *P. expansum* filtrate by exploitation 1,1-Diphenyl-2-picrylhydrazyl Penicillium (DPPH) free radical scavenging assay. The biofabricated SeNPs were formed in a spherical shape with dispersed narrow nanosized surrounding with active metabolites and an average size of 4 to 12.7 nm. The SeNPs depicted significant antioxidant activity as compared to ascorbic acid, where concentrations above 30 µg/L of Se-NPs had antioxidant activity above 50%. They also ran the RBC hemolytic assay to evaluate the biosafe nature of mycosynthesized SeNPs. Their results showed non-hemolytic activity on human RBCs at a concentration up to 250 µg/mL (Hashem et al. 2021). Furthermore, a study reported the biogenic development of AuNPs in the 20–40 nm range utilizing *P. rugulosum* in an extracellular approach. The AuNPs were applied to be conjugated with isolated genomic DNA from *E. coli* and *S. aureus*. Visual observation of agarose gel electrophoresis images confirmed the binding of very small concentration AuNPs with isolated DNA fragments in both organisms. The authors justified that the negatively charged phosphate backbone of the genomic DNA prevented AuNP aggregation. The ability of DNA to adhere to AuNPs promises a bright potential for the development of efficient nanostructure-based carriers for gene delivery systems (Mishra et al. 2012).

Additionally, a study described mycogenic fabrication of TeNPs via using *P. chrysogenum* with an average particulate size of 33.8 ± 6.833 nm and zeta potential of –17.4 mV. The authors evaluated the antioxidant activity and scavenging ability of fungus-based TeNPs on DPPH radicals and compared them to potassium tellurite. Their results indicated that mycogenerated TeNPs had a considerable antioxidant activity with an IC_{50} of 20.66 µg/mL of elemental Te and in a dose-dependent manner, whereas potassium tellurite had no IC_{50} at the highest tested concentration of 20 mM (equal to 2500 µg/mL of elemental Te). According to the authors, the outstanding DPPH radical scavenging activity of TeNPs can be due to their high surface-to-volume ratio, which allows TeNPs to easily react with DPPH radicals (Vahidi et al. 2021). Interestingly, extracellular biogenesis of AgNPs by *Penicillium* sp. with an average size of 18 nm was documented in a study. AgNPs showed high antibacterial, antioxidant, and anti-lipoxygenase activity, indicating that AgNPs might be used in a variety of biological functions. Furthermore, the AgNPs inhibited the tyrosinase enzyme at 98.86 ± 2.98% compared to arbutin as a standard (99.99 ± 1.28%). It also inhibited xanthine oxidase at 92.65 ± 1.81% when compared to standard (98.6%). Moreover, the release of lysosomal components from neutrophils at the inflammation site was prevented by AgNPs, as well as hemolysis of RBCs in the heat was inhibited by AgNPs (Govindappa et al. 2016).

As such, a study reported a green and efficient fabrication of extracellular AgNPs exploiting *P. oxalicum* cell filtrate at various pH levels under light radiation. When the pH was set to 12, the average size of AgNPs was 4 nm, with a spherical form, good dispersion, and high stability. At room temperature, AgNPs generated at pH 8 demonstrated a strong catalytic activity toward reducing methylene blue (MB) in the presence of $NaBH_4$. The authors proposed that the AgNPs acted as an electron relay system to overcome the kinetic barrier and catalyzed this reaction by facilitating the electron transfer between the BH_4^- donor and the MB acceptor. The authors also stated that because many organic dyes are detrimental to the environment and health, metallic NPs can play an important role as catalysts to lower the dyes catalytic performance due to their unique physicochemical characteristics and high surface area (Du et al. 2015). Moreover, a study reported the biogenic synthesis of AgNPs by using *P. citrinum* ranging from 2 to 20 nm in an extracellular approach. The authors explored the free radical scavenging activity of the AgNPs via the DPPH method. Their biomimetic AgNPs showed potent antioxidant potential, and the NPs radical scavenging ability increased with the increment in concentration. The authors proposed that the phenolic compounds from the aqueous extract of *P. citrinum* that capped the AgNPs might be responsible for the observed free radical scavenging activity (Danagoudar et al. 2020).

Besides, a study described the fungus-based biosynthesis of silver nanoparticles extracellularly using *P. italicum* with spherical and platelet-like structures in a size range between 32 and 100 nm. The biosynthesized AgNPs were proved to be potential antioxidants by showing effective and concentration-dependent radical scavenging activity against 2, 2-diphenyl-1-picrylhydrazyl, hydroxyl radical, and resazurin with 60%, 53%, and 50% antioxidant potential at 30 µg/mL against referenced antioxidant in each test, respectively. Based on their findings, the authors concluded that the synthesized AgNPs showed good antioxidant activity that allows AgNPs to be used as antioxidants and as ingredients of antioxidant formulations in the biomedical and pharmaceutical fields. In addition, antioxidants are beneficial for the management of many harmful diseases because of their scavenging ability (Taha et al. 2019).

Conclusion and future directions

Penicillium family from the fungal kingdom have demonstrated valuable properties in the biofabrication of nanostructures. In this chapter, the most recent utilization of *Penicillium* species in the biosynthesis of various MNPs and also their biological applications have been well discussed. The fabricated MNPs were mostly under 100 nm and the main production method was extracellular. The AgNPs and AuNPs have been respectively the highest proportion of the produced MNPs. The main structure of the MNPs was spherical. Overall, the significant antimicrobial, anticancer and other discussed activities represented the bioactivity of these green particles. Finally, it is believed that the use of *Penicillium* species for biofabrication of metal-based nanostructures would be an innovative approach in myco-nanotechnology.

76 *Mycosynthesis of Nanomaterials: Perspectives and Challenges*

Moreover, it is predictable that in the following years more activities and properties of the *Penicillium*-based biofabricated nanostructures will be reported.

References

Abd-Elsalam, K.A. 2021. Special Issue: Fungal Nanotechnology. J. Fungi. 7(8): 583.

Abd El Aty, A.A., Mohamed, A.A., Zohair, M.M., and Soliman, A.A. 2020. Statistically controlled biogenesis of silver nano-size by *Penicillium chrysogenum* MF318506 for biomedical application. Biocatal. Agric. Biotechnol. 25:101592.

Ali, F.T., El-Sheikh, H.H., El-Hady, M.M., Elaasser, M.M., and El-Agamy, D.M. 2014. Silver nanoparticles synthesized by *Penicillium Citreonigrum* and *Fusarium moniliforme* isolated from El-Sharkia, Egypt. Int. J. Sci. Eng Res. 5(4): 181–186.

Almaary, K.S., Sayed, S.R., Abd-Elkader, O.H., Dawoud, T.M., El Orabi, N.F., and Elgorban, A.M. 2020. Complete green synthesis of silver-nanoparticles applying seed-borne *Penicillium duclauxii*. Saudi J. Biol. Sci. 27(5): 1333–1339.

Ammar, H., and El-Desouky, T. 2016. Green synthesis of nanosilver particles by *Aspergillus terreus* HA 1N and *Penicillium expansum* HA 2N and its antifungal activity against mycotoxigenic fungi. J. Appl. Microbiol. 121(1): 89–100.

Asghar, M., Habib, S., Zaman, W., Hussain, S., Ali, H., and Saqib, S. 2020. Synthesis and characterization of microbial mediated cadmium oxide nanoparticles. Microsc. Res. Tech. 83(12): 1574–1584.

Barabadi, H. 2017. Nanobiotechnology: A promising scope of gold biotechnology. Cell. Mol. Biol. 63: 3-4. doi: 10.14715/cmb/2017.63.12.2.

Barabadi, H., Ovais, M., Shinwari, Z.K., and Saravanan, M. 2017. Anti-cancer green bionanomaterials: present status and future prospects. Green Chem. Lett. Rev. 10: 285–314. doi:10.1080/17518253.2 017.1385856.

Barabadi, H., Kobarfard, F., and Vahidi, H. 2018. Biosynthesis and characterization of biogenic tellurium nanoparticles by using *Penicillium chrysogenum* PTCC 5031: A novel approach in gold biotechnology. Iran. J. Pharm. Res. 17(Suppl2): 87.

Barabadi, H., Mohammadzadeh, A., Vahidi, H., Rashedi, M., Saravanan, M., Talank, N., and Alizadeh, A. 2021. *Penicillium chrysogenum*-derived silver nanoparticles: exploration of their antibacterial and biofilm inhibitory activity against the standard and pathogenic acinetobacter baumannii compared to tetracycline. J. Cluster Sci. in press, https://doi.org/10.1007/s10876-021-02121-5.

Boroumand Moghaddam, A., Namvar, F., Moniri, M., Md. Tahir, P., Azizi, S., and Mohamad, R. 2015. Nanoparticles biosynthesized by fungi and yeast: a review of their preparation, properties, and medical applications. Molecules. 20. doi: 10.3390/molecules200916540.

Chandrappa, C., Govindappa, M., Chandrasekar, N., Sarkar, S., Ooha, S., and Channabasava, R. 2016. Endophytic synthesis of silver chloride nanoparticles from *Penicillium* sp. of *Calophyllum apetalum*. Adv. Nat. Sci.: Nanosci. Nanotechnol. 7(2): 025016.

Chhipa, H. 2019. Mycosynthesis of nanoparticles for smart agricultural practice: a green and eco-friendly approach. In Green synthesis, characterization and applications of nanoparticles (pp. 87–109): Elsevier.

Crawford, L., Stepan, A.M., McAda, P.C., Rambosek, J.A., Confer, M.J., Vinci, V.A., and Reeves, C.D. 1995. Production of cephalosporin intermediates by feeding adipic acid to recombinant *Penicillium chrysogenum* strains expressing ring expansion activity. Bio/Technology. 13: 58–62. doi:10.1038/nbt0195-58.

Dahoumane, S.A., Jeffryes, C., Mechouet, M., and Agathos, S.N. 2017a. Biosynthesis of inorganic nanoparticles: a fresh look at the control of shape, size and composition. Bioengineering. 4(1): 14. doi:10.3390/bioengineering4010014.

Dahoumane, S.A., Mechouet, M., Wijesekera, K., Filipe, C.D.M., Sicard, C., Bazylinski, D.A., and Jeffryes, C. 2017b. Algae-mediated biosynthesis of inorganic nanomaterials as a promising route in nanobiotechnology—a review. Green Chem. 19: 552–587. doi:10.1039/C6GC02346K.

Danagoudar, A., Pratap, G., Shantaram, M., Chatterjee, B., Ghosh, K., Kanade, S.R., and Joshi, C.G. 2020. Cancer cell specific cytotoxic potential of the silver nanoparticles synthesized using the endophytic fungus, *Penicillium citrinum* CGJ-C2. Mater. Today Commun. 25: 101442.

Penicillium species-Mediated Biosynthesis of Nanomaterials 77

Dikshit, P.K., Kumar, J., Das, A.K., Sadhu, S., Sharma, S., Singh, S. et al. 2021. Green synthesis of metallic nanoparticles: applications and limitations. Catalysts. 11(8): 902.

Du, L., Xu, Q., Huang, M., Xian, L., and Feng, J.-X. 2015. Synthesis of small silver nanoparticles under light radiation by fungus *Penicillium oxalicum* and its application for the catalytic reduction of methylene blue. Mater. Chem. Phys. 160: 40–47.

Duran, N., Marcato P.D., Ingle A.P., Gade A.K., and M. Rai. 2010. Fungi mediated synthesis of silver nanoparticles: Characterization processes and applications. *In*: Rai, M., and Kövics, G. (eds.). Progress in Mycology. Springer, Dordrecht. https://doi.org/10.1007/978-90-481-3713-8_16.

El-Batal, A.I., El-Sayyad, G.S., Mosallam, F.M., and Fathy, R.M. 2020. *Penicillium chrysogenum*-mediated mycogenic synthesis of copper oxide nanoparticles using gamma rays for in vitro antimicrobial activity against some plant pathogens. J. Cluster Sci. 31(1): 79–90.

El-Sayed, A.S., and Ali, D. 2018. Biosynthesis and comparative bactericidal activity of silver nanoparticles synthesized by *Aspergillus flavus* and *Penicillium crustosum* against the multidrug-resistant bacteria. J. Microbiol. Biotechnol. in press, doi: 10.4014/jmb.1806.05089.

Fawcett, D., Verduin, J.J., Shah, M., Sharma, S.B., and Poinern, G.E.J. 2017. A review of current research into the biogenic synthesis of metal and metal oxide nanoparticles via marine algae and seagrasses. Int. J. Nanosci. 2017: 8013850. doi:10.1155/2017/8013850.

Feroze, N., Arshad, B., Younas, M., Afridi, M.I., Saqib, S., and Ayaz, A. 2020. Fungal mediated synthesis of silver nanoparticles and evaluation of antibacterial activity. Microsc. Res. Tech. 83(1): 72–80.

Fouda, A., Awad, M.A., Eid, A.M., Saied, E., Barghoth, M.G., Hamza, M.F. et al. 2021. An eco-friendly approach to the control of pathogenic microbes and anopheles stephensi malarial vector using magnesium oxide nanoparticles (Mg-NPs) fabricated by *Penicillium chrysogenum*. Int. J. Mol. Sci. 22(10): 5096.

Frisvad, J.C. 2014. *Penicillium/Penicillia* in food production. In Encyclopedia af food microbiology (pp. 14–18): Elsevier.

Frisvad, J.C., Smedsgaard, J., Larsen, T.O., and Samson, R.A. 2004. Mycotoxins, drugs and other extrolites produced by species in *Penicillium* subgenus *Penicillium*. Stud. Mycol. 49: 201–241.

Gade, A., Ingle, A., Whiteley, C., and Rai, M. 2010. Mycogenic metal nanoparticles: progress and applications. Biotechnol. Lett. 32: 593–600. doi:10.1007/s10529-009-0197-9.

Ghomi, A.R.G., Mohammadi-Khanaposhti, M., Vahidi, H., Kobarfard, F., Reza, M.A.S., and Barabadi, H. 2019. Fungus-mediated extracellular biosynthesis and characterization of zirconium nanoparticles using standard *Penicillium* species and their preliminary bactericidal potential: a novel biological approach to nanoparticle synthesis. Iran. J. Pharm. Res. 18(4): 2101.

Govindappa, M., Farheen, H., Chandrappa, C., Rai, R.V., and Raghavendra, V.B. 2016. Mycosynthesis of silver nanoparticles using extract of endophytic fungi, *Penicillium* species of *Glycosmis mauritiana*, and its antioxidant, antimicrobial, anti-inflammatory and tyrokinase inhibitory activity. Adv. Nat. Sci.: Nanosci. Nanotechnol. 7(3): 035014.

Grasso, G., Zane, D., and Dragone, R. 2020. Microbial nanotechnology: challenges and prospects for green biocatalytic synthesis of nanoscale materials for sensoristic and biomedical applications. Nanomaterials. 10(1): 11.

Hashem, A.H., Khalil, A.M.A., Reyad, A.M., and Salem, S.S. 2021. Biomedical applications of mycosynthesized selenium nanoparticles using *Penicillium expansum* ATTC 36200. Biol. Trace Elem. Res. 199: 3998–4008.

He, S., Zhang, Y., Guo, Z., and Gu, N. 2008. Biological synthesis of gold nanowires using extract of *Rhodopseudomonas capsulata*. Biotechnol. Progr. 24: 476–480. doi:10.1021/bp0703174.

Heera, P., and Shanmugam, S. 2015. Nanoparticle characterization and application: an overview. Int. J. Curr. Microbiol. App. Sci. 4: 379–386.

Hesham, A.E.L., Upadhyay, RS., Sharma, G.D., Manoharachary, C., and Gupta, V.K. 2020. Fungal Biotechnology and Bioengineering (pp. 482): Springer.

Houbraken, J., de Vries, R.P., and Samson, R.A. 2014. Modern taxonomy of biotechnologically important *Aspergillus* and *Penicillium* species. Adv. Appl. Microbiol. 86: 199–249.

Hulkoti, N.I., and Taranath, T.C. 2014. Biosynthesis of nanoparticles using microbes- a review. Colloids Surf., B. 121: 474–483. doi:10.1016/j.colsurfb.2014.05.027.

Hullmann, A., and Meyer, M. 2003. Publications and patents in nanotechnology. Scientometrics. 58: 507–527. doi: 10.1023/B:SCIE.0000006877.45467.a7.

78 *Mycosynthesis of Nanomaterials: Perspectives and Challenges*

Huston, M., DeBella, M., DiBella, M., and Gupta, A. 2021. Green synthesis of nanomaterials. Nanomaterials. 11(8): 2130.

Iravani, S., Korbekandi, H., Mirmohammadi, S.V., and Zolfaghari, B. 2014. Synthesis of silver nanoparticles: chemical, physical and biological methods. Res. Pharm. Sci. 9: 385–406.

Ismail, R.S., El-Sharkawy, R.M., Amin, B.H., and Swelim, M.A. 2021. Mycosynthesize of Ag-nanoparticles by *Penicillium expansum* and its antibacterial activity against bacterial pathogens. Plant Cell Biotechnol. Mol. Biol. 22(9-10): 13–25.

Khalil, N.M., Abd El-Ghany, M.N., and Rodríguez-Couto, S. 2019. Antifungal and anti-mycotoxin efficacy of biogenic silver nanoparticles produced by *Fusarium chlamydosporum* and *Penicillium chrysogenum* at non-cytotoxic doses. Chemosphere. 218: 477–486.

Khan, N., and Jameel, N. 2016. Antifungal activity of silver nanoparticles produced from fungus, *Penicillium fellutanum* at different pH. J. Microb. Biochem. Technol. 8: 5.

Khan, S.A., and Lee, C.-S. 2020. Green biological synthesis of nanoparticles and their biomedical applications. In Applications of nanotechnology for green synthesis (pp. 247–280): Springer.

Korcan, S.E., Kahraman, T., Acikbas, Y., Liman, R., Ciğerci, İ.H., Konuk, M., and Ocak, İ. 2021. Cyto–genotoxicity, antibacterial, and antibiofilm properties of green synthesized silver nanoparticles using *Penicillium toxicarium*. Microsc. Res. Tech. in preaa. https://doi.org/10.1002/jemt.23802.

Koul, B., Poonia, A.K., Yadav, D., and Jin, J.-O. 2021. Microbe-mediated biosynthesis of nanoparticles: applications and future prospects. Biomolecules. 11(6): 886.

Kozlovskiĭ, A., Zhelifonova, V., and Antipova, T. 2013. Fungi of the genus *Penicillium* as producers of physiologically active compounds (review). Prikl. Biokhim. Mikrobiol. 49(1): 5–16.

Kulkarni, N., and Muddapur, U. 2014. Biosynthesis of metal nanoparticles: a review. J. Nanotechnol. 2014:510246. doi: 10.1155/2014/510246.

Kumar, A., Asthana, M., Gupta, A., Nigam, D., and Mahajan, S. 2018. Secondary metabolism and antimicrobial metabolites of Penicillium in new and future developments In Microbial Biotechnology and Bioengineering (pp. 47–68). Elsevier.

Leitão, A.L. 2009. Potential of *Penicillium* species in the bioremediation field. Int. J. Environ. Res. Public Health 6(4): 1393–1417. doi:10.3390/ijerph6041393.

Lengke, M.F., Fleet, M.E., and Southam, G. 2006. Morphology of gold nanoparticles synthesized by filamentous cyanobacteria from gold(I)−thiosulfate and gold(III)−chloride complexes. Langmuir. 22: 2780–2787. doi:10.1021/la052652c.

Majeed, S., bin Abdullah, M.S., Dash, G.K., Ansari, M.T., and Nanda, A. 2016. Biochemical synthesis of silver nanoprticles using filamentous fungi *Penicillium decumbens* (MTCC-2494) and its efficacy against A-549 lung cancer cell line. Chin. J. Nat. Med. 14(8): 615–620.

Maliszewska, I., Lisiak, B., Popko, K., and Matczyszyn, K. 2017. Enhancement of the efficacy of photodynamic inactivation of *Candida albicans* with the use of biogenic gold nanoparticles. Photochem. Photobiol. 93(4): 1081–1090.

Manjunath, H.M., Joshi, C.G., and Raju, N.G. 2017. Biofabrication of gold nanoparticles using marine endophytic fungus–*Penicillium citrinum*. IET Nanobiotechnol. 11(1): 40–44.

Singh, M., Manikandan, S., and Kumaraguru, A.K. 2011. Nanoparticles: a new technology with wide applications. Res. J. Nanosci. Nanotechnol. 1(1): 1–11.

Mao, C., Flynn, C.E., Hayhurst, A., Sweeney, R., Qi, J., Georgiou, G. et al. 2003. Viral assembly of oriented quantum dot nanowires. Proc. Natl. Acad. Sci. 100(12): 6946–6951.

Mata, Y.N., Torres, E., Blázquez, M.L., Ballester, A., González, F., and Muñoz, J.A. 2009. Gold(III) biosorption and bioreduction with the brown alga *Fucus vesiculosus*. J. Hazard. Mater. 166: 612–618. doi:10.1016/j.jhazmat.2008.11.064.

Mishra, A., Tripathy, S.K., and Yun, S.-I. 2012. Fungus mediated synthesis of gold nanoparticles and their conjugation with genomic DNA isolated from *Escherichia coli* and *Staphylococcus aureus*. Process Biochem. 47(5): 701–711.

Mittal, A.K., Chisti, Y., and Banerjee, U.C. 2013. Synthesis of metallic nanoparticles using plant extracts. Biotechnol. Adv. 31: 346–356.

Molnár, Z., Bódai, V., Szakacs, G., Erdélyi, B., Fogarassy, Z., Sáfrán, G. et al. 2018. Green synthesis of gold nanoparticles by thermophilic filamentous fungi. Sci. Rep. 8: 3943. doi:10.1038/s41598-018-22112-3.

Penicillium species-Mediated Biosynthesis of Nanomaterials 79

Nabi, G., Khalid, N.R., and Khan, W. 2018. Synthesis of nanostructured based WO3 materials for photocatalytic applications. J. Inorg. Organomet. Polym Mater. 28(3): 777. doi:10.1007/s10904-017-0714-6.

Naveen, K.V., Sathiyaseelan, A., Mariadoss, A.V.A., Xiaowen, H., Saravanakumar, K., and Wang, M.-H. 2021. Fabrication of mycogenic silver nanoparticles using endophytic fungal extract and their characterization, antibacterial and cytotoxic activities. Inorg. Chem. Commun. 128: 108575. doi:10.1016/j.inoche.2021.108575.

Neethu, S., Midhun, S.J., Radhakrishnan, E., and Jyothis, M. 2020. Surface functionalization of central venous catheter with mycofabricated silver nanoparticles and its antibiofilm activity on multidrug resistant *Acinetobacter baumannii*. Microb. Pathogen. 138: 103832.

Neethu, S., Midhun, S.J., Sunil, M., Soumya, S., Radhakrishnan, E., and Jyothis, M. 2018. Efficient visible light induced synthesis of silver nanoparticles by *Penicillium polonicum* ARA 10 isolated from *Chetomorpha antennina* and its antibacterial efficacy against *Salmonella enterica* serovar *Typhimurium*. J. Photochem. Photobiol., B. 180: 175–185.

Ovais, M., Khalil, A.T., Raza, A., Khan, M.A., Ahmad, I., Islam, N.U. et al. 2016. Green synthesis of silver nanoparticles via plant extracts: beginning a new era in cancer theranostics. Nanomedicine. 11: 3157–3177. doi:10.2217/nnm-2016-0279.

Pal, G., Rai, P., and Pandey, A. 2019. Green synthesis of nanoparticles: A greener approach for a cleaner future. pp. 1–26. *In*: Green synthesis, Characterization and Applications of Nanoparticles. Elsevier.

Philip, D. 2009. Biosynthesis of Au, Ag and Au-Ag nanoparticles using edible mushroom extract. Spectrochim. Acta, Part A. 73: 374–381. doi:10.1016/j.saa.2009.02.037.

Philip, D. 2010. Green synthesis of gold and silver nanoparticles using *Hibiscus rosa sinensis*. Physica E. 42: 1417–1424. doi:https://doi.org/10.1016/j.physe.2009.11.081.

Pokropivny, V.V., and Skorokhod, V.V. 2008. New dimensionality classifications of nanostructures. Physica E. 40: 2521–2525. doi:10.1016/j.physe.2007.11.023.

Rai, M., Yadav, A., Bridge, P., and Gade, A.K. 2009. Myconanotechnology: a new and emerging science, pp. 258–267. *In*: Rai, M., and Bridge, P. (Eds.). Appl. Mycol.

Rafique, M., Nawaz, H., Rafique, M., Nabi, G., and Khalid, N.R. 2018. Material and method selection for efficient solid oxide fuel cell anode: recent advancements and reviews. Int. J. Energy Res. 43. doi:10.1002/er.4210.

Rajasekharreddy, P., Rani, P., and Bojja, S. 2010. Qualitative assessment of silver and gold nanoparticle synthesis in various plants: A photobiological approach. J. Nanopart. Res. 12: 1711–1721. doi:10.1007/s11051-010-9894-5.

Rana, A., Yadav, K., and Jagadevan, S. 2020. A comprehensive review on green synthesis of nature-inspired metal nanoparticles: Mechanism, application and toxicity. J. Cleaner Prod. 122880.

Raveendran, P., Fu, J., and Wallen, S.L. 2003. Completely "green" synthesis and stabilization of metal nanoparticles. J. Am. Chem. Soc. 125: 13940–13941. doi:10.1021/ja029267j.

Salem, S.S., Fouda, M.M., Fouda, A., Awad, M.A., Al-Olayan, E.M., Allam, A.A., and Shaheen, T.I. 2021. Antibacterial, cytotoxicity and larvicidal activity of green synthesized selenium nanoparticles using *Penicillium corylophilum*. J. Clust. Sci. 32(2): 351–361.

Sastry, M., Khan, M., and Kumar, R. 2003. Biosynthesis of metal nanoparticles using fungi and actinomycete. Curr. Sci. 85(2): 162–170.

Saxena, J., Sharma, P., and Singh, A. 2017. Biomimetic synthesis of AgNPs from *Penicillium chrysogenum* strain FGCC/BLS1 by optimising physico-cultural conditions and assessment of their antimicrobial potential. IET Nanobiotechnol. 11(5): 576–583.

Schuster, E., Dunn-Coleman, N., Frisvad, J., and van Dijck, P. 2002. On the safety of *Aspergillus niger*–a review. Appl. Microbiol. Biotechnol. 59: 426–435. doi:10.1007/s00253-002-1032-6.

Shankar, S.S., Ahmad, A., Pasricha, R., and Sastry, M. 2003. Bioreduction of chloroaurate ions by geranium leaves and its endophytic fungus yields gold nanoparticles of different shapes. J. Mater. Chem. 13: 1822–1826. doi:10.1039/B303808B.

Sheet, S., Sathishkumar, Y., Sivakumar, A.S., Shim, K.S., and Lee, Y.S. 2017. Low-shear-modeled microgravity-grown *Penicillium chrysogenum*-mediated biosynthesis of silver nanoparticles with enhanced antimicrobial activity and its anticancer effect in human liver cancer and fibroblast cells. Bioprocess. Biosyst. Eng. 40(10): 1529–1542.

80 Mycosynthesis of Nanomaterials: Perspectives and Challenges

Solanki, B., Ramani, H., Garaniya, N., and Parmar, D. 2016. Biosynthesis of silver nanoparticles using fungus *Penicillium brevicompactum* and evaluation of their anti-bacterial activity against some human pathogens. Res. J. Biotech. 11(8): 44–52.

Soliman, A.M., Abdel-Latif, W., Shehata, I.H., Fouda, A., Abdo, A.M., and Ahmed, Y.M. 2021. Green approach to overcome the resistance pattern of *Candida* spp. using biosynthesized silver nanoparticles fabricated by *Penicillium chrysogenum* F9. Biol. Trace Elem. Res. 199: 800–811.

Taha, Z.K., Hawar, S.N., and Sulaiman, G.M. 2019. Extracellular biosynthesis of silver nanoparticles from *Penicillium italicum* and its antioxidant, antimicrobial and cytotoxicity activities. Biotechnol. Lett. 41(8): 899–914.

Tahir, M.B., Nawaz, T., Nabi, G., Sagir, M., Khan, M.I., and Malik, N. 2020. Role of nanophotocatalysts for the treatment of hazardous organic and inorganic pollutants in wastewater. *Int. J. Environ. Anal. Chem.* in press. doi: 10.1080/03067319.2020.1723570.

Tyupa, D.V., Kalenov, S.V., Baurina, M.M., Yakubovich, L.M., Morozov, A.N., Zakalyukin, R.M., et al. 2016. Efficient continuous biosynthesis of silver nanoparticles by activated sludge micromycetes with enhanced tolerance to metal ion toxicity. Enzyme Microb. Technol. 95: 137–145.

Vahidi, H., Kobarfard, F., Alizadeh, A., Saravanan, M., and Barabadi, H. 2021. Green nanotechnology-based tellurium nanoparticles: Exploration of their antioxidant, antibacterial, antifungal and cytotoxic potentials against cancerous and normal cells compared to potassium tellurite. Inorg. Chem. Commun. 124: 108385.

Vahidi, H., Kobarfard, F., Kosar, Z., Mahjoub, M.A., Saravanan, M., and Barabadi, H. 2020. Mycosynthesis and characterization of selenium nanoparticles using standard *Penicillium chrysogenum* PTCC 5031 and their antibacterial activity: A novel approach in microbial nanotechnology. Nanomed. J. 7(4): 315–323.

Verma, S., Abirami, S., and Mahalakshmi, V. 2013. Anticancer and antibacterial activity of silver nanoparticles biosynthesized by *Penicillium* spp. and its synergistic effect with antibiotics. J. Microbiol. Res. 3(3): 54–71.

Visagie, C.M., Houbraken, J., Frisvad, J.C., Hong, S.-B., Klaassen, C.H.W., Perrone, G. et al. 2014. Identification and nomenclature of the genus *Penicillium*. *Stud. Mycol.* 78: 343–371. doi:10.1016/j.simyco.2014.09.001.

Wanarska, E., and Maliszewska, I. 2019. The possible mechanism of the formation of silver nanoparticles by *Penicillium cyclopium*. Bioorg. Chem. 93: 102803.

Wang, Y., and Xia, Y. 2004. Bottom-up and top-down approaches to the synthesis of monodispersed spherical colloids of low melting-point metals. Nano Lett. 4: 2047–2050. doi:10.1021/nl048689j.

Xie, J., Lee, J.Y., Wang, D.I.C., and Ting, Y.P. 2007. Identification of active biomolecules in the high-yield synthesis of single-crystalline gold nanoplates in algal solutions. Small. 3: 672–682. doi:10.1002/smll.200600612.

Yadav, A.N., Verma, P., Kumar, V., Sangwan, P., Mishra, S., Panjiar, N. et al. 2018. Biodiversity of the genus *Penicillium* in different habitats. In New and future developments in microbial biotechnology and bioengineering (pp. 3–18): Elsevier.

Yassin, M.A., El-Samawaty, A.E.-R.M., Dawoud, T.M., Abd-Elkader, O.H., Al Maary, K.S., Hatamleh, A.A., and Elgorban, A.M. 2017. Characterization and anti-*Aspergillus flavus* impact of nanoparticles synthesized by *Penicillium citrinum*. Saudi J. Biol. Sci. 24(6): 1243–1248.

Yassin, M.A., Elgorban, A.M., El-Samawaty, A.E.-R.M., and Almunqedhi, B.M. 2021. Biosynthesis of silver nanoparticles using *Penicillium verrucosum* and analysis of their antifungal activity. Saudi J. Biol. Sci. 28(4): 2123–2127.

Youssef, K., Hashim, A.F., Hussien, A., and Abd-Elsalam, K.A. 2017. Fungi as ecosynthesizers for nanoparticles and their application in agriculture. In Fungal Nanotechnology (pp. 55–75): Springer.

Zehetbauer, M.J., and Zhu, Y.T. 2009. Bulk Nanostructured Materials (pp. 736). John Wiley & Sons.

Zhang, X., Yan, S., Tyagi, R.D., and Surampalli, R.Y. 2011. Synthesis of nanoparticles by microorganisms and their application in enhancing microbiological reaction rates. Chemosphere. 82: 489–494. doi:10.1016/j.chemosphere.2010.10.023.

Zou, H., Wu, S., and Shen, J. 2008. Polymer/silica nanocomposites: preparation, characterization, properties, and applications. Chem. Rev. 108: 3893–3957. doi:10.1021/cr068035q.

CHAPTER 4

Biosynthesis of Iron Oxide Nanoparticles Using Fungi

*Amanpreet K. Sidhu** and *Priya Kaushal*

Introduction

Nanotechnology is a promising and rapidly growing discipline of research that has had a lot of success in the current era of technology. Nanoparticles are materials with distinctive size (usually ranging from 1 to 100 nanometers), structure, physio-chemical, electric, magnetic, thermal, mechanical, catalytic, and optical scattering (Chen et al. 2013; Rastar et al. 2013; Pantidos and Horsfall 2014; Arya et al. 2019; Hussain et al. 2019; Saleh et al. 2019; Turan et al. 2019; Vasantharaj et al. 2019). Nanoparticles properties and reactivity are mostly determined by their super tiny size and vast surface area (Campos et al. 2015; Srikar et al. 2016; Mourdikoudis et al. 2018). These are fascinating because they exist at the interface between bulk matter and atomic or molecular structures. A bulk material's physical and chemical properties remain constant regardless of its size, however, this is not always the case at the nanoscale. When investigated at the nanoscale, a number of bulk materials have revealed some amazing properties (Thakkar et al. 2010). The structure of nanoparticles is complex- they have three layers: the surface layer, the shell, and the core. The existence of functional groups like metal ions, surfactants, small molecules and polymers distinguishes one layer from the others. Nanoparticles are commonly symbolized as core (Khan et al. 2019). These have unique qualities such as size, shape, composition, and structural framework, which must be tuned during the synthesis process (Revati and Pandey 2011). Depending on their well-defined characteristics, nanoparticles are categorized into various classes like-carbon and lipid-based, metal, ceramics, semiconductor and polymeric (Bhatia 2016; Amoabediny et al. 2018; Ealias and Saravanakumar 2017; Khan et al. 2019). Iron,

---

Department of Biotechnology, Khalsa College, G.T. Road, Amritsar, Punjab 143002, India.
* Corresponding author: amanpreetkaursidhu@khalsacollege.edu.in

82   *Mycosynthesis of Nanomaterials: Perspectives and Challenges*

silver, copper, nickel, platinum, zinc, titanium, palladium, indium, silicon, tin, and gold oxides/dioxides have a wide range of applications in the environment (remediation, toxicology, water treatment, photodegradation), catalysis, textiles, electronics (optical limiting devices, batteries), mechanical industries, energy scavenging (nanogenerators) and pharmaceutical sector (drug delivery, cancer therapy, tissue repair) (Oskam 2006; Laurent et al. 2008; Kulkarni and Muddapur 2014; Aminabad et al. 2019; Kumar et al. 2018; Mei and Wu 2018; Odularu 2018; Khan et al. 2019; Malakootian et al. 2019; Massironi et al. 2019; Saleh et al. 2019; Shnoudeh et al. 2019; Vahed et al. 2019; Vasantharaj et al. 2019; Vinci and Rapa 2019; Yusof et al. 2019; Borah et al. 2020; Hernández et al. 2020; Vahidi et al. 2020). Iron oxides are one of the biologically acceptable nanoparticles because they have striking physical characteristics such as low curie temperature, a high surface area to volume ratio, superparamagnetism, stability in a liquid solution, and less prone to oxidation with broad applications in the biomedical field (gene, drug delivery, tissue repair, diagnosis), environment cleaning such as antibiotic degradation, dye adsorption, food-related processes, cosmetics, and controlling various pathogenic infections such as fungus, bacteria, and viruses (Khalil et al. 2017; Beheshtkhoo et al. 2018; Vasantharaj et al. 2019; Bhuiyan et al. 2020; Gao et al. 2020; Malhotra et al. 2020). The core and shell structure of iron oxide nanoparticles (IONPs) is also characteristic. Because of their core-shell structure, IONPs have properties that are similar to both hydrous iron oxides and metallic iron (Li and Zhang 2006). For their innate enzyme-like activity, these are now referred to as "Nanozymes" (Zhang et al. 2019; Gao et al. 2020). These particles have the ability to bind to peptides, nucleic acids, lipids, fatty acids, and numerous metabolites, among other biological substances (Singh et al. 2016; Jubran et al. 2020). IONPs have been synthesized using a variety of chemical, physical, and biological processes. Co-precipitation, micro-emulsion techniques, flow–injection, reverse micelles, sol-gel synthesis, and hydrothermal reactions are all examples of chemical synthesis (Laurent et al. 2008; Wu et al. 2008). The key stumbling block in these procedures is particle dispersion, clumping, and size equalization. Chemical-based approaches also include the use of solvents such as sodium borohydride, hydrazine, potassium bitartrate, and sodium dodecyl sulphate, all of which are hazardous to the environment since they produce noxious waste flows. Physical synthesis includes milling, grinding, and thermal ablation; all of these processes are time-consuming because of the high energy demand. Another significant disadvantage of these methods is incredibly time-consuming (Moghaddam et al. 2014; Rauwel et al. 2015; Khalil et al. 2017; Gahlawat and Choudhury 2019; Borah et al. 2020). Recently, there is a shift towards a greener, safer, biologically acceptable, stable, cost-effective, quick, and environmentally friendly process of synthesis (Salam et al. 2012; Senthilkumar and Sivakumar 2014; Aziz et al. 2015; Adelere and Lateef 2016; Ovais et al. 2016). Plants, bacteria, algae, actinomycetes, viruses, and fungi are used in the biological biosynthesis of iron oxide nanoparticles (Thakkar et al. 2010; Schröfel et al. 2014; Mirza et al. 2018; Sharma et al. 2019; Vasantharaj et al. 2019). This organic synthesis is based upon the use of a universal solvent, water, to produce nanoparticles free of poisonous chemical contaminants, which has led to their widespread acceptance in the biomedical sector (Pantidos and Horsfall 2014; Gholampoor et al. 2015).Green synthesis is based upon two basic stages, i.e., bioreduction and biosorption. Bioreduction is the chemical reduction of metal ions into stable forms, whereas biosorption is the process of attaching metal ions to the surface of organisms such as cell walls and peptides to form more stable complexes (Pantidos and Horsfall 2014). Furthermore, biological synthesis does not include the use of external capping agents for the fabrication of nanoparticles. Moreover, when compared to physicochemical approaches, this sort of synthesis takes less time (Gahlawat and Choudhury 2019). In this chapter, we focus on the biological synthesis of iron oxide nanoparticles by fungi as well as their benefits and drawbacks.

Magnetic Iron oxide nanoparticles

Iron oxides are widespread compounds in nature and can be easily synthesized in the laboratory. Eight different iron oxides are known of which magnetite (Fe_3O_4) and maghemite ($\gamma\ Fe_2O_3$) are quite popular (Wu et al. 2008). Apart from their strong magnetism, both magnetite and maghemite are shown to be biocompatible, being very attractive for different biomedical applications (Ajinkya et al. 2020).

Magnetite (Fe_3O_4): It contains both divalent and trivalent iron, it differs from other iron oxides. Its magnetic, electrical, and thermal properties are out of the norm. They are used in environmental contaminant removal and cell separation, magnetically guided medication delivery, sealing agents, damping and cooling mechanisms in loudspeakers, and contrast agents for magnetic resonance imaging (MRI) due to their outstanding characteristics (Blaney 2007).

Maghemite ($\gamma\ Fe_2O_3$): Moreover, γ-Fe_2O_3 is an ideal candidate for the fabrication of luminescent and magnetic dual-functional nano-composites due to its excellent transparent properties (Wu et al. 2010). Maghemite (γ-Fe_2O_3), a typical ferromagnetic mineral, is thermally unstable and is transformed to hematite at higher temperatures (Campos et al. 2015).

Methods for the synthesis of Iron oxide nanoparticles

Several approaches for the manufacture of iron oxide nanoparticles have been developed in recent years for use in various technological fields. Co-precipitation, microemulsion (Chin and Yaacob 2007), sol-gel (Woo et al. 2003), hydrothermal synthesis (Hayashi and Hakuta 2010; Alangari et al. 2022), thermal decomposition (Rane and Verenkar 2001), electrochemical deposition (Karimzadeh et al. 2017), and sonochemical synthetic route are some of the commonly used methods (Hassanjani et al. 2010). They can also be manufactured by processes such as laser pyrolysis (Alexandrescu et al. 2004), green synthesis using plant extracts (Latha and Gowri 2012), and microorganisms (Mazumdar and Haloi 2011; Mohamed et al. 2015).

The most common method for making magnetic nanoparticles is chemical co-precipitation. Nucleation, growth, and secondary processes such as agglomeration, attrition, and breakage are all part of the co-precipitation process (Sulistyaningsih et al. 2017). Using this method, one can synthesized nanoparticles with desired size, size distribution, and magnetic characteristics by selecting appropriate reactant

84 Mycosynthesis of Nanomaterials: Perspectives and Challenges

materials, reactant conditions, route, and experimental instruments. Factors that determine the precipitation process can influence the size and size distribution. Several research has been conducted to adjust parameters such as pH, temperature, ionic strength, salt nature, chelating agent addition, and Fe^{3+}/Fe^{2+} ratio, all of which affect the size and size distribution of manufactured particles (Lodhia et al. 2009). In the co-precipitation process, a number of variables influence the size and magnetic characteristics of nanoparticles. The characteristics of nanoparticles are also influenced by the type of alkali used. As a precipitant agent, some researchers utilized ammonia, while others used NaOH aqueous solution (Eivari and Rahdar 2013).

The size of the particles generally decreases as the Fe^{3+}/Fe^{2+} ratio, pH, and ionic strength of the medium increases. Fe^{3+}/Fe^{2+} ratio smaller than 0.3 results in the formation of Goethite. There are two phases for ratios between 0.3 and 0.5, consisting of tiny and big size nanoparticles. A ratio of 0.5, on the other hand, corresponds to magnetite stoichiometry, in which the produced particles are uniform in size and content (Jolivet et al. 1992). The higher the Fe^{3+}/Fe^{2+} ratio, the larger the particles that form. Monodisperse particles are produced when the ratio is between 0.4 and 0.6 (Babes et al. 1999). The size of the produced nanoparticles decreases as the pH rises. Particles shrink in size at higher pH and ionic strength, while they grow in size at lower pH and ionic strength due to the aging phase associated with Ostwald ripening (Vayssieres et al. 1998).

After adding alkali to observe the precipitation of Fe_3O_4 MNPs, two distinct reactions could occur during the precipitation of Fe_3O_4 from Fe^{2+} and Fe^{3+} salt combinations. The possible reaction for the formation of Fe_3O_4 magnetic nanoparticles is as follows:

$$Fe^{3+} + 3OH^- \rightarrow Fe(OH)_3$$

$$Fe(OH)_3 \rightarrow FeOOH + H_2O$$

$$Fe^{2+} + 2OH \rightarrow Fe(OH)_2$$

$$2FeOOH + Fe(OH)_2 \rightarrow Fe_3O_4\downarrow + 2H_2O$$

The reaction is quick and high-yielding and magnetite crystals appear almost immediately after the iron source is added (Mahdavi et al. 2013). Peternele et al. (2014) used co-precipitation to create iron oxide nanoparticles with limited size distribution and diameters between 5.05 and 7.21 nm. The reaction was carried out in an aqueous solution with a molar ratio of $Fe^{2+}/Fe^{3+} = 0.5$ and a pH of using both NaOH and NH_4OH separately in an aqueous solution with a molar ratio of $Fe^{2+}/Fe^{3+} = 0.5$. Khalil (2015) demonstrated the manufacture of monodispersed magnetite nanoparticles with a diameter of 7.84,0.05 nm at pH 9-11 using ammonium hydroxide and a single iron precursor (III). Eivari and Rahdar (2013) synthesized iron oxide nanoparticles with sizes ranging from 8 to 17 nm under various temperature and ammonium hydroxide concentration conditions, finding that increasing temperature affects different properties of nanoparticles while changing ammonium hydroxide concentration has no effect. Several studies have suggested that using surfactants like polyvinyl alcohol or dextran to create monodispersed NPs is an efficient method. These agents can be added to the reaction media or coated onto the particles after synthesis. Surfactants act as a capping agent, allowing particle size to be controlled and colloidal dispersions to be stabilized.

Characterization of the Iron oxide nanoparticles

Comprehensive surface characterization techniques are utilized to analyze the nanoparticles in order to have a deeper knowledge of their surface properties. The following are some of the most common approaches used to study iron oxide nanoparticles (Figure 1).

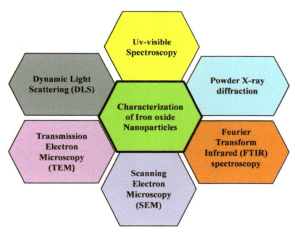

Figure 1. Characterization of Nanoparticles.

UV-visible absorption spectroscopy

Absorbance spectroscopy is a straightforward approach for determining a solution's optical characteristics. The concentration of molecules in a solution is measured by the amount of monochromatic light absorbed by the solution. Beer-Lamberts Law can be used to estimate the concentration of a solution. UV-visible absorption spectroscopy in the wavelength range of 200–700 nm is used to identify iron oxide nanoparticles. The collective oscillation of electrons in the conduction band at the surface of nanoparticles absorbs UV-visible electromagnetic radiation. The Surface Plasmon Resonance effect is the name for this phenomenon. Absorption bands for iron oxide nanoparticles are often observed in the 330–450 nm range (Vélez et al. 2016).

Powder X-ray diffraction (XRD) analysis

X-ray diffraction is a common method for determining important properties such as crystal structure, crystallite size, and strain. In nanoparticle research, X-ray diffraction patterns have been frequently used. Diffraction peaks in nanocrystalline materials are widened due to the randomly aligned crystals (Akbari et al. 2011). With the number of ingredients, the intensity increases or decreases. This technique is used to determine the metallic nature of particles and provides information on the unit cell's translational symmetry size and shape, as well as information on electron density inside the unit cell, i.e., where the atoms are located, from peak intensities (Akbari et al. 2011; Heera and Shanmugam 2015).

Fourier Transform Infrared (FTIR) spectroscopy

The functional groups associated with the nanoparticles are determined using FTIR spectroscopy. The vibration of Fe-O bonds in iron oxide is connected with the FTIR absorption bands at low wavenumbers below 700, i.e., between 500 and 700 cm^{-1}. The hydroxyl groups on the surface of the iron oxide nanoparticles are responsible for the unique absorption peak of about 3400 cm^{-1} (Mahdizadeh et al. 2012).

Dynamic Light Scattering (DLS) analysis

One of the most used methods for determining the size of MNPs is DLS, also known as photon correlation spectroscopy. The MNP suspension is exposed to a light beam (electromagnetic wave) during the DLS measurement, and the direction and intensity of the light beam are both altered as the incident light impinges on the MNP due to a process known as scattering. Because the MNPs are in constant random motion due to their kinetic energy, the fluctuation of intensity with time provides information about that random motion and may be utilized to calculate the particle diffusion coefficient. For spherical particles, the hydrodynamic radius of the particle RH may be derived from its diffusion coefficient using the Stokes-Einstein equation $D f = kBT/6\pi\eta RH$, where kB is the Boltzmann constant, T is the suspension temperature, RH is the hydrodynamic radius of the particle and η is the viscosity of the surrounding media.

Microscopic techniques

The morphology and size of nanoparticles are studied using microscopic techniques such as scanning electron microscopy (SEM) and transmission electron microscopy (TEM). In the realm of nanotechnology, these approaches are commonly used to characterize produced nanoparticles. Microscopy equipped with EDS (energy-dispersive X-ray spectroscopy) is used for analyzing the elements (elemental composition) present in a material giving information about the composition of nanoparticles alongwith capping existing on the surface of the nanoparticles (Tarafdar and Raliya 2013; Abdeen et al. 2016).

Fungi as a biological source for the green synthesis of the IONPs

Extracellular synthesis of iron oxide nanoparticles using fungal species is advantageous in terms of reproducibility of the synthesized nanoparticles, ease of scale-up, use of inexpensive raw materials for growth, high biomass forming capacity, simple downstream steps, low residue toxicity, and economic feasibility

(Tarafdar and Raliya 2013; Guilger-Casagrande and de Lima 2019; Chatterjee et al. 2020). Furthermore, fungal species have higher tolerance and bioaccumulation, which aids in the production of metal nanoparticles (Agarwal et al. 2017). Microbes ability to accumulate and/or extract metals has been used in various biological processes such as bioleaching, heavy metal removal, and bioremediation (Beveridge et al. 1996), and this capability of microbes has been used to accumulate and/or extract metals in various biological processes such as bioleaching, heavy metal removal, and bioremediation (Gericke and Pinches 2006). Fungi have been found to release a significant number of extracellular enzymes capable of hydrolyzing metals. The creation of numerous extracellularly generated nanoparticles is hypothesized to be caused by this enzymatic reduction of metal ions. Nitrate reductase is one such enzyme found in fungi (Abdeen et al. 2016). Although the overall mechanisms of metal nanoparticles (MNPs) synthesis employing fungus are still unknown.

Biosynthesis routes of Iron oxide nanoparticles using fungi

The ability of fungi to synthesize MNPs is thought to be dependent on the biological material utilized and the reaction conditions used. The strain of microorganism utilized, the growth parameters, and the method of sample preparation are all considered biological material, whereas the reaction circumstances include the metal precursor used, its concentration, pH, and reaction temperature (Silva et al. 2015). When reading the literature, one will notice a lot of variation in the beginning bio-material used for biosynthesis. To make the nanoparticles, some methods used fungal biomass, while others used the fungal culture's cell-free filtrate or fungal biomass homogenate, and yet others used fungal cell filtrate. The most extensively used approach required cultivating the fungi in the appropriate growth medium, then centrifuging the fungal mycelia to separate them. The mycelia is then combined with a metal precursor salt and cultured for 4–5 days at 28 degrees Celsius. This solution is filtered again, resulting in a suspension comprising manufactured nanoparticles (Kaul et al. 2012; Bhargava et al. 2013). In another method the fungi are cultivated in a suitable media and the mycelia biomass is isolated by filtration. The fungal biomass is dispersed in sterile distilled water and cultured on a rotary shaker for 1–5 days after being thoroughly washed with sterile distilled water. After incubation, cell-free filtrate (CFF) is collected by filtering out mycelium with Whatman filter paper1. The cell-free filtrate is then treated with an iron precursor salt mixture consisting of ferric chloride ($FeCl_3.6H_2O$) and ferrous sulfate ($FeSO_4.7H_2O$) in a 2:1 molar ratio (Chatterjee et al. 2020), $FeCl_3$ (103 M) (Tarafdar and Raliya 2013) and then incubated for a few hours on the rotary shaker to produce IONP production. For the manufacture of iron oxide nanoparticles, Abdeen et al. (2016) developed a method that combined microbial and physical processes. Using a cyclomixer, a fine homogenate of fungal mycelium was made in distilled water. The metal precursor salt $FeCl_3$ or $FeSO_4$ was added to this homogenate at a concentration of 2000 ppm. Under static conditions, the resultant mixture is incubated for 6 days. Centrifugation was used to collect the synthesized IONPs, which were then extensively cleaned with ethanol. The nanoparticles were then dried further in a supercritical pressure reactor at 300°C and 850 psi pressure. Fungal cell filtrate (FCF) was employed

88 *Mycosynthesis of Nanomaterials: Perspectives and Challenges*

to make iron oxide nanoparticles in a more dramatic variety. The selected fungal strain was cultivated in a suitable medium, and the culture medium was filtered through Whatman filter paper 1 using a vacuum pump after the incubation period was completed. To acquire the final fungal cell filtrate, the filtrate is centrifuged and re-filtered (FCF). At room temperature, an equal amount of FCF and a salt solution containing ferric trichloride ($FeCl_3$) and ferric dichloride ($FeCl_2$) (2:1 mM final concentration) are stirred for 5 minutes. The synthesis of IONPs is demonstrated by an abrupt shift in colour (Mahanty et al. 2019). This appears to be the simplest technique for fungal production of IONPs, eliminating the need for time-consuming filtration methods and costly equipment. Iron oxide nanoparticles were produced from the cell-free extract of *A. niger* BSC-1 in one study. Magnetite nanoparticles have a crystalline orthorhombic structure, according to XRD measurements. The average size of 17.29 nm was computed using the Debye-Scherrer equation, while the width of the IONPs evaluated by TEM was found to be between 20 and 40 nm. The particles were superparamagnetic, with a saturation magnetization of 55.19 emu/g, a coercivity of 0.0045 Oe, and a retentivity of 1.61 emu/g (Chatterjee et al. 2019). DLS analysis revealed that nanoparticles generated by *A. oryzae* TFR9 ranged in size from 10 to 24.6 nm. The production of spherical IONPs is depicted in a TEM micrograph (Tarafdar and Raliya 2013). The nanoparticles produced by *Aspergillus niger* mycelia homogenate YESM Pure iron and a large proportion of Fe_3O_4 spherical nanoparticles were obtained under supercritical ethanol conditions. For iron and Fe_3O_4 nanoparticles, the average size measured by SEM measurement is 18 and 50 nm, respectively. The magnetic characteristics of Fe and Fe_3O_4 nanoparticles indicate superparamagnetic and ferromagnetic-like behavior, respectively (Abdeen et al. 2016). The following Table 1 and Figure 2 provide concise information on the many reports encountered for the synthesis of IONP's by the fungus to date.

Mechanism of synthesis of IONPS by fungi

The biological synthesis of metal oxide nanoparticles has been extensively used as biological sources containing a wide range of biomolecules such as alkaloids, saponins, tannins, flavonoids, proteins, carbohydrates, steroids, enzymes (NADPH-dependent nitrate reductases, iron reductases), amino acids, cell wall components (alginate, laminarian), and exopolysaccharides (EPS) that aid in bioreduction, stabilisation, and capping of nanoparticles (Kumar et al. 2007; Ishak et al. 2019; Jacinto et al. 2020; AlKhattaf 2021). Although the underlying mechanism of bioreduction of metal ions by these major biomolecules is yet unknown, research findings suggest that surface chemistry (functional groups such as -CHO, -C=O, -COOH, -OH, -C-O-C, -C C) of these reducing agents plays a significant role in metal ion reduction (Gan and Li 2012; Akhtar et al. 2013; Jeevanandam et al. 2016; Ishak et al. 2019).

Iron oxide nanoparticles generated by chemical and physical processes are typically highly reactive in nature and can rapidly agglomerate when exposed to ambient oxygen. Furthermore, nanoparticles tend to agglomerate in order to reduce

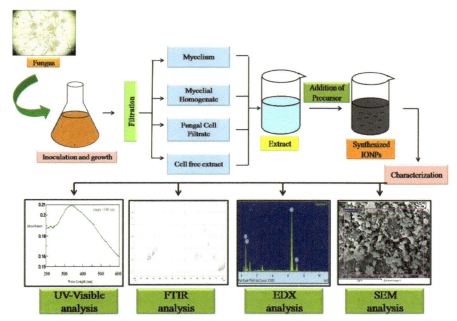

Figure 2. Biosynthesis routes of Iron Oxide Nanoparticles using fungi.

surface energy resulting from the high surface-area-to-volume ratio. To prevent agglomeration of the generated nanoparticles, a variety of stabilizers are added during the chemical and physical production processes. Biomolecules present in the extract act as capping agents and stable the generated nanoparticles, therefore no such stabilizing compounds are required in biological synthesis methods (Gudikandula and Charya Maringanti 2016). The size of iron oxide nanoparticles used in biomedical applications should be less than 20 nm, they should be uncapped, have strong magnetic characteristics, and have a critical transition temperature (Tc). Different reaction parameters such as temperature, pH, stirring, metal salt concentration, and extract volume are critical in the green synthesis strategy to create nanoparticles with the desired form, size, crystallinity, purity, stability, and morphology.

Use of fungal extract as capping agent

Due to the presence of huge amounts of proteins, high yields of nanoparticles, simplicity of handling, and creation of less harmful effluents, fungi are considered one of the most appealing sources for biological nanoparticle synthesis (Sidhu et al. 2022). In the creation of metal nanoparticles, fungi operate as both a reducing and stabilising agent. The biomolecules found in fungi bind nanoparticles, boosting their stability, inhibiting agglomeration, and aiding in the enhancement of the nanoparticles' biological activity and stability (Guilger-Casagrande and de Lima 2019; Priya et al. 2021). These biomolecules have a strong propensity for adhering

Table 1. Biosynthesis of Iron oxide nanoparticles by using fungi (Retracted from Priya et al. 2021 with upgradation).

Sr. No.	Fungal strain	Bio-material used	Iron Precursor used	Size (nm)	Shape	Magnetization	References
1.	*Fusarium oxysporum*	Fungal biomass	$K_3[Fe(CN)_6]$ and $K_4[Fe(CN)_6]$ (2:1 mM)	20–50 nm	Quasi-spherical	Ferrimagnetic	Bharde et al. 2006
2.	*Verticillium sp.*	Fungal biomass	$K_3[Fe(CN)_6]$ and $K_4[Fe(CN)_6]$ (2:1 mM)	100 to 400 nm	Cubo-octahedrally	Ferrimagnetic	Bharde et al. 2006
3.	*Curvularia lunata*	Fungal biomass	Fe_2O_3 (1000 ppm)	20.7 nm	-	-	Kaul et al. 2012
4.	*Chaetomium globosum*	Fungal biomass	Fe_2O_3 (1000 ppm)	25 nm	-	-	Kaul et al. 2012
5.	*A. fumigatus*	Fungal biomass	Fe_2O_3 (1000 ppm)	42.4 nm	-	-	Kaul et al. 2012
6.	*A. wentii*	Fungal biomass	Fe_2O_3 (1000 ppm)	46 nm	-	-	Kaul et al. 2012
7.	*Aspergillus oryzae* TFR9	Cell free filtrate	$FeCl_3$ (1mM)	10–24.6 nm	Spherical	-	Tarafdar and Raliya 2013
8.	*Aspergillus japonicus*	Fungal biomass	Potassium ferricyanide $K_3[Fe(CN)_6]$ and Potassium ferrocyanide $K_4[Fe(CN)_6]$ (1 mM:0.5 mM)	60–70 nm	Cubic	-	Bhargava et al. 2013
9.	*Aspergillus* sp.	Fungal biomass	$FeSO_4$ (3.0 mM)	5–200 nm	Spherical	-	Pavani and Kumar 2013
10.	*Alterneria alternate*	Fungal biomass	Iron(III) Nitrate (1 mM)	9 nm	Spherical	-	Mohamed et al. 2015
11.	*Aspergillus niger* YESM	fungal homogenate	$FeCl_3$, $FeSO_4$ (2000ppm each)	18 nm for Fe and 50 nm for Fe_3O_4	Spherical	Superparamagnetic for Fe and Ferromagnetic-like for Fe_3O_4	Abdeen et al. 2016

No.	Fungus	Source	Precursor	Size	Shape	Property	Reference
12.	*Trichoderma asperellum*	fungal cell filtrate	$FeCl_2$:$FeCl_3$ (1:2 mM)	18–32 nm	Spherical	-	Mahanty et al. 2019
13.	*Phialemoniopsis ocularis*	fungal cell filtrate	$FeCl_2$:$FeCl_3$ (1:2 mM)	6–22 nm	Spherical	-	Mahanty et al. 2019
14.	*Fusarium incarnatum*	fungal cell filtrate	$FeCl_2$:$FeCl_3$ (1:2 mM)	15–55 nm	Spherical	-	Mahanty et al. 2019
15.	*Aspergillus niger* BSC-1	Cell free filtrate	$FeCl_3$, $6H_2O$) and ferrous sulphate ($FeSO_4$.$7H_2O$) in 2 mM:1 Mm	20–40 nm	Orthorhombic	Superparamagnetic	Chatterjee et al. 2020
16.	*Rhizopus stolonifer*	Cell free filtrate	1M $FeCl_3$	-	-		Adeleye et al. 2020
17.	*Aspergillus flavus*	Cell free filtrate	1m M $FeSO_4$	28.6–33.8 nm	Spherical		Gouda et al. 2020
18.	*Penicillium oxalicum*	Cell free filtrate	100 mM$FeSO_4$	140.6 nm	spherical		Mathur et al. 2021
19.	*Tecomella undulata.*	Cell free filtrate	100 mM$FeSO_4$	140.6 nm	spherical		Mathur et al. 2021

to the surfaces of the nanoparticles, with proteins and amino acid residues coating the surfaces of the nanoparticles to provide stability and avoid agglomeration (Basavaraja and colleagues 2008). The free amino groups, often known as cysteine residues, are thought to interact with and bind the nanoparticle surfaces. Furthermore, cell wall enzymes are thought to supply negative carboxyl groups, which contribute to the electrostatic attraction between nanoparticles and biomolecules (Gole et al. 2001; Durán et al. 2011; Husseiny et al. 2015). Furthermore, the nucleophilic OH⁻ ions on the surfaces of the nanoparticles are considered to supply electrons for the reduction of metal ions, resulting in the creation of nanoparticles with a narrow size distribution (Gurunathan et al. 2009). Several studies have demonstrated the role of biomolecules from fungal extract, acting as a capping agent. Studies have used FTIR, XRD, EDS analysis or UV-visible spectroscopy to demonstrate the existence of a capping agent on the surface of nanoparticles. In most of the studies, proteins are found to be the major capping biomolecule. FTIR analysis reveals bands at 3385 represent the presence of primary amine (N-H) stretching whereas 1638 cm⁻¹ indicates amide (N-H) bending vibration. The bands attributed to C–N stretching vibrations of aromatic and aliphatic amines are found around 1383 and 1032 cm¹, supporting the presence of proteins on the surface of nanoparticles for their stability (Gudikandula et al. 2016) utilizing EDS data, workers have obtained the signal for C, N, and O signals representing biomolecules alongwith the strong signal obtained for the metal for which the nanoparticles are synthesized (Das et al. 2009; Elgorban et al. 2016). Recording fluorescence spectra at an excitation wavelength of 280 nm, which represents the transitions arising from aromatic amino acids tyrosine and tryptophan of proteins present on the surface of the particles, could also be used to assess the presence of capping protein on the surface of the nanoparticles (Kadam et al. 2019).

Conclusion

Out of bacteria, fungi, and algae, fungi are widely used due to their ability to release huge amounts of enzymes and tolerate flow pressure and agitation. Recent research has found that employing fungi to biogenically synthesized iron nanoparticles has various advantages and that these materials have great potential for a variety of applications. The biologically synthesized nanoparticles possess cappings derived from the fungi, which provide stability. Depending on the fungus, this capping may potentially have biological activity, working in tandem with the nanoparticle core's action. The ability to use different species of fungi and perform the synthesis under various physiological conditions, biomass quantity, biomass form, metal precursor type, and metal precursor concentration, among others, enables the production of nanoparticles with various physicochemical properties. However, there are a number of drawbacks that must be solved in order to properly use fungus for biogenic synthesis. These include knowing which fungus to employ, its growth parameters, the time it takes for the fungus to grow and complete the synthesis, and the parameters controlling the size, shape crystallinity and precise magnetic properties of the nanoparticles synthesized using fungi. Scaling up can also create challenges, such as the need for more research into the mechanisms that cause capping layers to form

Biosynthesis of Iron Oxide Nanoparticles Using Fungi   93

and the molecules present on them. Hence, further research is necessary to fill these gaps and come up with more tuned studies.

References

Abdeen, M., Sabry, S., Ghozlan, H., El-Gendy, A.A., and Carpenter, E.E. 2016. Microbial-physical synthesis of Fe and Fe_3O_4 magnetic nanoparticles using *Aspergillus niger* YESM1 and supercritical condition of ethanol. J. Nanomater., doi:10.1155/2016/9174891.

Adelere, I.A., and Lateef, A. 2016. A novel approach to the green synthesis of metallic nanoparticles: the use of agro-wastes, enzymes, and pigments. Nanotechnol. Rev. 5(6): 567–587. doi: 10.1515/ntrev-2016-0024.

Adeleye, T.M., Kareem, S.O., and Kekere-Ekun, A.A. 2020. Optimization studies on biosynthesis of iron nanoparticles using *Rhizopus stolonifer*. IOP Conf. Ser.: Mater. Sci. Eng. 805(1): 012037. doi: 10.1088/1757-899X/805/1/012037.

Agarwal, H., Kumar, S.V., and Rajeshkumar, S. 2017. A review on green synthesis of zinc oxide nanoparticles–An eco-friendly approach. Resource-Efficient Technologies 3(4): 406–413. doi: 10.1016/j.reffit.2017.03.002.

Ajinkya, N., Yu, X., Kaithal, P., Luo, H., Somani, P., and Ramakrishna, S. 2020. Magnetic Iron Oxide Nanoparticle (IONP) synthesis to applications: Present and future. Materials 13(20): 4644. doi: 10.3390/ma13204644.

Akbari, B., Tavandashti, M.P., and Zandrahimi, M. 2011. Particle size characterization of nanoparticles – a practical approach. Iran. J. Mater. Sci. Eng. 8(2): 48–56.

Akhtar, M.S., Panwar, J., and Yun, Y.S. 2013. Biogenic synthesis of metallic nanoparticles by plant extracts. ACS Sustain. Chem. Eng. 1(6): 591–602. doi:10.1021/sc300118u.

Alangari, A., Alqahtani, M.S., Mateen, A., Kalam, M.A., Alshememry, A., Ali, R., Kazi, M., Khalid, M. AlGhamdi, K.M., and Syed, R. 2022. Iron Oxide Nanoparticles: Preparation, Characterization, and Assessment of Antimicrobial and Anticancer Activity. Adsorption Science & Tech., Volume 2022, Article ID 1562051, 9 pages https://doi.org/10.1155/2022/1562051.

Alexandrescu, R.I., Morjan, I., Dumitrache, F., Voicu, I., Soare, I., Sandu, I., and Fleaca, C.T. 2004. Recent developments in the synthesis of iron-based nanostructures by laser pyrolysis: Integrating structural analysis with the experimental method. Solid State Phenom., 99-100,181–190.

Alkhattaf, F.S. 2021. Gold and silver nanoparticles: green synthesis, microbes, mechanism, factors, plant disease management and environmental risks. Saudi J. Biol. Sci. doi: 10.1016/j.sjbs.2021.03.078.

Aminabad, N.S., Farshbaf, M., and Akbarzadeh, A. 2019. Recent advances of gold nanoparticles in biomedical applications: State of the art. Cell Biochem. Biophys. 77(2): 123–137. doi: 10.1007/s12013-018-0863-4.

Amoabediny, G., Haghiralsadat, F., Naderinezhad, S., Helder, M.N., Akhoundi Kharanaghi, E., MohammadnejadArough, J., and Zandieh-Doulabi, B. 2018. Overview of preparation methods of polymeric and lipid-based (niosome, solid lipid, liposome) nanoparticles: A comprehensive review. Int. J. Polym. Mater. Polym. Biomater. 67(6): 383–400. doi: /10.1080/00914037.2017.1332623.

Arya, A., Mishra, V., and Chundawat, T.S. 2019. Green synthesis of silver nanoparticles from green algae (*Botryococcus braunii*) and its catalytic behavior for the synthesis of benzimidazoles. Chem. Data Collect. 20: 100190. doi: 10.1016/j.cdc.2019.100190.

Aziz, N., Faraz, M., Pandey, R., Shakir, M., Fatma, T., Varma, A., Barman, I., and Prasad, R. 2015. Facile algae-derived route to biogenic silver nanoparticles: synthesis, antibacterial, and photocatalytic properties. Langmuir 31(42): 11605–11612. doi: 10.1021/acs.langmuir.5b03081.

Babes, L., Denizot, B., and Tanguy, G. 1999. Synthesis of Iron oxide nanoparticles used as MRI contrast agents: A parametric study. J. Colloid Interface Sci. 212(2): 474–482.

Basavaraja, S., Balaji, S.D., Lagashetty, A., Rajasab, A.H., and Venkataraman, A. 2008. Extracellular biosynthesis of silver nanoparticles using the fungus *Fusarium semitectum*. Mater. Res. Bull. 43(5): 1164–1170. doi: 10.1016/j.materresbull.2007.06.020.

94 Mycosynthesis of Nanomaterials: Perspectives and Challenges

Beheshtkhoo, N., Kouhbanani, M.A.J., Savardashtaki, A., Amani, A.M., and Taghizadeh, S. 2018. Green synthesis of iron oxide nanoparticles by aqueous leaf extract of Daphne mezereum as a novel dye removing material. Appl. Phys. A 124(5): 363. doi: 10.1007/s00339-018-1782-3.

Beveridge, T.J., Hughes, M.N., Lee, H., Leung, K.T., Poole, R.K., Savvaidis, I., Silver, S., and Trevors, J.T. 1996. Metal-microbe interactions: contemporary approaches. Adv. Microb. Physiol. 38: 177–243. doi: 10.1016/S0065-2911(08)60158-7.

Bharde, A., Rautaray, D., Bansal, V., Ahmad, A., Sarkar, I., Yusuf, S.M., Milan, S., and Sastry, M. 2006. Extracellular biosynthesis of magnetite using fungi. Small 2(1): 135–141. doi: 10.1002/smll.200500180.

Bhargava, A., Jain, N., Barathi, M., Akhtar, M.S., Yun, Y.S., and Panwar, J. 2013. Synthesis, characterization and mechanistic insights of mycogenic iron oxide nanoparticles. pp. 337–348. *In*: Nanotechnology for sustainable development. Springer, Cham. doi: 10.1007/978-3-319-05041-6_27.

Bhuiyan, M.S.H., Miah, M.Y., Paul, S.C., Aka, T.D., Saha, O., Rahaman, M.M., Sharif, M.J.I., Habiba, O., and Ashaduzzaman, M. 2020. Green synthesis of iron oxide nanoparticle using *Carica papaya* leaf extract: application for photocatalytic degradation of remazol yellow RR dye and antibacterial activity. Heliyon, 6(8): e04603. doi: 10.1016/j.heliyon.2020.e04603.

Bhuiyan, M.S.H., Miah, M.Y., Paul, S.C., Aka, T.D., Saha, O., Rahaman, M.M., Sharif, M.J.I., Habiba, O., and Ashaduzzaman, M. 2020. Green synthesis of iron oxide nanoparticle using *Carica papaya* leaf extract: application for photocatalytic degradation of remazol yellow RR dye and antibacterial activity. Heliyon 6(8): e04603. doi: 10.1016/j.heliyon.2020.e04603.

Blaney, L. 2007. Magnetite (Fe3O4): properties, synthesis, and Applications. Lehigh Preserve 15: 33–81.

Borah, D., Das, N., Das, N., Bhattacharjee, A., Sarmah, P., Ghosh, K., Chandel, M., Rout, J., Pandey, P., Ghosh, N.N., and Bhattacharjee, C.R. 2020. Alga-mediated facile green synthesis of silver nanoparticles: Photophysical, catalytic and antibacterial activity. Appl. Organomet. Chem. 34(5): e5597. doi: 10.1002/aoc.5597.

Campos, E.A., Pinto, D.V.B.S., Oliveira, J.I.S., Mattos, E.C., and Dutra, R.C.L. 2015. Synthesis, Characterization and applications of iron oxide nanoparticles-a short review. J. Aerosp. Technol. Manag. 7(3): 267–276.

Chatterjee, S., Mahanty, S., Das, P., Chaudhuri, P., and Das, S. 2020. Biofabrication of iron oxide nanoparticles using manglicolous fungus *Aspergillus niger* BSC-1 and removal of Cr (VI) from aqueous solution. Chem. Eng. J. 385: 123790. doi: 10.1016/j.cej.2019.123790.

Chen, S., Yuan, R., Chai, Y., and Hu, F. 2013. Electrochemical sensing of hydrogen peroxide using metal nanoparticles: a review. Microchimica Acta, 180(1): 15–32. doi: 10.1007/s00604-012-0904-4.

Chin, A.B., and Yaacob, I.I. 2007. Synthesis and characterization of magnetic iron oxide nanoparticles via w/o microemulsion and Massart's procedure. J. Mater. Process. Technol. 191(1–3): 235–237.

Das, S.K., Das, A.R., and Guha, A.K. 2009. Gold nanoparticles: microbial synthesis and application in water hygiene management. Langmuir 25(14): 8192–8199. doi: 10.1021/la900585p.

Durán, N., Marcato, P.D., Durán, M., Yadav, A., Gade, A., and Rai, M. 2011. Mechanistic aspects in the biogenic synthesis of extracellular metal nanoparticles by peptides, bacteria, fungi, and plants. Appl. Microbiol. Biotechnol. 90(5): 1609–1624. doi: 10.1007/s00253-011-3249-8.

Ealias, A.M., and Saravanakumar, M.P. 2017. A review on the classification, characterisation, synthesis of nanoparticles and their application. *In*: IOP Conf. Ser. Mater. Sci. Eng. (Vol. 263, p. 032019). doi: 10.1088/1757-899X/263/3/032019 .

Eivari, H.A., and Rahdar, A. 2013. Some properties of iron oxide nanoparticles synthesized in different conditions. World Applied Programming 3(2): 52–55.

Elgorban, A.M., Aref, S.M., Seham, S.M., Elhindi, K.M., Bahkali, A.H., Sayed, S.R., and Manal, M.A. 2016. Extracellular synthesis of silver nanoparticles using *Aspergillus versicolor* and evaluation of their activity on plant pathogenic fungi. Mycosphere 7(6): 844–852. doi: 10.5943/mycosphere/7/6/15.

Gahlawat, G., and Choudhury, A.R. 2019. A review on the biosynthesis of metal and metal salt nanoparticles by microbes. RSC Adv. 9(23): 12944–12967. doi: 10.1039/C8RA10483B.

Gao, L., Fan, K., and Yan, X. 2020. Iron Oxide Nanozyme: A Multifunctional Enzyme Mimetics for Biomedical Application. *In*: Nanozymology (pp. 105–140). Springer, Singapore. doi: 10.1007/978-981-15-1490-6_5.

Gericke, M., and Pinches, A. 2006. Biological synthesis of metal nanoparticles. Hydrometallurgy 83(1-4): 132–140. doi: 10.1016/j.hydromet.2006.03.019.

Ghaffari-Moghaddam, M., Hadi-Dabanlou, R., Khajeh, M., Rakhshanipour, M., and Shameli, K. 2014. Green synthesis of silver nanoparticles using plant extracts. Korean J Chem Eng 31(4): 548–557. doi: 10.1007/s11814-014-0014-6.

Gholampoor, N., Emtiazi, G., and Emami, Z. 2015. The influence of Microbacteriumhominis and *Bacillus licheniformis* extracellular polymers on silver and iron oxide nanoparticles production; green biosynthesis and mechanism of bacterial nano production. J. Nanometer Mol. Nanotechnol. 4, 2. doi: 10.4172/2324-8777.1000160,

Gole, A., Dash, C., Ramakrishnan, V., Sainkar, S.R., Mandale, A. B., Rao, M., and Sastry, M. 2001. Pepsin–gold colloid conjugates: preparation, characterization, and enzymatic activity. Langmuir 17(5): 1674–1679. doi: 10.1021/la001164w.

Gouda, A.R., Sidkey, M.N., Shawky, A.H., and Abdel-hady, Y.A. 2020. Biosynthesis, characterization and antimicrobial activity of iron oxide nanoparticles synthesized by fungi. Al- Azhar J. Pharm. Sci. 62(2): 164–179. doi: 10.21608/ajps.2020.118382.

Guilger-Casagrande, M., and de Lima, R. 2019. Synthesis of silver nanoparticles mediated by fungi: A Review. Front. Bioeng. Biotechnol. 7. doi: 10.3389/fbioe.2019.00287.

Gurunathan, S., Kalishwaralal, K., Vaidyanathan, R., Venkataraman, D., Pandian, S.R.K., Muniyandi, J., Hariharan, N., and Eom, S.H. 2009. Biosynthesis, purification and characterization of silver nanoparticles using *Escherichia coli*. Colloids Surf. B: Biointerfaces,74(1): 328–335. doi: 10.1016/j.colsurfb.2009.07.048.

Hassanjani, A.R., Vaezi, M.R., Shokuhfar, A., and Rajabali, Z. 2010. Synthesis of iron oxide nanoparticles via sonochemical method and their characterization. Particuology 9(1): 95–99.

Hayashi, H., and Hakuta, Y. 2010. Hydrothermal synthesis of metal oxide nanoparticles in supercritical water. Mater, 3: 3794–3817.

Heera, P., and Shanmugam, S. 2015. Nanoparticle characterization and application: An overview. Int. J. Curr. Microbiol. Appl. Sci. 4(8): 379–386.

HernándezHernández, A.A., AguirreÁlvarez, G., CariñoCortés, R., MendozaHuizar, L.H., and JiménezAlvarado, R. 2020. Iron oxide nanoparticles: synthesis, functionalization, and applications in diagnosis and treatment of cancer. Chem. Pap. 74(11): 3809–3824. doi: 10.1007/s11696-020-01229-8.

Hussain, M., Raja, N.I., Iqbal, M., and Aslam, S. 2019. Applications of plant flavonoids in the green synthesis of colloidal silver nanoparticles and impacts on human health. Iran. J. Sci. Technol. Trans. A: Sci. 43(3): 1381–1392. doi: 10.1007/s40995-017-0431-6.

Husseiny, S.M., Salah, T.A., and Anter, H.A. 2015. Biosynthesis of size controlled silver nanoparticles by *Fusarium oxysporum*, their antibacterial and antitumor activities. Beni-Seuf Univ. J. Appl. Sci. 4(3): 225–231. doi: 10.1016/j.bjbas.2015.07.004.

Ishak, N.M., Kamarudin, S.K., and Timmiati, S.N. 2019. Green synthesis of metal and metal oxide nanoparticles via plant extracts: an overview. Mater. Res. Express 6(11): 112004. doi: 10.1088/2053-1591/ab4458.

Jacinto, M.J., Silva, V.C., Valladão, D.M.S., and Souto, R.S. 2020. Biosynthesis of magnetic iron oxide nanoparticles: a review. Biotechnol. Lett. 1–12. doi: 10.1007/s10529-020-03047-0.

Jeevanandam, J., Chan, Y.S., and Danquah, M.K. 2016. Biosynthesis of metal and metal oxide nanoparticles. ChemBioeng. Rev. 3(2): 55–67. doi: 10.1002/cben.201500018.

Jolivet, J.P., Belleville, P., and Trone, E. 1992. Influence of Fe (II) on the formation of the spinel iron oxide in alkaline medium. Clays Clay Miner 40(5): 531–539.

Jubran, A.S., Al-Zamely, O.M., and Al-Ammar, M.H. 2020. A Study of iron oxide nanoparticles synthesis by using bacteria. Int. J. Pharm. Qual. Assur., 11(01): 01–08.

Kadam, V.V., Ettiyappan, J.P., and Balakrishnan, R.M. 2019. Mechanistic insight into the endophytic fungus mediated synthesis of protein capped ZnO nanoparticles. Mater. Sci. Eng. B. 243: 214–221. doi: 10.1016/j.mseb.2019.04.017.

Karimzadeh, I., Aghazadeh, M., Doroudi, T., Ganjali, M.R., and Kolivand, P.H. 2017. Superparamagnetic iron oxide (Fe_3O_4) nanoparticles coated with PEG/PEI for biomedical applications: a facile and scalable preparation route based on the cathodic electrochemical deposition method. Adv. Phys. Chem. 1–7.

Kaul, R., Kumar, P., Burman, U., Joshi, P., Agrawal, A., Raliya, R., and Tarafdar, J. 2012. Magnesium and iron nanoparticles production using microorganisms and various salts. Mater. Sci.-Pol. 30(3): 254–258. doi: 10.2478/s13536-012-0028-x.

Khalil, M I. 2015. Co-precipitation in aqueous solution synthesis of magnetite nanoparticles using iron (III) salts as precursors. Arabian J. Chem. 8(2): 279–284. doi:10.1016/j.arabjc.2015.02.008.

Khalil, A.T., Ovais, M., Ullah, I., Ali, M., Shinwari, Z.K., and Maaza, M. 2017. Biosynthesis of iron oxide (Fe2O3) nanoparticles via aqueous extracts of *Sageretiathea* (Osbeck.) and their pharmacognostic properties. Green Chem. Lett. Rev., 10(4): 186–201. doi: 10.1080/17518253.2017.1339831.

Khan, I., Saeed, K., and Khan, I. 2019. Nanoparticles: Properties, applications and toxicities. Arab. J. Chem. 12(7): 908–931. doi: 10.1016/j.arabjc.2017.05.011.

Kulkarni, N., and Muddapur, U. 2014. Biosynthesis of metal nanoparticles: a review. J. Nanotechnol., doi: 10.1155/2014/510246.

Kumar, H., Venkatesh, N., Bhowmik, H., and Kuila, A. 2018. Metallic nanoparticle: a review. Biomed. J. Sci. Tech. Res. 4(2): 3765–3775. doi: 10.26717/BJSTR.2018.04.001011.

Kumar, S.A., Abyaneh, M.K., Gosavi, S.W., Kulkarni, S.K., Pasricha, R., Ahmad, A., and Khan, M.I. 2007. Nitrate reductase-mediated synthesis of silver nanoparticles from AgNO 3. Biotecnol. Lett. 29(3): 439–445. doi: 10.1007/s10529-006-9256-7.

Latha, N., and Gowri, M. 2012. Bio synthesis and characterization of Fe3O4 nanoparticles using *Caricaya Papaya* leaves extract. Int. J. Sci. Res. 3(11): 1551–1556.

Laurent, S., Forge, D., Port, M., Roch, A., Robic, C., Vander Elst, L., and Muller, R.N. 2008. Magnetic iron oxide nanoparticles: synthesis, stabilization, vectorization, physicochemical characterizations, and biological applications. Chem. Rev. 108(6): 2064–2110. doi: 10.1021/cr068445e.

Li, X.Q., and Zhang, W.X. 2006. Iron nanoparticles: The core−shell structure and unique properties for Ni (II) sequestration. Langmuir 22(10): 4638–4642. doi: 10.1021/la060057k.

Lodhia, J., Mandarano, G., Ferris, N.J., Eu, P., and Cowell, S.F. 2009. Development and use of iron oxide nanoparticles (part I): Synthesis of iron oxide nanoparticles for MRI. Biomed. Imaging Interv. J. 6408–6413.

Mahanty, S., Bakshi, M., Ghosh, S., Chatterjee, S., Bhattacharyya, S., Das, P., Das, S., and Chaudhuri, P. 2019. Green synthesis of iron oxide nanoparticles mediated by filamentous fungi isolated from Sundarban mangrove ecosystem, India. BioNanoScience 9(3): 637–651. doi: 10.1007/s12668-019-00644-w.

Mahdavi, M., Ahmad, M.B., Haron, M.J., Namvar. F., Nadi. B., Rahman, and M.Z, Amin, J. 2013. Synthesis, surface modification and characterisation of biocompatible magnetic iron oxide nanoparticles for biomedical applications. Molecules 278(7): 7533–7548. doi: 10.3390/molecules18077533.

Mahdizadeh, F., Karimi, A., and Ranjbarian, L. 2012. Immobilization of glucose oxidase on synthesized superparamagnetic Fe_3O_4 nanoparticles; Application for water deoxygenation. Int. J. Eng. Sci. 3(5): 1–6.

Malakootian, M., Yaseri, M., and Faraji, M. 2019. Removal of antibiotics from aqueous solutions by nanoparticles: a systematic review and meta-analysis. Environ. Sci. Pollut. Res., 26(9): 8444–8458. doi: 10.1007/s11356-019-04227-w.

Malhotra, N., Lee, J.S., Liman, R.A.D., Ruallo, J.M.S., Villaflores, O.B., Ger, T.R., and Hsiao, C.D. 2020. Potential toxicity of iron oxide magnetic nanoparticles: A review. Molecules 25(14): 3159. doi: 10.3390/molecules25143159.

Massironi, A., Morelli, A., Grassi, L., Puppi, D., Braccini, S., Maisetta, G., Esin, S., Batoni, G., Della Pina, C., and Chiellini, F. 2019. Ulvan as novel reducing and stabilizing agent from renewable algal biomass: Application to green synthesis of silver nanoparticles. Carbohydr. Polym. 203: 310–321. doi: 10.1016/j.carbpol.2018.09.066.

Mathur, P., Saini, S., Paul, E., Sharma, C., and Mehtani, P. 2021. Endophytic fungi mediated synthesis of iron nanoparticles: Characterization and application in methylene blue decolorization. Curr. Opin. Green Sustain. Chem. 4: 100053.doi: 10.1016/j.crgsc.2020.100053.

Mazumdar, H., and Haloi, N. 2011. A study on biosynthesis of iron nanoparticles by *Pleurotus* sp. J. Microbiol. Biotechnol. 1(3): 39–49.

Mei, W., and Wu, Q. 2018. Applications of metal nanoparticles in medicine/metal nanoparticles as anticancer agents. Metal Nanoparticles: Synthesis and Applications in Pharmaceutical Sciences, 169–190.

Mirza, A.U., Kareem, A., Nami, S.A., Khan, M.S., Rehman, S., Bhat, S.A., Mohammad, A., and Nishat, N. 2018. Biogenic synthesis of iron oxide nanoparticles using *Agrewia optiva* and *Prunus persica* phyto species: characterization, antibacterial and antioxidant activity. J. Photochem. Photobiol. B. Biol. 185: 262–274. doi: 10.1016/j.jphotobiol.2018.06.009.

Mohamed, Y.M., Azzam, A.M., Amin, B.H., and Safwat, N.A. 2015. Mycosynthesis of iron nanoparticles by *Alternaria alternata* and its antibacterial activity. Afr. J. Biotechnol., 14(14): 1234–1241.

Mourdikoudis, S., Pallares, R.M., and Thanh, N.T. 2018. Characterization techniques for nanoparticles: comparison and complementarity upon studying nanoparticle properties. Nanoscale, 10(27): 12871–12934. doi: 10.1039/C8NR02278J.

Odularu, A.T. 2018. Metal nanoparticles: thermal decomposition, biomedicinal applications to cancer treatment, and future perspectives. Bioinorg. Chem. Appl. doi: 10.1155/2018/9354708.

Oskam, G. 2006. Metal oxide nanoparticles: synthesis, characterization and application. J. Sol-Gel Sci. Technol. 37(3): 161–164. doi: 10.1007/s10971-005-6621-2.

Ovais, M., Khalil, A.T., Raza, A., Khan, M.A., Ahmad, I., Islam, N.U., Saravanan, M., Ubaid, M.F., Ali, M., and Shinwari, Z.K. 2016. Green synthesis of silver nanoparticles via plant extracts: beginning a new era in cancer theranostics. Nanomedicine, 12(23): 3157–3177. doi: 10.2217/nnm-2016-0279.

Pantidos, N., and Horsfall, L.E. 2014. Biological synthesis of metallic nanoparticles by bacteria, fungi and plants. J. Nanomed. Nanotechnol. 5(5): 1. doi: 10.4172/2157-7439.1000233.

Pavani, K.V., and Kumar, N.S. 2013. Adsorption of iron and synthesis of iron nanoparticles by *Aspergillus* species kvp 12. Am. J. Nanomater. 1(2): 24–26. doi: 10.12691/ajn-1-2–2.

Peternele, W.S., Fuentes, V.M., Fascineli, M.L., Silva, J.R., Silva, R.C., Lucci, C.M., and Azevedo, R.B. 2014. Experimental investigation of the coprecipitation method: An approach to obtain magnetite and maghemite nanoparticles with improved properties. J. Nanomater., Article ID 682985, 1–10.

Priya, Naveen, Kaur K., and Sidhu, A.K. 2021. Green synthesis: an eco-friendly route for the synthesis of iron oxide nanoparticles. Front. Nanotechnol. 3: 47. doi: 10.3389/fnano.2021.655062.

Rane, K.S., and Verenkar, V.M.S. 2001. Synthesis of ferrite grade γ-Fe_2O_3. Bull. Mat. Sci. 24(1): 39–45.

Rastar, A., Yazdanshenas, M.E., Rashidi, A., and Bidoki, S.M. 2013. Theoretical review of optical properties of nanoparticles. J. Eng. Fibers Fabr. 8(2): 155892501300800211. doi: 10.1177/155892501300800211.

Rauwel, P., Küünal, S., Ferdov, S., and Rauwel, E. 2015. A review on the green synthesis of silver nanoparticles and their morphologies studied via TEM. Adv. Mater. Sci. Eng. doi: 10.1155/2015/682749.

Salam, H.A., Rajiv, P., Kamaraj, M., Jagadeeswaran, P., Gunalan, S., and Sivaraj, R. 2012. Plants: green route for nanoparticle synthesis. Int. Res. J. Biol. Sci. 1(5): 85–90.

Saleh, T.A., Fadillah, G., and Saputra, O.A. 2019. Nanoparticles as components of electrochemical sensing platforms for the detection of petroleum pollutants: A review. TrAC- Trends Anal. Chem. 118: 194–206. doi: 10.1016/j.trac.2019.05.045.

Schröfel, A., Kratošová, G., Šafařík, I., Šafaříková, M., Raška, I., and Shor, L.M. 2014. Applications of biosynthesized metallic nanoparticles–a review. Acta Biomater. 10(10): 4023–4042. doi: 10.1016/j.actbio.2014.05.022.

Senthilkumar, S.R., and Sivakumar, T. 2014. Green tea (*Camellia sinensis*) mediated synthesis of zinc oxide (ZnO) nanoparticles and studies on their antimicrobial activities. Int. J. Pharm. Pharm. Sci. 6(6): 461–465.

Sharma, D., Kanchi, S., and Bisetty, K. 2019. Biogenic synthesis of nanoparticles: A review. Arab. J. Chem. 12(8): 3576–3600. doi: 10.1016/j.arabjc.2015.11.002.

Shnoudeh, A.J., Hamad, I., Abdo, R.W., Qadumii, L., Jaber, A.Y., Surchi, H.S., and Alkelany, S.Z. 2019. Synthesis, characterization, and applications of metal nanoparticles. In Biomater. Bionanotechnol. (pp. 527–612). Academic Press. doi: 10.1016/B978-0-12-814427-5.00015-9.

Sidhu, A.K., Verma, N., and Kaushal, P. 2022. Role of biogenic capping agents in the synthesis of metallic nanoparticles and evaluation of their therapeutic potential. Front. Nanotechnol. 3: 801620. doi: 10.3389/fnano.2021.801620.

Silva, L.P., Reis, I.G., and Bonatto, C.C. 2015. Green Synthesis of Metal Nanoparticles by Plants: Current Trends and Challenges. Green Processes for Nanotechnology, 259–275. doi: 10.1007/978-3-319-15461-9_9.

Singh, P., Kim, Y.J., Zhang, D., and Yang, D.C. 2016. Biological synthesis of nanoparticles from plants and microorganisms. Trends Biotechnol, 34(7): 588–599. doi: 10.1016/j.tibtech.2016.02.006.

Srikar, S.K., Giri, D.D., Pal, D.B., Mishra, P.K., and Upadhyay, S.N. 2016. Green synthesis of silver nanoparticles: a review. Green Sustain. Chem. 6(1): 34–56. doi: 10.4236/gsc.2016.61004.

Sulistyaningsih, T., Santosa, S.J., Siswanta, D., and Rusdiarso, B. 2017. Synthesis and characterization of magnetites obtained from mechanically and sonochemically assissted co-precipitation and reverse co-precipitation methods. Int. J. Manuf. Mater. Mech. Eng. 5(1).

Tarafdar, J.C., and Raliya, R. 2013. Rapid, low-cost, and ecofriendly approach for iron nanoparticle synthesis using *Aspergillus oryzae* TFR9. J. Nanopart. doi: 10.1155/2013/141274.

Thakkar, K.N., Mhatre, S.S., and Parikh, R.Y. 2010. Biological synthesis of metallic nanoparticles. Nanomed. : Nanotechnol. Biol. Med. 6(2): 257–262. doi: 10.1016/j.nano.2009.07.002.

Turan, N.B., Erkan, H.S., Engin, G.O., and Bilgili, M.S. 2019. Nanoparticles in the aquatic environment: Usage, properties, transformation and toxicity—A review. Process Saf. Environ. Prot. 130: 238–249. doi: 10.1016/j.psep.2019.08.014.

Vahidi, H., Barabadi, H., and Saravanan, M. 2020. Emerging selenium nanoparticles to combat cancer: a systematic review. J. Clust. Sci. 31(2): 301–309. doi: 10.1007/s10876-019-01671-z.

Vasantharaj, S., Sathiyavimal, S., Senthilkumar, P., Lewis Oscar, F., and Pugazhendhi, A. 2019. Biosynthesis of iron oxide nanoparticles using leaf extract of Ruelliatuberosa: antimicrobial properties and their applications in photocatalytic degradation. J. Photochem. Photobiol. B, Biol. 192: 74–82. doi: 10.1016/j.jphotobiol.2018.12.025.

Vayssieres, L., Chaneac, C., and Trone, E. 1998. Size tailoring of magnetite particles formed by aqueous precipitation: An example of thermodynamic stability of nanometric oxide particles. J. Colloid Interface Sci. 205(2): 205–212.

Vélez, E., Campillo, G.E., Morales, G., Hincapié, C., Osorio, J., Arnache, O., Uribe, J.I., and Jaramillo, F. 2016. Mercury removal in wastewater by iron oxide nanoparticles. In J. Phys. Conf. Ser. 687(1): 012050. doi: 10.1088/1742-6596/687/1/012050.

Vinci, G., and Rapa, M. 2019. Noble metal nanoparticles applications: Recent trends in food control. Bioengineering 6(1): 10. doi: 10.3390/bioengineering6010010.

Woo, K., Lee, H.J., Ahn, J.P. and Park, Y.S. 2003. Sol-gel mediated synthesis of Fe_2O_3 nanorods. Adv. Mater. 15(20): 1761–1764.

Wu, W., He, Q., and Jiang, C. 2008. Magnetic iron oxide nanoparticles: synthesis and surface functionalization strategies. Nanoscale Res. Lett. 3(11): 397. doi: 10.1007/s11671-008-9174-9.

Wu, W., Xiao, X.H., Zhang, S.F. Peng, T.C., Zhou, J., Ren, F. and Jiang, C.Z. 2010. Synthesis and magnetic properties of maghemite (γ-Fe_2O_3) Short-Nanotubes. Nanoscale Res. Lett. 5(9): 1474–1479.

Yusof, H.M., Mohamad, R., and Zaidan, U.H. 2019. Microbial synthesis of zinc oxide nanoparticles and their potential application as an antimicrobial agent and a feed supplement in animal industry: a review. J. Anim. Sci. Biotechnol. 10(1): 57. doi: 10.1186/s40104-019-0368-z.

Zhang, R., Zhou, Y., Yan, X., and Fan, K. 2019. Advances in chiral nanozymes: a review. Microchimica Acta 186(12): 782. doi: 10.1007/s00604-019-3922-7.

Zununi Vahed, S., Fathi, N., Samiei, M., Maleki Dizaj, S., and Sharifi, S. 2019. Targeted cancer drug delivery with aptamer-functionalized polymeric nanoparticles. Journal of Drug Targeting 27(3): 292–299. doi: 10.1080/1061186X.2018.1491978.

CHAPTER 5

Mycosynthesis of Chitosan Nanoparticles

Mayuri Napagoda[1,*] *and Sanjeeva Witharana*[2]

Introduction

Chitin is a polysaccharide polymer found in the exoskeletons of insects, crustaceans and mollusks and in the cell walls of some fungi. After cellulose, chitin is the second most abundant polysaccharide in nature. It mainly consists of the aminosugar N-acetylglucosamine and is usually complex with other polysaccharides and proteins (Yeul and Rayalu 2013). The deacetylated form of chitin also naturally exists in some fungi and is known as chitosan (Figure 1). However, chitosan is less prevalent in nature compared to chitin. On the other hand, the deacetylation of chitin is seldomly complete, and in consequence, acetamide groups are present at a certain percentage. The typical commercial chitosan has around 85% degree of deacetylation (Yeul and Rayalu 2013; Islam et al. 2017).

At present, the most widely available raw materials for chitin production are the shells of crustaceans (crabs, shrimps, and prawns). However, due to the multifaceted organization of chitin with other components in the crustacean shells, harsh treatments are essential to obtain chitin and then chitosan on a commercial scale. Moreover, the chitosan obtained from these sources suffers some inconsistencies like high molecular weight, protein contamination, and inconsistent levels of deacetylation thus limiting its applications. Under these circumstances, chitosan extracted from fungal sources is emerging as an alternative to crustacean-derived chitosan. The enzymatic deacetylation takes place naturally in various classes of fungi thus resulting in the formation of chitosan. Therefore, by varying the fermentation conditions, it is possible to obtain fungal-chitosan with desirable molecular weight and degree of deacetylation (Dhillon et al. 2013).

---

[1] Faculty of Medicine, University of Ruhuna, Galle 80 000, Sri Lanka.
[2] Faculty of Engineering, University of Moratuwa, Moratuwa 10 400, Sri Lanka.
* Corresponding author: mayurinapagoda@yahoo.com

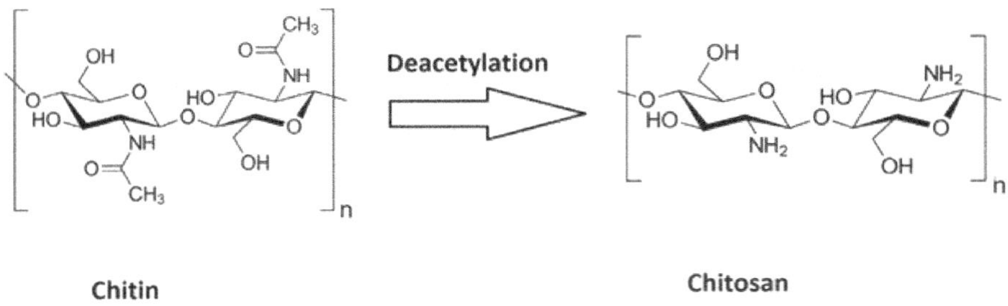

Figure 1. Deacetylation of chitin results in chitosan.

Both chitin and chitosan exhibit desirable features like non-toxicity, biodegradability, biocompatibility, bioresorptivity, and hence emerging as alternatives to synthetic polymers. However, chitosan is chemically more active than chitin due to the presence of hydroxyl and amine groups on each deacetylated unit. Once these functional groups are subjected to chemical modifications, the physical and mechanical properties of chitosan also get altered producing a wide range of derivatives with diverse applications (Alves and Mano 2008; Islam et al. 2017). Interestingly, chitosan has become one of the commonly used materials for polymeric nanoparticle synthesis. For example, nanoparticles prepared with chitosan or its derivatives inherit properties like positive surface charge, mucoadhesive, and adherence ability, which make them strong candidates for drug delivery applications (Mohammed et al. 2017). Chitosan nanoparticles have been employed in the delivery of various drugs with anticancer, anti-inflammatory, and antibiotics activities through mucosal, nasal, topical, oral, vaginal, or parenteral routes. Moreover, chitosan nanoparticles have been exploited as carriers of other therapeutic agents like peptides, proteins, vaccines, and DNA via parenteral and non-parenteral routes. Further, these nanoparticles are widely used in regenerative medicine, mostly for cartilage and bone tissue engineering and cell therapy (Jayasuriya 2017). Like in the field of medicine, chitosan nanoparticles have many applications in agriculture where they are used in herbicide, fungicide, insecticide, and fertilizer delivery. Once the active ingredients are encapsulated into chitosan nanocarriers, the controlled release properties and high bioavailability of these nanoformulations help to minimize the wastage and leaching of the active ingredients. Further, the small size of the nanoformulations enhances the uptake of the active ingredients of the agrochemicals (Maluin and Hussein 2020). Moreover, chitosan nanoparticles are employed in wastewater treatment, used as a filler material as well in food packaging (Yanat and Schroën 2021).

The properties of chitosan nanoparticles may show considerable variation with respect to the preparation methods, and surface modification techniques, etc. The conventional preparative methods often involved the use of hazardous chemicals, high temperatures, or high shear force (Yanat and Schroën 2021). As a result, the preparation of nano-fungal chitosan has gained increased attention over recent years mainly due to the sustainability, simplicity and efficacy of this technique (Alsaggaf et al. 2020). The aim of this chapter is to enlighten the readers on the fungal-based synthesis of chitosan nanoparticles and their potential applications.

Preparation of chitosan nanoparticles: The conventional approaches

Chitosan nanoparticles were developed by different methods such as emulsification, precipitation or ionic/covalent crosslinking (Yanat and Schroën 2021).

Chitosan nanoparticles were first described by Ohya et al. (1994) who applied the glutaraldehyde crosslinking technique and the water-in-oil (W/O) emulsion method to prepare small-sized chitosan-gel nanospheres for the delivery of the anticancer drug 5-fluorouracil (5-FU) and its derivatives. The amino group of chitosan and the aldehyde group of the crosslinking agent were important in the preparation of the nanoparticles. In this method, an aqueous solution containing 5-FU derivatives and the oil phase comprised of toluene and Span 80 as the stabilizer were mixed and sonicated. Thereafter glutaraldehyde solution saturated with toluene in the presence of Span 80 was slowly added to the above emulsion and stirred well to obtain chitosan nanoparticles. Centrifugation was used to separate the nanoparticles, which were subsequently subjected to multiple washing steps and drying under a vacuum (Ohya et al. 1994). Jameela et al. (1998) utilized the glutaraldehyde crosslinking approach to preparing chitosan microspheres loaded with progesterone. Covalent cross-linking was also employed in reversed micelles (microemulsion) approach for the preparation of chitosan nanoparticles. Here, an aqueous phase comprised of chitosan and glutaraldehyde was mixed with the organic phase that contained an organic solvent (e.g., n-hexane, isooctane) and a lipophilic surfactant (e.g., cetyltrimethylammonium bromide, sodium bis (2-ethylhexyl) sulfosuccinate). This technique produced ultrafine nanoparticles of less than 100 nm in size (Kafshgari et al. 2012). The sieving method also involved the cross-linking of an aqueous acidic solution of chitosan using glutaraldehyde. Thereafter the cross-linked chitosan was passed through a sieve and then washed with NaOH solution to remove the unreacted glutaraldehyde and dried at 40°C (Kashyap et al. 2015). However, due to the toxicity and drug integrity issues associated with glutaraldehyde, the classical methods that are based on glutaraldehyde crosslinking are no longer in use (Yanat and Schroën 2021).

Precipitation-based approaches like phase inversion precipitation and emulsion-droplet coalescence are also used in the preparation of chitosan nanoparticles. The phase inversion precipitation technique involves emulsification combined with precipitation. For example, El-Shabouri (2002) prepared an oil-in-water emulsion with an organic phase (methylene chloride and acetone) and an aqueous solution containing chitosan, gelatin-A, sodium glycocholate in the presence of poloxamer as a stabilizer. High-pressure homogenization was employed to obtain emulsion droplets and the evaporation of solvents resulted in the precipitation of nanoparticles. The encapsulation of highly lipophilic, poorly absorbable drug cyclosporin-A with chitosan nanoparticles had improved the oral bioavailability of cyclosporin-A. On the other hand, the emulsion-droplet coalescence technique is based on the coalescence of chitosan droplets with NaOH which severs as a precipitation agent (Tokumitsu et al. 1999). Tokumitsu et al. (1999) prepared gadopentetic acid-loaded chitosan nanoparticles by this approach. In this method two separate emulsions were prepared; one containing gadopentetic acid and chitosan in liquid paraffin and the other containing NaOH in liquid paraffin. Once the two emulsions were combined, chitosan gets deposited as nanoparticles, as a result of the coalescence of droplets. Interestingly, the releasing properties and the ability for long-term retention of gadopentetic acid in the tumor tissues indicated the suitability of the prepared nanoparticles to be developed into intra-tumoral injectable devices for gadolinium neutron-capture therapy of cancer (Tokumitsu et al. 1999). However, the popularity of the aforementioned precipitation-based approaches in the synthesis of chitosan nanoparticles has decreased due to the involvement of harmful chemicals and high shear force (Yanat and Schroën 2021).

Ionic gelation is another preparation method that involves the crosslinking of oppositely charged groups; i.e., protonated amino groups in chitosan and negatively charged low-molecular-weight compounds like sodium tripolyphosphate (TPP), sodium sulfate. This is a simple technique in which chitosan in an aqueous acidic solution is mixed with the aqueous solution of TPP under vigorous stirring. The electrostatic interaction between chitosan and TPP lead to the formation of spherical-shaped nanoparticles (Liu and Gao 2009). This method does not require any harmful crosslinkers and the size of the nanoparticles can be adjusted easily (Yanat and Schroën 2021).

Ionic gelation with radical polymerization can also be used in the preparation of chitosan nanoparticles for drug delivery. Hu et al. (2002) prepared chitosan–poly(acrylic acid) complex nanoparticles by polymerizing acrylic acid into a chitosan template (Hu et al. 2002). The resulting nanoparticles possessed several features like positive surface charges and small particle size and were successfully used to deliver insulin, silk peptide, and serum albumin through the oral route. However, the preparation of chitosan nanoparticles by this method is time-consuming (Yanat and Schroën 2021).

Self-assembly is identified as a feasible and cost-effective route to prepare chitosan nanoparticles, particularly for drug delivery applications. Self-assembly is a spontaneous process and it can be used to obtain structures with particular functions and properties (Yang et al. 2014). The preparation of chitosan nanoparticles by this method is based on multiple simultaneous interactions between chitosan and other molecules (Quiñones et al. 2018). The ability of chitosan polyelectrolyte to form complexes with anionic materials like hyaluronic acid or alginate is used in this process. Further, the introduction of hydrophobic moieties by grafting is also useful to modify the hydrophobicity of chitosan and thereby promote self-assembly. The self-assembly process can produce highly stable chitosan nanoparticles that are capable of encapsulating hydrophilic and lipophilic drugs (Quiñones et al. 2018; Yanat and Schroën 2021).

The top-down approach is also employed in the preparation of chitosan nanoparticles. This involves acid hydrolysis of chitin to form chitin nanoparticles followed by deacetylation using alkaline treatment. However, it is a highly time-consuming process and requires an extra step for drug loading (Yanat and Schroën 2021).

The more recent methods used for the preparation of chitosan nanoparticles include the spray drying technique and supercritical-CO_2-assisted solubilization and atomization (SCASA). In the spray drying method, chitosan solution is passed through a nanospray dryer at an air temperature of 120°C–150°C (Ngan et al. 2014). Likewise, Huang et al. (2010) have prepared chitosan/iron (II, III) oxide nanoparticles by a spray-drying method. Although spray drying is a simple and fast process that does not require another separation or drying step, it also has some limitations like large particle size and unsuitability to use with temperature-sensitive substances (Yanat and Schroën 2021). SCASA too does not require another separation or drying step and it is acidic or harmful solvent-free method. This process requires only water and CO_2 during preparation. Here, pressurized CO_2 is dissolved in water to lower the pH and allow the dissolution of chitosan as well. The resulting solution is fed through a spraying nozzle that leads to atomization and generation of dried chitosan nanoparticles (Hijazi et al. 2014). Although this is a greener approach, the large particle size, need of a specially designed system, and the time-intensiveness are some of the drawbacks of this method. In addition to the aforementioned preparative methods, there were some attempts to prepare chitosan nanoparticles employing plant extracts or bacterial cultures; for example, the use of leaf extract of *Pelargonium graveolens* (El-Naggar et al. 2022) and the isolates of *Klebsiella pneumoniae* (Hussein and Aldujaili 2022).

Figure 2 illustrates different methods involved in the synthesis of chitosan nanoparticles.

Mycosynthesis of chitosan nanoparticles

Over the recent years, there has been increased attention to the biological synthesis of nanomaterials. The conventional physico-chemical methods employed in the preparation of nanomaterials often use hazardous chemicals and the generation of toxic waste. Therefore, environmentally benign approaches like biogenic synthesis are emerging as an alternative. Among the various biological agents employed in nanomaterial synthesis, fungi are preferred by many scientists. In comparison to bacteria, fungi are more efficient in producing a large quantity of nanoparticles mainly due to the secretion of enzymes that can be directly involved in the bioreduction and stabilization of the particles (Khandel and Shahi 2018). The synthesis of nanoparticles using fungi is known as "mycosynthesis" and this has contributed to the development of a new area named "myconanotechnology" (Rai et al. 2021).

A number of nanomaterials have been synthesized using fungi so far, most of those were metal nanoparticles. *Saccharomyces cerevisiae*, *Fusarium* spp. *Aspergillus* spp., and *Penicillium* spp. have played a significant role in mycosynthesis of various nanomaterials (Dhillon et al. 2012; Nag et al. 2021; Shelke et al. 2022). Table 1 summarizes some examples of mycosynthesized nanomaterials.

Although the mycosynthesis of metal nanoparticles has been practiced for many years, mycosynthesis of chitosan nanoparticles is a relatively new development. In addition to the toxicity issues related to most of the physico-chemical methods employed in the synthesis of chitosan nanoparticles, their polydispersity (most often 250–400 nm in size) is a considerable drawback. The size variation may affect their physico-chemical characteristics like charge, loading efficiency, and controlled drug release (Kamat et al. 2015; Sathiyabama and Parthasarathy 2016). However,

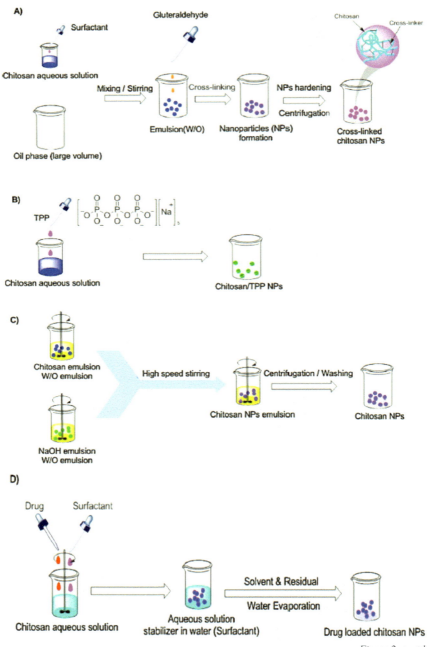

Figure 2. contd. ...

...Figure 2. contd.

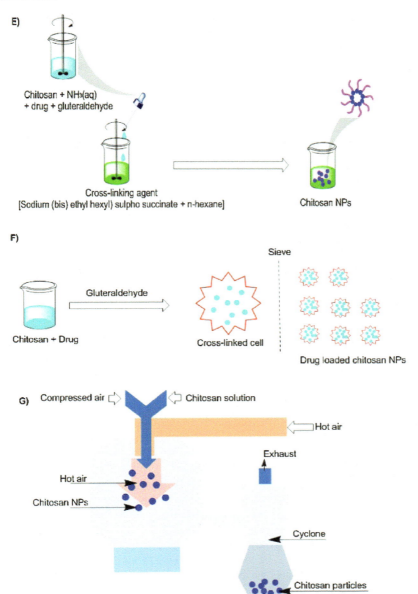

Figure 2. Schematic representation of various methods for the synthesis of chitosan nanoparticles. (A) Emulsion cross-linking; (B) ionotropic gelation; (C) emulsion-droplet coalescence; (D) precipitation; (E) reverse micelles; (F) sieving; and (G) spray drying.
[Figure adopted from Kashyap et al. (2015) with permission from Elsevier]

106 *Mycosynthesis of Nanomaterials: Perspectives and Challenges*

Table 1. Examples of some mycosynthesized nanomaterials.

Nanomaterial	Fungus species	Size and shape of the nanomaterial	References
Ag	*Fusarium oxysporum*	20–50 nm, Spherical	Duran et al. (2005)
	Aspergillus niger	3–30 nm, Spherical	Jaidev and Narasimha (2010)
	Phoma glomerata	60–80 nm, Spherical	Birla et al. (2009)
	Trichoderma viride	5–40 nm, Spherical, rod-like	Fayaz et al. (2009)
	Volvariella volvaceae	15 nm, Spherical	Thakkar et al. (2010)
TiO$_2$	*Aspergillus flavus*	62–74 nm, Spherical	Rajakumar et al. (2012)
ZnO	*Aspergillus fumigatus*	1.2–6.8 nm, Spherical and hexagonal	Raliya et al. (2013)
Au	*Helminthosporium solani*	2–70 nm, Polydispersed	Kumar et al. (2008)
	Penicillium brevicompactum	10–50 nm, Spherical	Mishra et al. (2011)
	Candida albicans	20–40 nm, 60–80 nm, Spherical & non-spherical	Chauhan et al. (2011)
	Fusarium oxysporum	2–50 nm, Spherical, monodispersity	Zhang et al. (2011)
Pt	*Fusarium oxyporum*	70–180 nm, Rectangular, triangular, spherical and aggregates	Govender et al. (2009)
Fe	*Aspergillus oryzae*	10–24.6 nm, Spherical	Tarafdar and Raliya (2013)

biogenic synthesis has yielded stable chitosan nanoparticles of less than 100 nm in size (Sathiyabama and Parthasarathy 2016)

One of the early attempts at mycogenic synthesis of chitosan nanoparticles involved the addition of anionic proteins isolated from the endophytic fungus *Penicillium oxalicum* to chitosan solution at pH 4.8. The reaction mixture was stirred and incubated overnight at room temperature. Thereafter, the colloidal suspension was centrifuged and the resulting precipitate was freeze-dried. The formation of nanoparticles was confirmed by UV-Visible spectrophotometric analysis that indicated a single peak at 285 nm. Fourier-transform infrared spectroscopy (FTIR), high-resolution transmission electron microscopy (HRTEM), X-ray diffraction (XRD), and zeta potential analysis revealed physico-chemical properties of the synthesized nanoparticles. The bands at 1602.8, 1564.18, and 1403.5 cm^{-1} in the FTIR spectrum indicated the binding of protein to chitosan. HRTEM images confirmed the well-dispersed and spherical-shaped chitosan nanoparticles. The zeta potential was determined as -37mV specifying the colloidal stability of the synthesized nanoparticles while the absence of a peak in the diffractogram was interpreted as a characteristic feature of an amorphous structure. These mycogenic chitosan nanoparticles were subjected to antimicrobial activity studies against some phytopathogens; *Pyricularia grisea, Alternaria solani, Fusarium oxysporum*. Further, enhanced germination percentage, length of root and shoot, seed vigor index, and vegetative biomass of seedlings were observed in the chickpea seeds treated with chitosan nanoparticles signifying its' role as a plant growth promoter (Sathiyabama and Parthasarathy 2016).

Similarly, Saravanakumar et al. (2018) produced chitosan nanoparticles using the crude enzymes isolated from the fungus *Trichoderma harzianum*. The crude enzyme was slowly added to a chitosan solution in which the pH was adjusted to 4.8. After a magnetic agitation of 30 min, the solution was incubated overnight at room temperature, and thereafter, colloidal chitosan was obtained. This was subjected to centrifugation and then freeze-dried to make a powder. The presence of chitosan nanoparticles was confirmed by the maximum plasmon resonance peak at 280 nm and the field emission transmission electron microscopic (FETEM) analysis indicated that those nanoparticles were polydispersed and spherical (Figure 3). The particle size analyzer (PSA) revealed that the size of the nanoparticles was in a range of 0.01–0.314 μm while the amorphous structure of the nanoparticles was confirmed through XRD analysis. Moreover, a high water solubility was observed in these nanomaterials. In the FTIR spectrum, modifications and changes in the peaks due to carboxylic acid, aromatic amine, amide, and aliphatic primary amine groups were observed and these were attributed to the involvement or binding of the crude enzyme with chitosan. The prepared nanoparticles were capable of damaging the bacterial cell wall and hence displayed bactericidal activity against *Staphylococcus aureus* and *Salmonella enterica Typhimurium*. The biocompatibility of the nanoparticles was confirmed by a cell viability assay performed with a water-soluble tetrazolium (WST)-1 assay kit. In the above assay, the mycosynthesized chitosan nanoparticles did not show any cytotoxic effect on murine fibroblast NIH-3T3 cells. Further, the mycosynthezied nanoparticles exhibited higher antioxidant activity in comparison to the nanoparticles synthesized using tripolyphosphate (Saravanakumar et al. 2018).

Apart from the fungal enzyme-mediated synthesis of chitosan nanoparticles, there are several reports on the production of chitosan nanoparticles using chitin or chitosan extracted from various fungi. Since these protocols involve the use of chemical agents such as sodium tripolyphosphate, the claim for their mycogenicity is

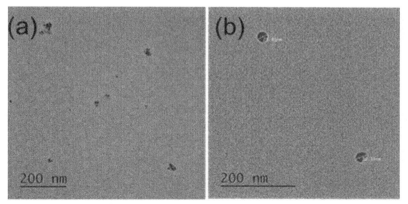

Figure 3. Characterization of *T. harzianum* based T-CSNP: Transmission electron microscopic analysis (a.b).
[Figure adopted from Saravanakumar et al. (2018) with permission from Elsevier]

questionable. Yet most of those preparative strategies yielded chitosan nanoparticles with relatively a small size and desirable bioactivities. Alsharari et al. (2018) extracted fungal chitosan from the mycelia of *Cunninghamella elegans* to produce chitosan nanoparticles for the remediation and biosorption of contaminant heavy metals, Pb^{2+} and Cu^{2+}. Sodium tripolyphosphate was applied to synthesize chitosan nanoparticles from fungal chitosan and the particle size of the nanoparticles was found to be in the range of 5–45 nm. Encouragingly, the synthesized chitosan nanoparticles were highly effective than bulk chitosan for the remediation and biosorption of contaminant metals (Alsharari et al. 2018). In another exercise, chitosan extracted from dried mycelia of *Rhizopus stolonifer* was used to prepare chitosan nanoparticles by ionic gelation of tri-sodium polyphosphate. These nanoparticles exhibited high antimicrobial activity, yet devoid of toxic effects in both *in vitro* and *in vivo* assays (Darwesh et al. 2018).

El Rabey et al. (2019) designed an experiment to obtain fungal chitosan from the filamentous fungus *Amylomyces rouxii* and transform it into nano-form by gelation with ionotropic method. Thereafter, the attempts were made to load fluconazole into the nanoparticles and then to evaluate the potential antimycotic activity of the synthesized nanoconjugates against drug-resistant *Candida* spp. Apart from the small particle size, high stability, and distribution, the nanoconjugates exhibited strong anti-Candidal activity. Cotton textiles with these nanoconjugates were fabricated and the potent antimycotic exhibited by these materials indicated their suitability in the health care discipline (El Rabey et al. 2019).

Boruah and Dutta (2020) more recently explored the fungi *Fusarium oxysporum*, *Metarhizium anisopliae*, *Beauveria bassiana* and *Trichoderma viride* as sources of chitin, which then converted to chitosan and later to chitosan nanoparticles with the use of sodium tripolyphosphate. Electron microscopic analysis revealed that synthesized chitosan nanoparticles are of near-spherical shape. In dynamic light scattering (DLS) analysis, the chitosan nanoparticles based on *B. bassiana* and *T. viride* showed less than 100 nm of average size (78.36, 89.03 nm respectively) while the nanoparticles based on *F. oxysporum* and *M. anisopliae* were above 150 nm in average size (Boruah and Dutta 2020).

In another study, chitosan nanoparticles were synthesized using extracted chitosan from *Aspergillus niger* mycelia and sodium tripolyphosphate. This led to the production of nanoparticles in the size range 55.2–118.9 nm with a mean diameter of 68.4 nm. Later, these nanoparticles were used to produce nanocomposites with silver nanoparticles synthesized from the algal biomass *Codium capitatum*. The antibacterial activity of individual nanoparticles, as well as the chitosan-silver nanocomposite, was determined against drug-resistant pathogens of *Salmonella typhimurium* and *Staphylococcus aureus*. Although all nanoparticles exhibited antibacterial activity, the activity was much more prominent in the chitosan-silver nanocomposite (Alsaggaf et al. 2020).

Based on the aforementioned examples it is obvious that mycosynthezied nanoparticles display a range of bioactivities that can be utilized in different fields like medicine, agriculture, food industries, and environmental chemistry.

Conclusion

Mycosynthesis of chitosan nanoparticles is relatively a novel area of research that can lead to the production of stable, non-toxic, and bioactive nanoparticles. Nevertheless, only a handful of studies have been conducted hitherto on this topic. Apart from the fungal enzyme-mediated synthesis of chitosan nanoparticles, several research groups focused on the synthesis of chitosan nanoparticles from chitin or chitosan extracted from various fungi in the presence of some chemical agents. The mycosynthesized chitosan nanoparticles have displayed different bioactivities. However, further investigations are required to establish the suitability and efficacy of different fungal species in mycosynthesis of nanochitosan and the possible utility of the mycogenic nanoparticles in different fields.

References

Alsharari, S.F., Tayel, A.A., and Moussa, S.H. 2018. Soil emendation with nano-fungal chitosan for heavy metals biosorption. Int. J. Biol. Macromol. 118(Pt B): 2265–2268.

Alsaggaf, M.S., Tayel, A.A., Alghuthaymi, M.A., and Moussa, S.H. 2020. Synergistic antimicrobial action of phyco-synthesized silver nanoparticles and nano-fungal chitosan composites against drug resistant bacterial pathogens, Biotechnol. Biotechnol. Equip. 34(1): 631–639.

Alves, N.M., and Mano, J.F. 2008. Chitosan derivatives obtained by chemical modifications for biomedical and environmental applications. Int. J. Biol. Macromol. 43(5): 401–414.

Birla, S.S., Tiwari, V.V., Gade, A.K., Ingle, A.P., Yadav, A.P., and Rai, M.K. 2009. Fabrication of silver nanoparticles by *Phoma glomerata* and its combined effect against *Escherichia coli*, *Pseudomonas aeruginosa* and *Staphylococcus aureus*. Lett. Appl. Microbiol. 48: 173–179.

Boruah, S., and Dutta, P. 2020. Fungus mediated biogenic synthesis and characterization of chitosan nanoparticles and its combine effect with *Trichoderma asperellum* against *Fusarium oxysporum*, *Sclerotium rolfsii* and *Rhizoctonia solani*. Indian Phytopathol. 74(1): 81–93.

Chauhan, A., Zubair, S., Tufail, S., Sherwani, A., Sajid, M., Raman, S.C., Azam, A., and Owais, M. 2011. Fungus-mediated biological synthesis of gold nanoparticles: potential in detection of liver cancer. Int. J. Nanomed. 6: 2305–2319.

Darwesh, O.M., Sultan, Y.Y., Seif, M.M., and Marrez, D.A. 2018. Bio-evaluation of crustacean and fungal nano-chitosan for applying as food ingredient. Toxicol. Rep. 5: 348–356.

Dhillon, G.S., Brar, S.K., Kaur, S., and Verma, M. 2012. Green approach for nanoparticle biosynthesis by fungi: Current trends and applications. Crit. Rev. Biotechnol. 32(1): 49–73.

Dhillon, G.S., Kaur, S., Brar, S.K., and Verma, M. 2013. Green synthesis approach: extraction of chitosan from fungus mycelia. Crit. Rev. Biotechnol. 33(4): 379–403.

Duran, N., Marcato, P.D., Alves, O.L., de Souza, G.I.H., and Esposito, E. 2005. Mechanistic aspects of biosynthesis of silver nanoparticles by several *Fusarium oxysporum* strains. J. Nanobiotechnol. 3: 1–7.

Fayaz, A.M., Balaji, K., Girilal, M., Kalaichelvan, P.T., and Venkatesan, R. 2009. Mycobased synthesis of silver nanoparticles and their incorporation into sodium alginate films for vegetable and fruit preservation. J. Agric. Food Chem. 57: 6246–6252.

El-Naggar, N.EA., Saber, W.I.A., Zweil, A.M., and Bashir, S.I. 2022. An innovative green synthesis approach of chitosan nanoparticles and their inhibitory activity against phytopathogenic *Botrytis cinerea* on strawberry leaves. Sci. Rep. 12: 3515. DOI: 10.1038/s41598-022-07073-y.

El Rabey, H.A., Almutairi, F.M., Alalawy, A.I., Al-Duais, M.A., Sakran, M.I., Zidan, N.S., and Tayel, A.A. 2019. Augmented control of drug-resistant *Candida* spp. via fluconazole loading into fungal chitosan nanoparticles. Int. J. Biol. Macromol. 141: 511–516.

El-Shabouri, M.H. 2002. Positively charged nanoparticles for improving the oral bioavailability of cyclosporin-A. Int. J. Pharm. 249: 101–108.

110 *Mycosynthesis of Nanomaterials: Perspectives and Challenges*

Govender, Y., Riddin, T., Gericke, M., and Whiteley, C.G. 2009. Bioreduction of platinum salts into nanoparticles: A mechanistic perspective. Biotechnol. Lett. 31(1): 95–100.

Hussein, A.A., and Aldujaili, N.H. 2022. Biological preparation of chitosan nanoparticles using *Klebsiella pneumonia*. AIP Conference Proceedings 2386; 020002. DOI: 10.1063/5.0067233.

Hu, Y., Jiang, X., Ding, Y., Ge, H., Yuan, Y., and Yang, C. 2002. Synthesis and characterization of chitosan-poly(acrylic acid) nanoparticles. Biomaterials. 23: 3193–3201.

Huang, H.Y., Shieh, Y.T., Shih, C.M., and Twu, Y.K. 2010. Magnetic chitosan/iron (II, III) oxide nanoparticles prepared by spray-drying. Carbohydr. Polym. 81: 906–910.

Hijazi, N., Rodier, E., Letourneau, J.J., Louati, H., Sauceau, M., Le Moigne, N., Benezet, J.C., and Fages, F. 2014. Chitosan nanoparticles generation using CO_2 assisted processes. J. Supercrit. Fluids. 95: 118–128.

Islam, S., Bhuiyan, M.A.R., and Islam, M.N. 2017. Chitin and chitosan: Structure, properties and applications in biomedical engineering. J. Polym. Environ. 25: 854–866.

Jaidev, L.R., and Narasimha, G. 2010. Fungal mediated biosynthesis of silver nanoparticles, characterization and antimicrobial activity. Colloids Surf. B Biointerfaces. 81(2): 430–433.

Jameela, S.R., Kumary, T.V., Lal, A.V., and Jayakrishnan, A. 1998. Progesterone-loaded chitosan microspheres: A long acting biodegradable controlled delivery system. J. Control Release 52: 17–24.

Jayasuriya, A.C. 2017. Production of micro- and nanoscale chitosan particles for biomedical applications. pp. 185–209. *In*: Jennings, J.A., and Bumgardner, J.D. (eds.). Chitosan Based Biomaterials Volume 1. Woodhead Publishing, Sawston, UK.

Kafshgari, M.H., Khorram, M., Mansouri, M., Samimi, A., and Osfouri, S. 2012. Preparation of alginate and chitosan nanoparticles using a new reverse micellar system. Iran Polym. J. 21: 99–107.

Kamat, V., Marathe, I., Ghormade, V., Bodas, D., and Paknikar, K. 2015. Synthesis of monodisperse chitosan nanoparticles and *in situ* drug loading using active microreactor. ACS Appl. Mater Interfaces. 7(41): 22839–22847.

Kashyap, P.L., Xiang, X., and Heiden, P. 2015. Chitosan nanoparticle based delivery systems for sustainable agriculture. Int. J. Biol. Macromol. 77: 36–51.

Khandel, P., and Shahi, S.K. 2018. Mycogenic nanoparticles and their bio-prospective applications: Current status and future challenges. J. Nanostruct. Chem. 8: 369–391.

Kumar, S.A., Peter, Y.A., and Nadeau, J.L. 2008. Facile biosynthesis, separation and conjugation of gold nanoparticles to doxorubicin. Nanotechnology 19(49): 495101. doi:10.1088/0957-4484/19/49/49510.

Liu, H., and Gao, C. 2009. Preparation and properties of ionically cross-linked chitosan nanoparticles. Polym. Adv. Technol. 20: 613–619.

Maluin, F.N., and Hussein, M.Z. 2020. Chitosan-based agronanochemicals as a sustainable alternative in crop protection. Molecules 25(7): 1611. doi:10.3390/molecules25071611.

Mishra, A., Tripathy, S., Wahab, R., Jeong, S.H., Hwang, I., Yang, Y.B., Kim, Y.S., Shin, H.S., and Yun, S.I. 2011. Microbial synthesis of gold nanoparticles using the fungus *Penicillium brevicompactum* and their cytotoxic effects against mouse mayo blast cancer C2C12 cells. Appl. Microbiol. Biotechnol. 929(3): 617–630.

Mohammed, M.A., Syeda, J.T.M., Wasan, K.M., and Wasan, E.K. 2017. An overview of chitosan nanoparticles and its application in non-parenteral drug delivery. Pharmaceutics. 9(4): 53. doi: 10.3390/pharmaceutics9040053.

Nag, M., Lahiri, D., Sarkar, T., Ghosh, S., Dey, A., Edinur, H.A., Pati, S., and Ray, R.R. 2021. Microbial fabrication of nanomaterial and its role in disintegration of exopolymeric matrices of biofilm. Front Chem. 9: 690590. doi: 10.3389/fchem.2021.690590.

Ngan, L.T.K., Wang, SL., Hiep, Ð.M., Luong, P.M., Vui, N.T., Ðinh, T.M., and Dzung, N.A. 2014. Preparation of chitosan nanoparticles by spray drying, and their antibacterial activity. Res. Chem. Intermed. 40: 2165–2175.

Ohya, Y., Shiratani, M., Kobayashi, H., and Ouchi, T. 1994. Release behavior of 5-fluorouracil from chitosan-gel nanospheres immobilizing 5-fluorouracil coated with polysaccharides and their cell specific cytotoxicity. J. Macromol. Sci. A. 31: 629–642.

Quiñones, J.P., Peniche, H., and Peniche, C. 2018. Chitosan based self-assembled nanoparticles in drug delivery. Polymers. 10(3): 235. doi: 10.3390/polym10030235.

Rai, M., Bonde, S., Golinska, P., Trzcińska-Wencel, J., Gade, A., Abd-Elsalam, K.A., Shende, S., Gaikwad, S., and Ingle, A.P. 2021. *Fusarium* as a novel fungus for the synthesis of nanoparticles: Mechanism and applications. J. Fungi. 7(2): 139. doi: 10.3390/jof7020139.

Rajakumar, G., Rahuman, A., Roopan, S.M., Khanna, V.G., Elango, G., Kamaraj, C., Zahir, A.A., and Velayutham, K. 2012. Fungus-mediated biosynthesis and characterization of TiO_2 nanoparticles and their activity against pathogenic bacteria. Spectrochim Acta Mol. Biomol. Spectrosc. 91: 23–29.

Raliya, R., Rathore, I., and Tarafdar, J.C. 2013. Developmental of microbial nanofactory for zinc, magnesium and titanium nanoparticles production using soil fungi. J. Bionanoscience 7(5): 590–596.

Shelke, D., Chambhare, M., and Sonawane, H. 2022. Mycosynthesis of nanoparticles and their potential application in pharmaceutical bioprocessing. *In*: Sarma, H., Gupta, S., Narayan, M., Prasad, R. and Krishnan, A. (eds.). Engineered Nanomaterials for Innovative Therapies and Biomedicine. Nanotechnology in the Life Sciences. Springer, Cham. DOI: 10.1007/978-3-030-82918-6_17.

Saravanakumar, K., Chelliah, R., MubarakAli, D., Jeevithan, E., Oh, D.H., Kathiresan, K., and Wang, M.H. 2018. Fungal enzyme-mediated synthesis of chitosan nanoparticles and its biocompatibility, antioxidant and bactericidal properties. Int. J. Biol. Macromol. 118(Pt B): 1542–1549.

Sathiyabama, M., and Parthasarathy, R. 2016. Biological preparation of chitosan nanoparticles and its *in vitro* antifungal efficacy against some phytopathogenic fungi. Carbohydr. Polym. 151: 321–325.

Tarafdar, J.C., and Raliya, R. 2013. Rapid, low-cost, and ecofriendly approach for iron nanoparticle synthesis using *Aspergillus oryzae* TFR9. J. Nanoparticles. 2013: 141274. doi: 10.1155/2013/141274.

Thakkar, K.N., Mhatre, S.S., and Parikh, R.Y. 2010. Biological synthesis of metallicnanoparticles. Nanomedicine 6(2): 257–262.

Tokumitsu, H., Ichikawa, H., and Fukumori, Y. 1999. Chitosan-gadopentetic acid complex nanoparticles for gadolinium neutron-capture therapy of cancer: Preparation by novel emulsion-droplet coalescence technique and characterization. Pharm. Res. 16: 1830–1835.

Yanat, M., and Schroën, K. 2021. Preparation methods and applications of chitosan nanoparticles; with an outlook toward reinforcement of biodegradable packaging, React Funct. Polym. 161: 104849. doi: 10.1016/j.reactfunctpolym.2021.104849.

Yang, Y., Wang, S., Wang, Y., Wang, X., Wang, Q., and Chen, M. 2014. Advances in self-assembled chitosan nanomaterials for drug delivery. Biotechnol. Adv. 32: 1301–1316.

Yeul, V.S., and Rayalu, S.S. 2013. Unprecedented chitin and chitosan: A chemical overview. J. Polym. Environ. 21: 606–614.

Zhang, X., He, X., Wang, K., and Yang, X. 2011. Different active biomolecules involved in biosynthesis of gold nanoparticles by three fungus species. J. Biomed. Nanotech. 7(2): 245–254.

CHAPTER 6

Fungi-Mediated Fabrication of Copper Nanoparticles and Copper Oxide Nanoparticles, Physical Characterization and Antimicrobial Activity

Ishita Saha,[1] Parimal Karmakar[1] and Debalina Bhattacharya[2,]*

Introduction

Nanoparticles have gained widespread attention in the last few years, due to their distinctive characteristics such as optical, thermal, mechanical, magnetic, and chemical properties compared to bulk materials. They have diameters ranging from 1–100 nm. Nanoparticles are mostly used in modern biomedical applications such as therapeutics, diagnostics, catalysis, electronics, bioimaging, and optical sensing (Zikalala et al. 2018). The biomedically used nanoparticles are commonly known as nanomedicine. Nanoparticles can interrelate closely with microorganisms due to their large surface area to volume ratio and small size which can interact with contaminants for degradation (Riehemann et al. 2009). Various nano metals for example silver, gold, copper, zinc, titanium, chromium, etc., have gained widespread interest due to their properties against bacteria (Akhavan et al. 2011). Cu is the most cost-effective metal among Au and Ag. In ancient times copper gained attention for industrial purposes, electrical equipment, construction materials, antimicrobial agents, and alloy formation with other metals. Copper also plays an important role

---

[1] Department of Life science & Biotechnology, Jadavpur University, Kolkata-700032.
[2] Department of Microbiology, Maulana Azad College, Kolkata-700013.
* Corresponding author: debalina.bhattacharya13@gmail.com

as a fungicide, fertilizer, and essential trace micronutrient for human, animal, and aquatic species (Malhotra et al. 2020). Copper in the bulk form is less available in the body, and much of its amount excreted out with feces causing environmental pollution and economic loss. In the nineteenth century, the application of nanotechnology offers a novel approach to synthesis copper nanoparticles (Cu NPs). Due to their small size and high surface area to volume ratio, copper nanoparticles (Cu NPs) have received attention as powerful antibacterial and antimicrobial agents and have been used in a variety of industries including glass colour and the ceramic industry (El-Batal et al. 2018). But the stability of copper nanoparticle is very low, so metal oxides nanoparticles, i.e., copper oxide nanoparticles (CuO NPs) has gained attention. Copper oxide (CuOs), which are inorganic NPs, can be easily oxidised from Cu NPs. Cu and CuO NPs are widely employed as anticancer, antibacterial, and antioxidant agents (Gawande et al. 2016; Khan et al. 2019). This is because both NPs can interact with the biological system at the cellular level for a variety of reactions and functions. As Cu is a transition metal, Cu/CuO NPs have been used to generate free radicals, causing DNA damage of tumor cells, and programmed cell death with no adverse effects on other nearby cells (Pramanik et al. 2012). Cu/CuO NPs can be synthesized by physical, chemical, and biological methods.

The use of living organisms in green chemistry leads to the production of nanoparticles. This procedure uses plant extracts, animal extracts, enzymes, and microbes, making it dependable, simple, inexpensive, biocompatible, and environmental-friendly. During the synthesis of nanoparticles in the biological approach, the biological system acts as reducing and capping agents. Fungi are the preferred microbes because of their fast growth rate, bio-properties, and simple structures that are easy to operate. Furthermore, these organisms create maximum biomolecules, which in turn drive the synthesis of NPs. Apart from physical and chemical synthesis, myconanotechnology is constantly evolving and has influenced many thrusts in NP manufacturing, including a wide range of biological applications for human health (Ingle et al. 2014). The problem of nanoparticle agglomeration during green synthesis necessitates the use of stabilizers such as ligands, polymers, and surfactants. The size of nanoparticles is highly influenced by the pH of the solution. Cu NPs produced under acidic (pH 5.5) conditions had greater solubility and zeta potential (Cho et al. 2012). CuO NPs form instead of Cu NPs at higher pH levels due to an excess of sodium hydroxide, and a further increase in pH generates Cu hydroxide without any Cu NPs formation (Soomro et al. 2014). Cu NPs' biological activity is determined by their physical and chemical properties, such as size, structure, concentration, and surface charge. Fungal-derived NPs exhibited the best anticancer and antimicrobial activity. Therefore, the focus of this chapter is to provide information on the fabrication of fungi-mediated copper nanoparticles (Cu NPs) and copper oxide nanoparticles (CuO NPs), their physical characterizations and their application as antimicrobial agents.

Synthesis process of copper nanoparticles and copper oxide nanoparticles from different fungal sources

The synthesis process is widely separated into two main categories, i.e., the bottom-up approach (constructive approach) and the top-down approach (destructive approach) (Khan et al. 2019). In bottom-up approach, the nano-sized particles are formed into small atomic size particles wherein top-down approach, the large molecules are converted into smaller ones, and the smaller molecules lead to the formation of suitable nanomaterials. Cu NPs and CuO NPs have been produced by physical, chemical, and biological approaches. The common physical methods of the nanoparticle's synthesis process are laser ablation, aerosol techniques, microwave-assisted, and radiolysis techniques; chemical methods and microemulsion techniques, etc. (Abou El-Nour et al. 2010; Thakkar et al. 2010). O'Sullivan et al. observed that CuO NPs were readily synthesized and stabilized by high surfactant concentrations such as polysorbate 40 and polysorbate 60 (O'Sullivan et al. 2004). Strong reducing agents can be used for controlling the shape and size of the nanoparticles. However, these nanoparticles synthesized by physical or chemical methods lead to generating by-products that have unfavorable effects on the environment, high cost, low product efficiency, and high energy consumption. Therefore, fungi-mediated synthesis of Cu/CuO NPs are gaining attention day by day. These fungi-synthesized Cu/CuO NPs are more cost-effective, biocompatible, biodegradable, robust, stable, and longer shelf life. Here in this chapter, we have summarized the fungi-mediated fabrication process of Cu NPs and CuO NPs.

Fungi-mediated synthesis of Cu NPs and CuO NPs are almost similar. It mainly depends on pH, temperature, variations in incubation time and fungal biomass concentration (Alghuthaymi et al. 2015). The metal ions transform into metal oxide nanoparticles with the help of fungal biomolecules. The fungus hydrolyzes the metal ion complexes formed by the positively charged metal ions to produce metal oxide nanoparticles. The fungal proteins also act as capping agents to metal oxide nanoparticles, which is eco-friendly and more stable (Zikalala et al. 2018). In several studies Cu NPs are synthesized from various fungal sources like *Agaricus bisporus, Trichoderma atroviride, Trichoderma koningiopsis, Hypocrea lixii, Fusarium proliferatum,* etc. (Salvadori et al. 2014a; Kasana et al. 2016). Kalaimurugan et al. isolated the whole-cell biomass of *Fusarium proliferatum* from a soil sample and used the biomass for Cu NP synthesis (Kalaimurugan et al. 2019). They used 10 mM copper sulfate solution in 100 ml fungal biomass solution and after 30 mins the color was changed from blue to dark brown (Kalaimurugan et al. 2019). Whereas, CuO NPs are synthesized from different species of fungi, such as *Alternaria alternata, Aspergillus terreus, Botrytis cinerea, Penicillium chrysogenum, Pleurotus ostreatus, Rhodotorula mucilaginosa, Stereum hirsutum, Trichoderma asperellum,* etc. (Waris et al. 2021). Hassabo et al. synthesized different sizes CuO NPs from 64.7 nm to 51.6 nm using yeast (Hassabo et al. 2021). Yeast solution was treated with copper salt and the color was changed from light sky blue to light green after 24 h incubation (Hassabo et al. 2021). However, Cuevas et al. synthesized Cu/CuO NPs from the same fungi *Stereum hirsutum* and used three types of copper salt-like, copper sulfate, copper chloride, and copper nitrate respectively (Cuevas et al. 2015). A list of fungi

Fungi-Mediated Fabrication of Copper and Copper Oxide Nanoparticles 115

for the synthesis of Cu NPs and CuO NPs along with their sizes have been shown in Tables 1(a) and 1(b) respectively.

Table 1 (a). List of fungal sources used for the synthesis of copper nanoparticles along with their sizes.

Source of Cu NPs (fungi)	Color	UV range (nm)	Size (nm)	References
Agaricus bisporus	Blue to Green	370, 690	2–10	Sriramulu et al. 2020
Stereum hirsutum	Blue	670	5–20	Cuevas et al. 2015
Trichoderma atroviride	-	580–600	5–25	Natesan et al. 2021
Trichoderma koningiopsis	-	-	87.5	Salvadori et al. 2014a
Hypocrea lixii	-	-	24.5	Salvadori et al. 2013
Fusarium proliferatum	Blue to dark brown	575	12.5 ± 0.21	Kalaimurugan et al. 2019
Penicillium ochrochloron	-	-	-	Lacerda et al. 2019

Table 1 (b). List of fungal sources used for the synthesis of copper oxide nanoparticles along with their sizes.

Source of CuO NPs (fungi)	Color	UV range (nm)	Size (nm)	Reference
Aspergillus terreus	Blue to Green	260–270	15.75	Mousa et al. 2020; Mani et al. 2021
Trichoderma harzianum	Pale yellow brown to colloidal dark brown	-	38–77	Consolo et al. 2020
Botrytis cinerea	Dark brown	-	60–80	Kovačec et al. 2017
Penicillium chrysogenum	Dark reddish brown	410	9.70	El-Batal et al. 2020
Pleurotus ostreatus	-	-	25–36	El-Batal et al. 2018
Rhodotorula mucilaginosa	-	-	10.5	Salvadori et al. 2013
Trichoderma asperellum	Light yellow to dark brown	285-295	110	Saravanakumar et al. 2019
Aspergillus fumigatus	Light blue to light green	335	48	Ghareib et al. 2019
Penicillium aurantiogriseum	Pale yellow to brown	265	91–97	Honary et al. 2012
Penicillium citrinum	Pale yellow to brown	265	91–116	Honary et al. 2012
Penicillium waksmanii	Pale yellow to brown	265	80–87	Honary et al. 2012
Aspergillus niger	Blue to greenish black	-	30–40	Hasanin et al. 2021
Yeast extract	Blue to green	730	64.7–51.6	Hassabo et al. 2022

Isolation, purification, screening, and molecular characterization of fungal strains

The collected samples were placed in Petri dishes containing various kinds of media, including potato agar (PDA), malt extract agar (MEA), and Sabouraud dextrose agar (SDA) media. The culture media was supplemented with the antibiotic streptomycin and incubated at $28 \pm 2°C$ for 11–14 days (Zhong et al. 2011). The isolated fungus was then identified by observing the various characteristics of the mycelia like mycelium morphology, types of conidia/spore mass color, distinctive dorsal and reverse colony color, and diffusible pigment, sporophore, and spore chain morphology (Mani et al. 2018). Then the PCR primers for rDNA, 18S rRNA fragment identification that belongs to different mycelia were used. Finally, the strains were mounted on the glass slides for staining and were then examined under a light microscope at 40X and 100X magnification.

Fungal biomass and filtrate preparation for Cu NPs and CuO NPs synthesis

Fungal biomass was then prepared by inoculating with fungi spores in various broth media like Sabouraud Dextrose broth, Czapek-Dox's broth media, etc. The biomass was then incubated at room temperature at 35°C in a shaking condition for a few days. The biomass was filtered by Whatman filter paper (Abdeen et al. 2016) and it was separated from the culture medium by washing with deionized water. After the incubation period, biomass was separated from the medium by centrifugation. Finally, the cell-free supernatant portion of this medium was utilized for nanoparticle synthesis (Mousa et al. 2020).

Synthesis of Cu NPs and CuO NPs from fungal filtrate

Next, the supernatant was mixed with a copper salt (Copper sulfate, copper chloride, copper nitrate) solution at Vol./Vol. ratio and incubated at 30°C. The resultant mixture exhibited a change in its color after the reduction of sulfate/chloride/nitrate ions from Cu^{+2} to Cu^0 and it is the earliest signal for the synthesis of nanoparticles. According to El Sayed et al., the centrifuged nanoparticles were collected, re-dispersed in sterilized deionized water, air-dried, and finally resuspended in distilled water for further studies (El Sayed and El-Sayed 2020). Figure 1 represents the overview of the fungi-mediated synthesis of CuO NPs.

Physical characterization of fungi-mediated Cu NPs and CuO NPs

Cu NPs and CuO NPs were characterized by different techniques like Ultraviolet-visible spectroscopy (UV-Vis), Fourier transform infrared spectroscopy (FT-IR), X-ray diffraction (XRD), Scanning electron microscopy (SEM), Energy dispersive x-ray analysis (EDX), Transmission electron microscopy (TEM), Dynamic light scattering (DLS), Zeta potential, Thermo gravimetric analysis (TGA) and Differential thermal analysis (DTA), etc.

Ultraviolet-visible spectroscopy

Ultraviolet-visible spectroscopy is a characterizing tool used to detect the optical properties of a synthesized material. The synthesized Cu NPs and CuO NPs were

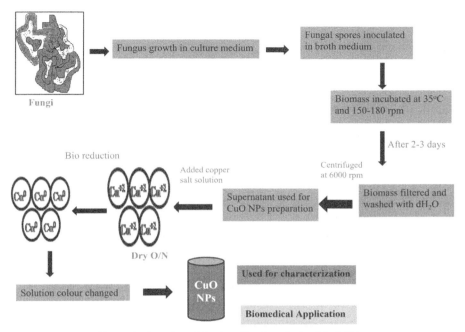

Figure 1. Overview of fungi-mediated synthesis of CuO NPs.

diffused in an aqueous solution and the absorbance was measured from 250 nm to 800 nm to obtain the absorption maxima (λ_{max}). Sriramulu et al. synthesized Cu NPs from *Agaricus bisporus* extract and observed the absorption peak at 551 nm (Sriramulu et al. 2020). Cu NPs were produced from *Trichoderma atroviride* extract and were treated with copper sulfate salt (Natesan et al. 2021). *Stereum hirsutum* fungus was used for the synthesis of Cu NPs and CuO NPs with the help of three different copper salts such as copper chloride, copper sulfate and copper nitrate respectively. The maximum absorbance range on surface plasmon resonance absorption (SPR) of Cu NPs or CuO NPs were 620–710 nm (Cuevas et al. 2015). Honary et al. synthesized CuO NPs from extracellular metabolites of three molds *Penicillium aurantiogriseum*, *Penicillium citrinum*, and *Penicillium waksmanii*, respectively, and copper sulfate as salt (Honary et al. 2012). The UV absorbance of the synthesized nanoparticles at 265 nm and the fluorescence spectrum showed a broad emission peak at 448 nm when excited at 433 nm (Honary et al. 2012). The absorption spectra of Cu NPs and CuO NPs lie at different positions because the NPs were synthesized from different sources of fungal extract solution as shown in Table 1 (a) and (b).

X-ray diffraction (XRD)

X-ray diffractometer is used for recognizing the crystal phase of materials with CuKα radiation (θ =1.5406 Å), operating at a voltage of 40 kV and a current of 40 mA at room temperature. The intensity data were collected at a range of 20° to

118 *Mycosynthesis of Nanomaterials: Perspectives and Challenges*

90°. Cu NPs and CuO NPs peaks were analyzed by JCPDS software. The absorbance peak varied from one another because nanoparticles were synthesized from the various fungal solutions. Fungi-mediated Cu NPs showed the diffraction peaks (2θ) at 43.90°, 50.18°, and 74.67° and the corresponding planes are at (111), (200) and (220) according to Sriramulu et al. (2020). The fabricated Cu NPs synthesized from *Trichoderma atroviride* fungi have four diffraction peaks at 36.15°, 42.91°, 49.91°, and 73.61°, which corresponds to index of (111), (200), and (220) respectively (Natesan et al. 2021). Salvadori et al synthesized Cu NPs from dead biomass of *Hypocrea lixii* and compared the XRD pattern of Cu NPs and CuO NPs (Salvadori et al. 2013). Other results showed the XRD patterns of *Fusarium proliferatum* biomass impregnated copper. The nanoparticles have shown sharp and clear peaks at 45.1037°, 37. 8541°, 32.4132°, 27.1053°, 65.4325° and 77.0722° relating to (111), (220), (200), and (311) planes (Kalaimurugan et al. 2019). In the case of CuO NPs four strong absorption peak angles (2θ) were normally shown at 29°, 37°, 44°, and 62° respectively, with crystalline planes (110), (111), (200), and (220) according to Khatami et al. (2019). Mani et al., reported the 2θ diffraction peaks at 28.20°, 30.58°, 31.51°, 34°, 45.38°, 46.52°, and 75.20° confirming the presence of CuO NPs (Mani et al. 2021). Mousa et al. synthesized CuO NPs from *Aspergillus terreus* and they have reported the presence of (110), (002), (111), (202), (020), (113), (113), and (222) planes which confirm CuO NPs formation (Mousa et al. 2021). Hassabo et al. synthesized CuO NPs from marine yeast and observed different diffraction patterns due to the degradation of a yeast cell (Hassabo et al. 2021). Cuevas et al. used *Stereum hirsutum* to synthesize Cu NPs and CuO NPs, and revealed an X-ray diffraction pattern with peaks at 43.6, 50.7, and 74.45, which represent the (111), (200), (220), and (222) planes of Cu NPs, and peaks at 29.4, 36.8, 42.1, 61.9, and 77.6 which represent the (110), (111), (200), (220), and (222) planes of the CuO NPs respectively (Cuevas et al. 2015).

Fourier transform infrared spectroscopy (FTIR)

FTIR spectrophotometer was utilized to analyze the various functional groups or biomolecules that are responsible for the bioreduction of materials. The observed intense bands were compared with the standard values of various functional groups for the necessary identification. Cu NPs are transformed to the oxide form of CuO NPs with the help of proteins and enzymes in the endophytic fungus. Several studies have observed various stretching at different positions for the synthesis process of Cu NPs and CuO NPs. Salvadori et al. synthesized Cu NPs from *Hypocrea lixii* fungus and observed two bands at 1649 cm^{-1} and 1532 cm^{-1} that correspond to the bending vibrations of amide I and amide II, respectively (Salvadori et al. 2013). The researchers reported the proteins/peptides are bounded to the surface of Cu NPs and these agents act as a capping agent (Salvadori et al. 2013). Mani et al. synthesized CuO NPs from endophytic fungus *Aspergillus terreus* and observed different peaks at 3321.41cm^{-1} to 3450cm^{-1} (-OH), 3200 cm^{-1} (-OH), 1682.1 cm^{-1} (C=C), 1600 cm^{-1}, 1100 cm^{-1} to 1000 cm^{-1} (C-O), 1016.14 cm^{-1} to 644.70 cm^{-1} (CuO), 700 cm^{-1} to 400 cm^{-1} (CuO nano-structures vibrations) and 900 cm^{-1} to 700 cm^{-1} (C-H) (Mani et al. 2021). Cuevas et al. synthesized Cu/CuO NPs from *Stereum hirsutum* fungus and observed the stretching at different positions like 3280 cm^{-1} and 2924 cm^{-1} (primary and secondary amines), 1029 cm^{-1} (C-N), 1243 cm^{-1} and 1244 cm^{-1} (amides I and amides III), 1076 cm^{-1} (amides II), and 900–1105 cm^{-1} (C–O–C pyranose ring stretching, beta-glycosidic linkages, and glycosidic ethers bonds of some polycarbohydrates) (Cuevas et al. 2015).

Energy-dispersive X-ray spectroscopy (EDX)

EDX is an x-ray technique that was performed to detect the presence of elemental composition of materials. EDX systems are attached to scanning electron microscopy or transmission electron microscopy instruments. The data generated by EDX analysis consist of spectra showing peaks corresponding to the elements making up the true composition of the sample being analyzed. Cu NPs were synthesized from *Agaricus bisporus* and the copper percentage was confirmed by EDX analysis (Sriramulu et al. 2020). CuO NPs were synthesized from *Aspergillus terreus* and other fungal sources where the elemental composition of fungal mediated CuO NPs show copper 68.7% and oxides 31.2% respectively (Zhao et al. 2020).

Scanning electron microscopy (SEM)

A scanning electron microscope works by focusing on a sample's surface with a focused electron beam. The size and shape of the synthesized nanoparticles were analyzed by SEM. *Agaricus bisporus* mediated copper nanoparticles have appeared as a spherical-shaped structure with agglomeration. Sriramulu et al. reported that the agglomeration nature of synthesized Cu NPs was present due to the hydroxyl groups of fungus (Sriramulu et al. 2020). Kandasamy et al. synthesized fungal-mediated CuO NPs from *Trichoderma asperellum* and reported the synthesized particles are spherical and evenly agglomerated (Kayalvizhi et al. 2020).

Transmission electron microscopy (TEM)

Transmission electron microscopy is a technique to examine a specimen by sending a beam of electrons through it. The goal of this procedure is to determine the size of the electrons on the surface. Salvadori et al. synthesized Cu NPs from dead biomass of fungus *Hypocrea lixii* and observed that the size range of spherical-shaped Cu NPs was 24.5 nm in TEM and was produced extracellularly (Salvadori et al. 2013). The fraction of dead biomass of the fungus *Trichoderma koningiopsis* impregnated with copper examined by TEM showed Cu NPs were present in the cell wall but not in the cytoplasm and cytoplasmic membrane (Salvadori et al. 2014b). In addition, it was observed that NPs are produced intracellularly and are uniformly distributed when analyzed in TEM (Salvadori et al. 2014b). Mostly the green synthesized CuO NPs are spherical in shape and are appeared by agglomeration of small particles with a size range of 20–35 nm. (Prakash et al. 2018). Whereas in another report it was observed that Cu NPs or CuO NPs synthesized from the mycelium-free extract of *Stereum hirsutum*, exhibit a spherical shape, monodispersity with diameters between 5 to 20 nm (Cuevas et al. 2015). Therefore, the source is a very important factor for

120 *Mycosynthesis of Nanomaterials: Perspectives and Challenges*

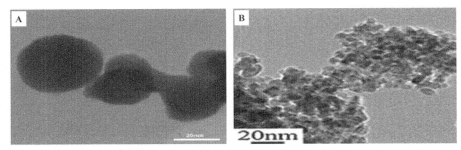

Figure 2. TEM images of (A) Cu NPs and (B) CuO NPs. This image has been modified and reprinted from the source Chalandar et al. 2017 from an open-access journal (Biosci Biotechnol Res Asia) under the Creative Commons Attribution license".

nanoparticle size and shape determination. Table 1 (a) and 1 (b) summarizes the different fungal sources and different sizes of Cu/CuO NPs and Figure 2 represents the TEM images of (A) Cu NPs and (B) CuO NPs.

Dynamic light scattering (DLS) and zeta potential

The characterization of nanoparticles is of paramount importance to the development of effective nanoformulations for the therapeutic approach. The size and surface charge of nanoparticles are known to play a crucial role in the proper characterization of these components. DLS (dynamic light scattering) and ZP (zeta potential) measurements are very useful techniques for ascertaining particle size and surface charge. The hydrodynamic sizes of nanoparticles were measured by DLS. The large positive or large negative zeta potential value depends upon the aggregation of synthesized nanoparticles. It was reported that the hydrodynamic size of nanoparticles was significantly larger than their TEM sizes because the sample measured in TEM was in dry state, whereas the DLS measures the size of the hydrated state of particles (Pramanik et al. 2012). Sriramulu et al. synthesized Cu NPs from *Agaricus bisporus* and observed the zeta potential value of this nanoparticle was –40 mv, and the size range was 10–60 nm based on DLS technique. Whereas the TEM images reported the size of this nanoparticle was 2–10 nm (Sriramulu et al. 2020). El-Batal et al. used *Penicillium chrysogenum* fungus for mycogenic synthesis of CuO NPs. They have observed the particle size distribution on DLS was 10.7 nm whereas the size distribution of CuO NPs was 9.7 nm on TEM (El-Batal et al. 2020). Therefore, these results indicated that the particle size measured by DLS, and TEM varies.

Antimicrobial activity of fungi-mediated Cu NPs and CuO NPs

In recent years, various microorganisms are increasingly resistant to numerous antibiotics. Therefore, copper-based antiseptics take an important part to overcome this situation. CuO NPs are strong antimicrobial agents that can be used against a wide range of bacteria. Nanoparticles interact with phosphorous moieties of bacterial DNA, inactivate replication, and suppress the functions of the enzymes. Furthermore,

Fungi-Mediated Fabrication of Copper and Copper Oxide Nanoparticles   121

bacterial cells do not produce respiratory enzymes and stop ATP production which leads to cell death. The positively charged nanoparticle interacts with the negatively charged bacterial cell membrane and as a result, several changes were observed such as rupture, cytoplasmic shrinkage, detachment, etc. (Borkow and Gabbay 2009). CuO NPs can modify or inhibit bacterial enzyme synthesis by binding to $-SH$ groups of bacterial proteins (Bhattacharya et al. 2020). These particles were found to be more effective against Gram-positive and Gram-negative microorganisms like *Escherichia coli, Bacillus subtilis, Pseudomonas aeruginosa, Staphylococcus aureus* and show extraordinary performance against pathogenic bacteria or broad-spectrum bacterial species. Here, we have mainly focused on the antibacterial and antifungal activity of fungi-mediated Cu NPs and CuO NPs. *Agaricus bisporus* mediated Cu NPs have applied against a wide variety of bacterial growth and the increasing order of zone of inhibition in *Proteus vulgaris > Escherichia coli > Pseudomonas aeruginosa > Klebsiella pneumoniae > Enterobacter aerogenes* were observed (Sriramulu et al. 2020). *Fusarium proliferatum* mediated synthesis of Cu NPs inhibits the growth of many bacterial species such as *Bacillus* sp., *Staphylococcus* sp., *Klebsiella* sp. and *E. coli* (Kalaimurugan et al. 2019). Whereas Natesan et al., applied *Trichoderma atroviride* mediated Cu NPs against two pathogenic fungi *Poria hypolateritia* and *Pestalotiopsis theae*, and showed strong antifungal activity (Natesan et al. 2021). Several studies have reported that fungi-mediated CuO NPs can inhibit the growth of various fungal cells like *Candida albicans, Candida krusei, Candida glabrata* and *Saccharomyces cerevisiae* (Amiri et al. 2017). Ghareib et al. synthesized CuO NPs from *Aspergillus fumigatus* fungus and it was applied against Gram-positive *Staphylococcus aureus* and Gram-negative *Klebsiella pneumoniae* to inhibit their growth (Ghareib et al. 2019). *Penicillium chrysogenum* fungi were used for the synthesis of CuO NPs and ZnO NPs and both are applied for antibacterial activity. The result showed the better inhibitory effects of CuO NPs on Gram-negative and Gram-positive pathogenic bacteria compared to ZnO NPs. Also, CuO NPs inhibit the growth of Gram-positive bacteria at a very low concentration compared to the ZnO NPs (Mohamed et al. 2021). *P. chrysogenum* filtrate mediated synthesis of CuO NPs showed a promising antibacterial and antifungal activity. Cu^{2+} ion and *P. chrysogenum* filtrate did not stimulate bacterial growth. These CuO NPs suspension was also applied against different fungal strains such as *Fusarium oxysporum, Aspergillus niger, Erysiphae cichoracearum* and it was observed that the synthesized nanoparticle showed a significant effect on pathogenic fungi compared to the *P. chrysogenum* filtrate and the standard antifungal agent itself (El-Batal et al. 2020). CuO NPs are found to be a better antibacterial agent compared to the antifungal agent (Mani et al. 2021) because the fungal cell wall is very rigid. It is composed of N-acetylglucose amine, so the metallic nanoparticle was not invaded easily into the fungal cell wall. The small-sized nanoparticles can penetrate easily on the surface of the pathogen's vegetative part and the toxicity of copper ions highly favored the nanometric size (Muñoz-Escobar and Reyes-López 2020). Therefore, the nano form of metallic copper has a more potent ability to prevent the pathogen hyphal growth than its oxide form (CuO) (Oussou-Azo et al. 2020).

Fungi-mediated CuO NPs are also applied in various fields like food processing, packaging, cultivation purposes, photocatalytic degradation, and as antioxidant, anticancer, and antidiabetic agent (Noor et al. 2020). The myco-synthesized CuO NPs are small and have a high surface area to volume ratio which allows them to interact with the microbial cell membrane (Usman et al. 2013). Furthermore, it inhibits the DNA helical structure, damages the essential protein through binding to -SH or carboxyl groups of an essential enzyme, and leads to destroying selective permeability (Yoon et al. 2007).

Conclusion and future prospects

Copper and Copper oxide nanoparticles can be produced from organic and inorganic materials to have their unique physical, chemical, and biological properties. They can interact with different tissues and cells in the body. Currently, the delivery of copper oxide nanoparticles to the body is most important due to their biomedical applications like transformed diagnostics and therapeutic strategies. The synthesis of copper and copper oxide nanoparticles is carried out by three different mechanisms-physical, chemical, and biological. Biological synthesis is more potent, eco-friendly, and biocompatible when compared to physical and chemical methods. Traditional medicines are potent alternatives for allopathy drugs which generally exhibit many side effects. Chinese and Indian traditional medicinal practices used various plant materials to treat numerous ailments without causing side effects. Therefore, we have utilized nanotechnology to formulate phyto-based drugs to treat wound healing, infections, and diseases like cancer, and diabetes. Fungi-mediated synthesis of copper nanoparticles and copper oxide nanoparticles has proven to be effective in targeting various types of cancer cells and consequently inducing apoptosis. They are used as photocatalysts and promote a wide range of antibacterial, antifungal, and antidiabetic activity. Copper nanoparticles and copper oxide nanoparticles are cost-effective and are potent alternatives for other metal-based nanoparticles such as gold, silver, platinum, and lead. So, many researchers have fabricated fungus-mediated Cu/CuO nanoparticles from various fungal sources. Moreover, the fungi-mediated synthesis of copper nanoparticles and copper oxide nanoparticles coated with biopolymer has boosted their physicochemical properties and their evaluation for biological activities like anticancer activity, antioxidant activity, anti-diabetic activity, etc. All biological activities are found to be greater in capped Cu/CuO nanoparticles when compared to the uncapped ones. Furthermore, capped Cu/CuO NPs can be efficiently used as drugs and as a carrier of diagnostic molecules. Therefore, more efficient, cheap, non-toxic polymeric coated fungi-mediated copper nanoparticles or copper oxide nanoparticles can be used as a potent therapeutic agent in near future.

References

Abdeen, M., Sabry, S., Ghozlan, H., El-Gendy, A.A., and Carpenter, E.E. 2016. Microbial-physical synthesis of Fe and Fe_3O_4 magnetic nanoparticles using *Aspergillus niger* YESM1 and supercritical condition of ethanol. J. Nanomater.

Abou El-Nour, K.M., Eftaiha, A.A., Al-Warthan, A., and Ammar, R.A. 2010. Synthesis and applications of silver nanoparticles. Arab. J. Chem. 3(3): 135–140.

Akhavan, O., Azimirad, R., Safa, S. and Hasani, E. 2011. CuO/Cu (OH)$_2$ hierarchical nanostructures as bactericidal photocatalysts. J. Mater. Chem. 21(26): 9634–9640.

Alghuthaymi, M.A., Almoammar, H., Rai, M., Said-Galiev, E., and Abd-Elsalam, K.A. 2015. Myconanoparticles: synthesis and their role in phytopathogens management. Biotechnol. Biotechnol. Equip. 29(2): 221–236.

Amiri, M., Etemadifar, Z., Daneshkazemi, A., and Nateghi, M. 2017. Antimicrobial effect of copper oxide nanoparticles on some oral bacteria and Candida species. J. Dent. Biomater. 4(1): 347.

Bhattacharya, D., Saha, R., and Mukhopadhyay, M. 2020. Combination therapy using metal nanoparticles for skin infections. Nanotechnology in Skin, Soft Tissue, and Bone Infections, 49–69, Springer.

Borkow, G., and Gabbay, J. 2009. Copper, an ancient remedy returning to fight microbial, fungal and viral infections. Curr. Chem. Biol. 3(3): 272–278.

Chalandar, H.E, Ghorbani, H.R, Attar, H., and Alavi, S.A. 2017. Antifungal effect of copper and copper oxide nanoparticles against *Penicillium* on orange fruit. Biosci. Biotechnol. Res. Asia 14(1): 279–284.

Cho, W.S., Duffin, R., Poland, C.A., Duschl, A., Oostingh, G.J., MacNee, W., Bradley, M., Megson, I.L., and Donaldson, K. 2012. Differential pro-inflammatory effects of metal oxide nanoparticles and their soluble ions in vitro and in vivo; zinc and copper nanoparticles, but not their ions, recruit eosinophils to the lungs. Nanotoxicology 6(1): 22–35.

Consolo, V.F., Torres-Nicolini, A., and Alvarez, V.A. 2020. Mycosinthetized Ag, CuO and ZnO nanoparticles from a promising *Trichoderma harzianum* strain and their antifungal potential against important phytopathogens. Sci. Rep. 10(1): 1–9.

Cuevas, R., Durán, N., Diez, M.C., Tortella, G.R., and Rubilar, O. 2015. Extracellular biosynthesis of copper and copper oxide nanoparticles by *Stereum hirsutum*, a native white-rot fungus from chilean forests. J. Nanomater..

El-Batal, A.I., Al-Hazmi, N.E., Mosallam, F.M., and El-Sayyad, G.S. 2018. Biogenic synthesis of copper nanoparticles by natural polysaccharides and *Pleurotus ostreatus* fermented fenugreek using gamma rays with antioxidant and antimicrobial potential towards some wound pathogens. Microb. Pathog. 118: 159–169.

El-Batal, A.I., El-Sayyad, G.S., Mosallam, F.M., and Fathy, R.M. 2020. *Penicillium chrysogenum*-mediated mycogenic synthesis of copper oxide nanoparticles using gamma rays for in vitro antimicrobial activity against some plant pathogens. J. Clust. Sci. 31(1): 79–90.

El Sayed, M.T., and El-Sayed, A.S. 2020. Biocidal activity of metal nanoparticles synthesized by *Fusarium solani* against multidrug-resistant bacteria and mycotoxigenic fungi. J. Microbiol. Biotechnol. 30(2): 226–236.

Gawande, M.B., Goswami, A., Felpin, F.X., Asefa, T., Huang, X., Silva, R., Zou, X., Zboril, R., and Varma, R.S. 2016. Cu and Cu-based nanoparticles: synthesis and applications in catalysis. Chem. Revi. 116(6): 3722–3811.

Ghareib, M., Abdallah, W., Tahon, M., and Tallima, A. 2019. Biosynthesis of copper oxide nanoparticles using the preformed biomass of *Aspergillus fumigatus* and their antibacterial and photocatalytic activities. Dig. J. Nanomater. Biostructures 14(2).

Hasanin, M., Hashem, A.H., Lashin, I., and Hassan, S.A. 2021. *In vitro* improvement and rooting of banana plantlets using antifungal nanocomposite based on myco-synthesized copper oxide nanoparticles and starch. Biomass Convers. Biorefin., 1–11.

Hassabo, A.A., Ibrahim, E., Ali, B., and Emam, H.E. 2021. Anticancer effects of biosynthesized Cu_2O nanoparticles using marine yeast. Biocatal. Agric. Biotechnol. 102261.

Honary, S., Barabadi, H., Gharaei-Fathabad, E., and Naghibi, F. 2012. Green synthesis of copper oxide nanoparticles using *Penicillium aurantiogriseum, Penicillium citrinum and Penicillium waksmanii*. Dig. J. Nanomater Bios. 7(3): 999–1005.

Ingle, A.P., Duran, N., and Rai, M. 2014. Bioactivity, mechanism of action, and cytotoxicity of copper-based nanoparticles: a review. Appl. Microbiol. Biotechnol. 98(3): 1001–1009.

Kalaimurugan, D., Sivasankar, P., Lavanya, K., Shivakumar, M.S., and Venkatesan, S. 2019. Antibacterial and larvicidal activity of *Fusarium proliferatum* (YNS2) whole cell biomass mediated copper nanoparticles. J. Clust. Sci. 30(4): 1071–1080.

Kasana, R.C., Panwar, N.R., Kaul, R.K., and Kumar, P. 2016. Copper nanoparticles in agriculture: biological synthesis and antimicrobial activity. In Nanoscience in Food and Agriculture 3: 129–143.

Kayalvizhi, S., Sengottaiyan, A., Selvankumar, T., Senthilkumar, B., Sudhakar, C., and Selvam, K. 2020. Eco-friendly cost-effective approach for synthesis of copper oxide nanoparticles for enhanced photocatalytic performance. Optik, 202: 163507.

Khan, I., Saeed, K., and Khan, I. 2019. Nanoparticles: Properties, applications, and toxicities. Arab. J. Chem., 12(7): 908–931.

Khatami, M., Varma, R.S., Heydari, M., Peydayesh, M., Sedighi, A., Agha Askari, H., Rohani, M., Baniasadi, M., Arkia, S., Seyedi, F., and Khatami, S. 2019. Copper oxide nanoparticles greener synthesis using tea and its antifungal efficiency on *Fusarium solani*. Geomicrobiol. J. 36(9): 777–781.

Kovačec, E., Regvar, M., van Elteren, J.T., Arčon, I., Papp, T., Makovec, D., and Vogel-Mikuš, K. 2017. Biotransformation of copper oxide nanoparticles by the pathogenic fungus *Botrytis cinerea*. Chemosphere 180: 178–185.

Lacerda, E.C.M., dos Passos Galluzzi Baltazar, M., Dos Reis, T.A., do Nascimento, C.A.O., Côrrea, B., and Gimenes, L.J. 2019. Copper biosorption from an aqueous solution by the dead biomass of *Penicillium ochrochloron*. Environ. Monit. Assess. 191(4): 1–8.

Malhotra, N., Ger, T.R., Uapipatanakul, B., Huang, J.C., Chen, K.H.C., and Hsiao, C.D. 2020. Review of copper and copper nanoparticle toxicity in fish. Nanomaterials 10(6): 1126.

Mani, V.M., Soundari, A.J.P.G., and Tamilarasi, S. 2018. Determination of *in vitro* cytotoxicity and anti-angiogenesis for a bioactive compound from *Aspergillus terreus* FC36AY1 Isolated from Aegle marmelos around Western Ghats, India. In Medicinal Chemistry. IntechOpen.

Mani, V.M., Kalaivani, S., Sabarathinam, S., Vasuki, M., Soundari, A.J.P.G., Das, M.A., Elfasakhany, A., and Pugazhendhi, A. 2021. Copper oxide nanoparticles synthesized from an endophytic fungus *Aspergillus terreus*: Bioactivity and anti-cancer evaluations. Environ. Res. 201: 111502.

Mohamed, A.A., Abu-Elghait, M., Ahmed, N.E., and Salem, S.S. 2021. Eco-friendly mycogenic synthesis of ZnO and CuO nanoparticles for *in vitro* antibacterial, antibiofilm, and antifungal applications. Biol. Trace Elem. Res. 199(7): 2788–2799.

Mousa, A.M., Aziz, O.A.A., Al-Hagar, O.E., Gizawy, M.A., Allan, K.F., and Attallah, M.F. 2020. Biosynthetic new composite material containing CuO nanoparticles produced by *Aspergillus terreus* for 47Sc separation of cancer theranostics application from irradiated Ca target. Appl. Radiat. Isot. 166: 109389.

Mousa, S.A., El-Sayed, E.S.R., Mohamed, S.S., El-Seoud, M.A.A., Elmehlawy, A.A., and Abdou, D.A. 2021. Novel mycosynthesis of Co_3O_4, CuO, Fe_3O_4, NiO, and ZnO nanoparticles by the endophytic *Aspergillus terreus* and evaluation of their antioxidant and antimicrobial activities. Appl. Microbiol. Biotechnol., 105(2): 741–753.

Muñoz-Escobar, A., and Reyes-López, S.Y. 2020. Antifungal susceptibility of *Candida* species to copper oxide nanoparticles on polycaprolactone fibers (PCL-CuONPs). PLoS One, 15(2): e0228864.

Natesan, K., Ponmurugan, P., Gnanamangai, B.M., Manigandan, V., Joy, S.P.J., Jayakumar, C., and Amsaveni, G. 2021. Biosynthesis of silica and copper nanoparticles from *Trichoderma, Streptomyces* and *Pseudomonas* spp. evaluated against collar canker and red root-rot disease of tea plants. Arch. Phytopathol. Pflanzenschutz, 54(1-2): 56–85.

Noor, S., Shah, Z., Javed, A., Ali, A., Hussain, S.B., Zafar, S., Ali, H., and Muhammad, S.A. 2020. A fungal based synthesis method for copper nanoparticles with the determination of anticancer, antidiabetic and antibacterial activities. J. Microbiol. Methods 174: 105966.

O'Sullivan, S.M., Woods, J.A., and O'Brien, N.M. 2004. Use of Tween 40 and Tween 80 to deliver a mixture of phytochemicals to human colonic adenocarcinoma cell (CaCo-2) monolayers. Br. J. Nutr. 91(5): 757–764.

Oussou-Azo, A.F., Nakama, T., Nakamura, M., Futagami, T., and Vestergaard, M.D.C.M. 2020. Antifungal potential of nanostructured crystalline copper and its oxide forms. Nanomaterials 10(5): 1003.

Prakash, S., Elavarasan, N., Venkatesan, A., Subashini, K., Sowndharya, M., and Sujatha, V. 2018. Green synthesis of copper oxide nanoparticles and its effective applications in Biginelli reaction, BTB photodegradation and antibacterial activity. Adv Powder Technol. 29(12): 3315–3326.

Pramanik, A., Laha, D., Bhattacharya, D., Pramanik, P., and Karmakar, P. 2012. A novel study of antibacterial activity of copper iodide nanoparticle mediated by DNA and membrane damage. Colloid Surf B: Biointerfaces 96: 50–55.

Riehemann, K., Schneider, S.W., Luger, T.A., Godin, B., Ferrari, M., and Fuchs, H. 2009. Nanomedicine—challenge and perspectives. Angew. Chem. Int. Ed., 48(5): 872–897.

Salvadori, M.R., Lepre, L.F., Ando, R.A., Oller do Nascimento, C.A., and Corrêa, B. 2013. Biosynthesis and uptake of copper nanoparticles by dead biomass of *Hypocrea lixii* isolated from the metal mine in the Brazilian Amazon region. PLoS One, 8(11): e80519.

Salvadori, M.R., Ando, R.A., Oller do Nascimento, C.A., and Corrêa, B. 2014a. Intracellular biosynthesis and removal of copper nanoparticles by dead biomass of yeast isolated from the wastewater of a mine in the Brazilian Amazonia. PLoS One 9(1): 87968.

Salvadori, M.R., Ando, R.A., Oller Do Nascimento, C.A., and Correa, B. 2014b. Bioremediation from wastewater and extracellular synthesis of copper nanoparticles by the fungus *Trichoderma koningiopsis*. J. Environ. Sci. Health A 49(11): 1286–1295.

Saravanakumar, K., Shanmugam, S., Varukattu, N.B., MubarakAli, D., Kathiresan, K., and Wang, M.H. 2019. Biosynthesis and characterization of copper oxide nanoparticles from indigenous fungi and its effect of photothermolysis on human lung carcinoma. J. Photochem. Photobiol. B: Biol. 190: 103–109.

Soomro, R.A., Sherazi, S.H., Memon, N., Shah, M., Kalwar, N., Hallam, K.R., and Shah, A. 2014. Synthesis of air stable copper nanoparticles and their use in catalysis. Adv. Mater Lett. 5: 191–198.

Sriramulu, M., Shanmugam, S., and Ponnusamy, V.K. 2020. *Agaricus bisporus* mediated biosynthesis of copper nanoparticles and its biological effects: An *in-vitro* study. Colloid and Interface Science Communications, 35: 100254.

Thakkar, K.N., Mhatre, S.S., and Parikh, R.Y. 2010. Biological synthesis of metallic nanoparticles. Nanomed.: Nanotechnol. Biol. Med. 6(2): 257–262.

Usman, M.S., El Zowalaty, M.E., Shameli, K., Zainuddin, N., Salama, M., and Ibrahim, N.A. 2013. Synthesis, characterization, and antimicrobial properties of copper nanoparticles. Int. J. Nanomedicine, 8: 4467.

Waris, A., Din, M., Ali, A., Ali, M., Afridi, S., Baset, A., and Khan, A.U. 2021. A comprehensive review of green synthesis of copper oxide nanoparticles and their diverse biomedical applications. Inorgan. Chem. Commun. 123: 1387–7003.

Yoon, K.Y., Byeon, J.H., Park, J.H., and Hwang, J. 2007. Susceptibility constants of *Escherichia coli* and *Bacillus subtilis* to silver and copper nanoparticles. Sci. Total Environ. 373(2-3): 572–575.

Zhao, S., Su, X., Wang, Y., Yang, X., Bi, M., He, Q., and Chen, Y. 2020. Copper oxide nanoparticles inhibited denitrifying enzymes and electron transport system activities to influence soil denitrification and N2O emission. Chemosphere 245: 125394.

Zhong, K., Gao, X., Xu, Z., Gao, H., Fan, S., Yamaguchi, I., Li, L., and Chen, R. 2011. Antioxidant activity of a novel Streptomyces strain Eri12 isolated from the rhizosphere of Rhizoma curcumae Longae. Curr. Res. Bacteriol. 4(2): 63–72.

Zikalala, N., Matshetshe, K., Parani, S., and Oluwafemi, O.S. 2018. Biosynthesis protocols for colloidal metal oxide nanoparticles. Nano-Struct. Nano-Objects, 16: 288–29.

CHAPTER 7

Mycogenic Synthesis of Silver Nanoparticles and its Optimization

Joanna Trzcińska-Wencel, Magdalena Wypij and
*Patrycja Golińska**

Introduction

Today, over 1.5 million species of fungi are believed to thrive and survive in various habitats on Earth. However, only 70,000 species have been well identified. Fungi being predominately decomposer organisms secrete many enzymes extracellularly to break-down compounds of great complexity into simpler forms and use diverse energy sources (Blackwell 2011; Adebayo et al. 2021). Therefore, their wide range and diversity makes them an important tool for synthesis, applications, and developing new products in nanotechnology, including silver nanoparticles (AgNPs) (Rai et al. 2009; Guilger-Casagrande and Lima 2019; Adebayo et al. 2021). Moreover, fungal-mediated synthesis of AgNPs offers various advantages over biosynthesis by bacteria, plants, or algae. In comparison to the other biological systems, it is believed that the fungi may have greater potential for NPs synthesis because they release a large number of metabolites, especially proteins, demonstrate bioaccumulation ability and high binding capacity, tolerance to metals, and intracellular uptake competencies for metals (Mandal et al. 2006; Narayanan and Sakthivel 2010; Alghuthaymi et al. 2015; Rauwel et al. 2015; Singh et al. 2016; Madakka et al. 2018). It is suggested that biomolecules such as peptides, proteins, or enzymes secreted by these microorganisms are responsible for the reduction of silver ions and induction of nanoparticle synthesis and encapsulation (Konappa et al. 2021). In the contaminated

---

Department of Microbiology, Faculty of Biological and Veterinary Sciences, Nicolaus Copernicus University, Lwowska 1, 87 100 Toruń, Poland.
* Corresponding author: golinska@umk.pl

environment, fungi as a biological system, attempt to reduce the number of silver ions to prevent silver toxicity and environmental pollution (Pandiarajan et al. 2011; Adebayo et al. 2021). Furthermore, the fungal growth is easy to perform, fast, and provides a huge amount of biomass used for nanoparticle synthesis that can take place intra- and/or extracellularly (Boroumand Moghaddam et al. 2015; Adebayo et al. 2021). In addition, fungal biomass does not require additional steps to extract the filtrate which is used for the biosynthesis of AgNPs (Gade et al. 2008). It should be also emphasized that the biosynthesis of silver nanoparticles is an eco-friendly process when compared to chemical synthesis that involves the use of reagents that are toxic and hazardous to health and the environment (Guilger-Casagrande and Lima 2019). Although mycosynthesis of AgNPs is important for process efficiency, the classical strategy for optimization of biosynthesis parameters through fixing all process parameters except one variable is time and labor-consuming as well as not consider the interactions between different bioprocess factors (Othman et al. 2017). Moreover, the genetic manipulation to overexpress specific enzymes in order to intensify synthesis is much more difficult among eukaryotes than prokaryotes (Rauwel et al. 2015).

Fungi as nanofactories—an overview

Many fungal strains have been reported as a biological system for the synthesis of AgNPs including *Fusarium oxysporum, Aspergillus niger, Alternaria alternata, Penicillium oxalicum, Fusarium solani, Trichoderma asperellum, Rhizopus stolonifer, Cladosporium cladosporioides, Candida albicans, Phoma* spp. and many others (Gade et al. 2008; Mukherjee et al. 2008; Abd El-Aziz et al. 2015; AbdelRahim et al. 2017; Hamzah et al. 2018; Feroze et al. 2019; Hulikere and Joshi 2019; Hikmet and Hussein 2021; Govindappa et al. 2022).

The biosynthesis of AgNPs is preliminarily confirmed by visual observation of the color change of the reaction mixture consisting of fungal extract and silver nitrate ($AgNO_3$) and then by measuring the absorbance of the AgNPs suspension using UV-Vis spectroscopy. Generally, UV-Vis spectra of AgNPs are monitored in the range of 200–700 nm and the presence of a maximum absorption peak around 380–450 nm is related to the AgNPs characteristic surface plasmon resonance (SPR) (Desai et al. 2012; Mistry et al. 2021). Basically, physico-chemical properties of the biosynthesized AgNPs including shape, size, size distribution, crystallinity, stability, and capping agents are investigated by using analytical techniques such as Transmission Electron Microscopy (TEM), High-Resolution TEM (HRTEM), Scanning Electron Microscopy (SEM), Field-Emission SEM (FESEM), Atomic Force Microscopy (AFM), Dynamic Light Scattering (DLS), X-ray diffraction (XRD), Zeta potential and Fourier Transform Infrared Spectroscopy (FTIR) (Verma et al. 2010; Mishra et al. 2011; Kumar et al. 2012; Costa Silva et al. 2017; Elamawi et al. 2018).

These physical and chemical properties, but also biological activity of biosynthesized AgNPs depend on the fungal strains used for the synthesis. Therefore, various fungal nanofactories that have the ability to synthesize AgNPs are described below and in Table 1. The majority of the studies have involved fungi

128 *Mycosynthesis of Nanomaterials: Perspectives and Challenges*

Table 1. An overview of fungi and yeasts used for intra- and extra-cellular synthesis of AgNPs with various shapes and sizes.

Fungi	Type of synthesis	A max	Morphology	Size [nm]	References
Acremonium borodinense	extracellular	420	spherical		Lawrance et al. 2021
Alternaria alternata	extracellular	440	spherical	8–48	Govindappa et al. 2022
Aspergillus brunneoviolaceus	extracellular	411	spherical	0.7–15.2	Mistry et al. 2021
Aspergillus caespitosus	Exo-filtrates	410	spherical, hexagonal, irregular shapes	10–50	El-Bendary et al. 2021
Aspergillus caespitosus	Endo-filtrates	410	spherical, hexagonal, irregular shapes	3–100	El-Bendary et al. 2021
Aspergillus clavatus	extracellular	415	Spherical, hexagonal, polydispersed	10–25	Verma et al. 2010
Aspergillus fumigatus	extracellular	420	spherical, triangular	5–25	Bhainsa and D'souza 2006
Aspergillus niger	extracellular	420	spherical	20	Gade et al. 2008
Aspergillus oryzae NRRL447	exogenous proteins	410	spherical	15–109.6	Elshafei et al. 2021
Aspergillus terreus	extracellular	440	spherical, nearly spherical	1–20	Li et al. 2012
Aspergillus terreus BA6	extracellular	420	spherical	7–23	Lotfy et al. 2021
Candida albicans	extracellular	420	different morphologies	20–80	Rahimi et al. 2016
Candida albicans	extracellular	429	spherical	40.19	Hikmet and Hussein 2021
Candida glabrata	extracellular	460	spherical and oval	2–15	Jalal et al. 2018
Candida guilliermondii	extracellular	425	near spherical	10–20	Mishra et al. 2011
Cladosporium cladosporioides	extracellular	440	spherical	30–60	Hulikere and Joshi 2019
Cladosporium cladosporioides	extracellular	415	spherical, polydisperse	10–100	Balaji et al. 2009
Cladosporium halotolerans	extracellular	415	spherical	20	Ameen et al. 2021
Colletotrichum sp. ALF2-6	extracellular	420	spherical, near to spherical, triangular and hexagonal	5–60	Azmath et al. 2016
Duddingtonia flagrans	extracellular	413	quasispherical	10.3 ± 7.2	Costa Silva et al. 2017

Table 1 contd. ...

...Table 1 contd.

Fungi	Type of synthesis	A max	Morphology	Size [nm]	References
Fusarium acuminatum	Extracellular	420	spherical	5-40	Ingle et al. 2008
Fusarium equiseti	extracellular	300	spherical, with smooth surface	2–50	Haji Basheerudeen et al. 2021
Fusarium mangiferae	extracellular	416	spherical, oval	25–52	Hamzah et al. 2018
Fusarium oxysporum	extracellular	440	Spherical, needle-shaped, monodispersed	3.4–26.8	Ishida et al. 2014
Fusarium oxysporum	extracellular	413	spherical, monodispersed	40 ± 5.0	Salaheldin et al. 2016
Fusarium oxysporum	extracellular	413	highly variable, spherical, triangular, aggregated	5–15	Ahmad et al. 2003
Fusarium oxysporum	intracellular	430	spherical, aggregated	25–50	Korbekandi et al. 2013
*Fusarium oxysporum-*NFW16	extracellular	423	spherical, monodispersed	30–36.1	Ilahi et al. 2021
Fusarium semitectum	extracellular	420	spherical	10–60	Basavaraja et al. 2008
Fusarium solani	extracellular	415	spherical	5–30	Abd El-Aziz at al. 2015
Guignardia mangiferae	extracellular	417	spherical	5–30	Balakumaran et al. 2015
Penicillium brevicompactum WA 2315	extracellular	420	different morphologies	23–105	Shaligram et al. 2009
Penicillium chrysogenum	extracellular	400	spherical, monodispersed	48.2	Barabadi et al. 2021
Penicillium fellutanum	extracellular	430	spherical	5–25	Kathiresan et al. 2009
Penicillium oxalicum	extracellular		spherical	60–80	Feroze et al. 2019
Penicillium oxalicum GRS-1	extracellular	420	spherical, monodispersed	10–40	Rose et al. 2019
Penicillium radiatolobatum	extracellular	410	spherical, triangle, hexagonal, polydispersed	5.09–24.85	Naveen et al. 2021
Penicillium toxicarium	extracellular	396	spherical	48.03–73.04	Korcan et al. 2021
Phaenerochaete chrysosporium	extracellular	470	pyramidal	50–200	Vigneshwaran et al. 2006

Table 1 contd. ...

130 *Mycosynthesis of Nanomaterials: Perspectives and Challenges*

...Table 1 contd.

Fungi	Type of synthesis	A max	Morphology	Size [nm]	References
Phoma glomerata	extracellular	440	spherical	60–80	Birla et al. 2009
Rhizopus stolonifer	extracellular	420	spherical	6.04	AbdelRahim et al. 2017
Saccharomyces cerevisiae	extracellular	410	spherical	> 70	Niknejad et al. 2015
Trichoderma asperellum	extracellular	410	spherical	13–18	Mukherjee et al. 2008
Trichoderma harzianum	extracellular	438	cubic	72	Konappa et al. 2021
Trichoderma longibrachiatum	extracellular	385	spherical, polydispersed	5–25	Elamawi et al. 2018
Trichoderma viride	extracellular		polydispersed, globular	1–50	Elgorban et al. 2016
Verticillium sp.	intracellular	450	spherical	25	Mukherjee et al. 2001

Key: A, absorbance

Fusarium spp., *Aspergillus* spp., *Trichoderma* spp., and *Penicillium* spp. Among the presented studies, extracellular synthesis predominated and the synthesized nanoparticles were mainly spherical in shape. However, pyramidal-shaped ones were synthesized from *Phaenerochaete chrysosporium* (Vigneshwaran et al. 2006) while highly heterogeneous nanoparticles were biofabricated using *Fusarium oxysporum* (Ahmad et al. 2003), *Penicillium brevicompactum* WA 2315 (Shaligram et al. 2009), *Penicillium radiatolobatum* (Naveen et al. 2021) and *Colletotrichum* sp. ALF2-6 (Azmath et al. 2016). In fact, the smallest nanoparticles were synthetized from *Candida glabrata* and *Aspergillus brunneoviolaceus* with sizes in the range 2–15 nm and 0.7–15.2 nm, respectively (Jalal et al. 2018; Mistry et al. 2021). Maximum absorbance peaks ranged at a wavelength from 300 nm (with NP size 2–50 nm) to 470 nm (with NP size 50–200 nm) which may be associated with the size distribution of the mycogenic nanoparticles (Vigneshwaran et al. 2006; Haji Basheerudeen et al. 2021).

Aspergillus spp.

Several studies investigated the use of biomass or cellular filtrate of *Aspergillus* species for the synthesis of AgNPs (Bhainsa and D'souza 2006; Gade et al. 2008; Verma et al. 2010; Yari et al. 2022). The ability to synthesize nanoparticles by *Aspergillus* spp. was demonstrated with absorbance maximum peaks in the 410–440 nm wavelength range and spherical or irregular shapes (Table 1). Therefore, Wang and coworkers (2021) synthesized polydispersed spherical silver nanoparticles with sizes between 1 and 24 nm using the culture supernatants of *Aspergillus sydowii* strain isolated from soil. TEM-based size distribution of bio-AgNPs showed that 38%

of the AgNPs were in size from 1 to 5 nm, 45% in a range of 5–10 nm, and 12% between 10 and 15 nm (Wang et al. 2021). El-Bendary and coworkers (2021) compared the biosynthesis of AgNPs carried out using the extracellular and intracellular filtrates of the *Aspergillus caespitosus*. Both exo- and endocellularly synthesized AgNPs showed an absorbance peak at 410 nm. Nanoparticles were spherical and hexagonal, had anisotropic crystalline structure, and ranged in size 10–50 nm and 3–100 nm, respectively. Moreover, UV–visible spectroscopy confirmed, that biosynthesized AgNPs (bioAgNPs) were stable for up to 5 months of storage at 4°C (El-Bendary et al. 2021). Recently, Yari et al. (2022) synthetized spherical and agglomerated AgNPs with a size > 40 nm using mycelial extract of *Aspergillus terreus*. In addition, the small zeta potential value (+ 0.7 mV) of synthesized nanoparticles resulted in a high agglomeration propensity. Moreover, they proposed that the hydroxyl and amide I functional groups detected in the fungal extract from *A. terreus* may assist in forming and stabilizing biosynthesized nanoparticles.

Fusarium spp.

Many scientists used different fungi of the genus *Fusarium* to prepare AgNPs with different sizes and shapes depending upon species (Ishida et al. 2014; Korbekandi et al. 2013; Abd El-Aziz et al. 2015; Rai et al. 2021). *Fusarium* spp. are filamentous fungi that are widely distributed in soil, water and associated with plants (Nelson et al. 1994). *F. oxysporum* is the most frequently reported fungus for AgNPs synthesis (Durán et al. 2005; Salaheldin et al. 2016; Korbekandi et al. 2013; Ilahi et al. 2021; Allend et al. 2021). Ahmad et al. (2003) reported for the first time on the extracellular biosynthesis of AgNPs by a eukaryotic system, namely *Fusarium oxysporum*. Transmission electron microscopy (TEM) showed spherical, triangular, and small particles in the range of 5–15 nm and their aggregates in the size range 5–50 nm. In addition, the authors proposed that the NADH-dependent reductase released by the fungus may be responsible for the reduction of Ag$^+$ ions and the formation of silver nanoparticles (Ahmed et al. 2003). Furthermore, Anil Kumar et al. (2007) elucidated this mechanism by *in vitro* synthesis of AgNPs with nitrate reductase purified from *Fusarium oxysporum*, phytochelatin, and, α-NADPH as a cofactor. The enzymatic synthesis provided spherical nanoparticles in the size of 10–25 nm in diameter and stabilized by capping peptides. UV-visible spectroscopy showed an absorption peak at 413 nm. Surface plasmon resonance (SPR) absorption peak for silver nanoparticles was not observed in the absence of enzyme or phytochelatin or α-NADPH cofactor in the reaction mixture. This indicated that the reduction of Ag$^+$ ions involves enzymatic reduction of nitrate to nitrite. Moreover, the enzymatic process is simple and easy for downstream processing and may be suitable for mass-scale production (Anil Kumar et al. 2007). Ishida and coworkers (2014) also reported extracellular biosynthesis of AgNPs by using *F. oxysporum*, but these AgNPs were slightly different in morphology. They were in the size range 3.4–26.8, spherical and needle-shaped (Ishida et al. 2014). Otherwise, intracellular synthesis of AgNPs by *F. oxysporum* was reported by Korbekandi et al. (2013). However, intracellular synthesis required additional processing steps to release synthesized nanoparticles from cells and is less efficient compared to extracellular process

132 *Mycosynthesis of Nanomaterials: Perspectives and Challenges*

(Durán et al. 2005). Recently, marine endophytic fungi *Fusarium equiseti* isolated from seaweed were used for the synthesis of AgNPs. Analysis of morphological features of these AgNPs proved their small size (2–50 nm), spherical shape, and smooth surface (Haji Basheerudeen et al. 2021). Other reports provide mycosynthesis of silver nanoparticles by different *Fusarium* species, such as *F. acuminatum, F. mangiferae, F. semitectum, F. solani,* and *F. pallidoroseum* (Ingle et al. 2008; Haji Basheerudeen et al. 2021; Basavaraja et al. 2008; Abd El-Aziz et al. 2015; Shukla et al. 2021), as showed in Table 1.

Penicillium spp.

Various species of the genus *Penicillium* were capable of producing AgNPs with different sizes and morphologies in a green approach (Shaligram et al. 2009; Kathiresan et al. 2009; Rose et al. 2019; Korcan et al. 2021). Silver nanoparticles synthesized from *Penicillum* spp., showed a remarkable variety of shapes (Table 1), including spherical (Kathiresan et al. 2009; Feroze et al. 2019; Rose et al. 2019), hexagonal, and triangle ones (Naveen et al. 2021). Hence, AgNPs with sizes ranging from 5.09 to 24.85 nm were prepared from the endophytic fungus *Penicillium radiatolobatum*. The morphology of these NPs was highly variable, namely mostly spherical, but some triangles and hexagonal were also recorded (Naveen et al. 2021). Similar results were presented by Kathiresan et al. (2009). They proved an extracellular biosynthesis of spherical and small (5–25 nm) AgNPs by synthesized from *P. fellutanum* isolated from mangrove root soil. As mentioned previously, the size, shape, and dispersion of the biosynthesized nanoparticles vary depending on the fungal strains and methods used in the synthesis process. For example, spherical, but different sizes (60–80 nm and 10–40 nm) AgNPs from *Penicillium oxalicum* were obtained by Feroze et al. (2019) and Rose et al. (2019), respectively.

Trichoderma spp.

Several non-pathogenic, species of the *Trichoderma* were used for the extracellular biosynthesis of spherical AgNPs including *T. asperellum, T. longibrachiatum*, and *T. viride* (Mukherjee et al. 2008; Elgorban et al. 2016; Elamawi et al. 2018; Qu et al. 2021). UV-visible spectroscopy provided maximum absorption peaks for all silver nanoparticles synthesized from *Trichoderma* in the range from 385 nm for *T. longibrachiatum* (size 5–25 nm) to 438 nm for *T. harzanium* (size 72 nm) (Table 1). Size analysis of AgNPs from *T. viride* using TEM showed a formation of spherical and irregularly formed, polydispersed particles in a range 1–50 nm. TEM analysis proved also a thin layer of organic material on the surface of nanoparticles, which provided stability of AgNPs (Elgorban et al. 2016). Silver nanoparticles synthesized from *T. asperellum* were spherical in shape and size ranging from 13–18 nm (Mukherjee et al. 2008). Konappa and coworkers (2021) synthesized cubic, polydispersed AgNPs using *T. harzianum* cell filtrate. This cell filtrate was rich in the alkaloids, flavanones, steroids, and phospholipids that were believed to be the ones responsible for the reduction of ions and/or capping silver nanoparticles (Konappa et al. 2021). In addition, other *Trichoderma* strains namely,

T. atroviride, T. crissum, T. longibrachiatum, T. spirale, T. virens, T. afroharzianum, T. hamatum, T. citrinoviride, T. koningiopsis and *T. velutinum* were successfully used for the synthesis of spherical AgNPs in size ranged from 5 to 35 nm. However, the results confirmed that AgNPs were most efficiently synthesized by the species of *T. longibrachiatum* (155.2 ± 15.02 mg of AgNPs per 100 mL of filtrate) (Qu et al. 2021).

Yeasts

Recently, many researchers have focused on the synthesis of silver nanoparticles using yeasts, which are non-toxic, and exhibit bio-accumulative potential for the synthesis of nanoparticles with high efficiency and low cost. Moreover, yeasts adopt different mechanisms for the formation and stabilization of AgNPs that display differences in size and shape (Boroumand Moghaddam et al. 2015; Skalickova et al. 2017; Shu et al. 2020). Thus, the presence of silver ions in the culture medium triggers a stress response, including the production of biomolecules responsible for internal stress elimination such as phytochelatin synthase and glutathione. Bioreduction of silver ions into nanoparticles is enabled by the redox and nucleophilic properties of these molecules (Kowshik et al. 2003; Skalickova et al. 2017). In addition, Shu et al. (2020) studied the formation of nanoparticles in yeast cell-free extract and suggested that reducing agents such as amino acids, vitamins, and carbohydrates present in yeast extract were responsible for Ag^+ ions reduction (Shu et al. 2020). Moreover, another study indicated a potential role of pigments in nanoparticle formation, specifically cell-associated melanin produced by *Yarrowia lipolytica*. Silver ions may be reduced to silver nanoparticles during the transition of quinone residues of melanin to alternate between the hydroxyl and quinone forms (Apte et al. 2013a; 2013b; Roy et al. 2019).

A number of studies have confirmed the ability of yeasts to the synthesis of AgNPs including *Candida guilliermondii, Candida utilis, Saccharomyces cerevisiae, Saccharomyces uvarum, Pichia kudriavzevii,* and marine yeast of *Candida* sp. VITDKGB (Mishra et al. 2011; Niknejad et al. 2015; Waghmare et al. 2015; Korbekandi et al. 2016; Dhabalia et al. 2020; Kumar et al. 2011; Ammar et al. 2021). Jalal et al. (2018) reported the synthesis of spherical and oval AgNPs by *C. glabrata*. The produced nanoparticles ranged in size from 2 to 15 nm and were smaller than those synthesized by other *Candida* species, especially those belonging to *C. albicans* (Table 1). Size distribution, shape, state of aggregation, and surrounding dielectric medium of nanoparticles affect the optical absorption spectrum of silver nanoparticles (Smitha et al. 2008). Ag NPs synthesized from *C. glabrata* showed a maximum absorbance peak at wavelength 460 nm. This SPR wavelength shifted towards a longer wavelength region indicating polydisperse and agglomerated nanoparticles (Vanaja et al. 2013; Jalal et al. 2018).

Other fungi

A number of scientists have developed a variety of fungi for the synthesis of metal nanoparticles. These include fungal extract from endophytic fungi, namely

Alternaria alternata isolated from Dendrophthoe falcata (Govindappa et al. 2022), Colletotrichum sp. ALF2-6 isolated from Andrographis paniculata (Azmath et al. 2016), Guignardia mangiferae isolated from the leaves of Citrus sp. (Balakumaran et al. 2015). Verticillium sp. isolated from the Taxus plant was employed for intracellular synthesis of AgNPs. Interestingly, a number of silver nanoparticles were observed in TEM micrographs associated with the cell wall of the Verticillium cells after the synthesis process. Obtained nanoparticles were spherical, with an average size of 25 nm ± 12 nm and fairly good monodispersity (Mukherjee et al. 2001).

Otherwise, spherical, monodispersed AgNPs with an average size of 9.46 ± 2.64 nm were synthesized by a mycelial aqueous extract of Rhizopus stolonifer isolated from tomato fruits (AbdelRahim et al. 2017). In addition, Phoma glomerata was studied for the rapid synthesis of spherical, small size (average size 19 nm), and very stable (zeta potential value of −30.7 mV) AgNPs in the presence of bright sunlight as an inducing agent. Furthermore, the authors suggested that photosensitizated aromatic molecules from fungal extract generate free electrons, which may be involved in the rapid synthesis of AgNPs under sunlight (Gade et al. 2014). Besides, some other fungi have been found for AgNPs synthesis, namely nematophagous fungus Duddingtonia flagrans (Costa Silva et al. 2017), white-rot fungus, Phaenerochaete chrysosporium (Vigneshwaran et al. 2006), soilborne pathogen Rhizoctonia solani isolated from Rosmarinus officinalis (Ashrafi et al. 2013) and many others (Table 1).

Influence of physico-chemical parameters on mycosynthesis of AgNPs

Control of biosynthesis conditions of nanoparticles is an important area of research in nanoscience (AbdelRahim et al. 2017). The optimization of these conditions is the preliminary investigation for their large-scale production and industrial applications (Nayak et al. 2011). According to Srikar and coauthors (2016), the major physical and chemical parameters that affect the synthesis of AgNPs include temperature, metal ion concentration, pH of the reaction mixture, duration of reaction, and agitation. It should be emphasized that these parameters affect both the quality and quantity of the synthesized silver nanoparticles, as well as, their properties for further applications (Das et al. 2020). Therefore, by adjusting synthesis conditions it is possible to manipulate the metabolism of fungi and subsequently biosynthesis of nanoparticles with the desired characteristics, such as specific size and morphology (Zielonka and Klimek-Ochab 2017). However, according to Birla et al. (2013), the physico-chemical conditions during the synthesis with the use of the fungus depends on the selected strain, its growth requirements, and metabolic activity. Table 2 shows selected studies on the synthesis of AgNPs using different fungal species and synthesis conditions.

Effect of temperature

Temperature is one of the key controlling factors affecting the formation of AgNPs and their properties (Figure 1) (Birla et al. 2013; Rose et al. 2019) that can also influence the reaction rate of the synthesis process (Phanjom and Ahmed 2017). For

Table 2. Optimization of the synthesis of silver nanoparticles by fungi.

Fungus	Type of parameters	Optimized conditions	Nanoparticle characteristics	References
Arthroderma fulvum	temperature, pH, concentration of $AgNO_3$ and time of reaction	55°C; pH 10; 1.5 mM $AgNO_3$; 12 h	20.56 nm, spherical	Xue et al. 2016
Aspergillus fumigatus BTCB10	temperature, fungal culture age, substrate concentration and biomass weight, pH	25°C; 7 days of culture; 7 g of biomass; 1 mM $AgNO_3$; pH 6	322.8 nm, spherical	Shahzad et al. 2019
Aspergillus niger NRC1731	pH, time incubation and concentration of $AgNO_3$	1.82 mM silver nitrate for 34 h at pH 7.0	3–20 nm, spherical	Elsayed et al. 2018
Aspergillus oryzae MTCC1846	temperature, pH and concentration of $AgNO_3$	90°C; pH 10; 1 mM $AgNO_3$	4–9 nm, spherical	Phanjom and Ahmed 2017
Aspergillus sydowii	temperature, pH and $AgNO_3$ concentrations	50°C; 8.0 pH; 1.5 mM $AgNO_3$	1–21 nm, spherical	Wang et al. 2021
Aspergillus terreus BA6	reaction pH, dextrose (g/l), peptone (g/l), and $AgNO_3$ concentration (mM)	pH 6.25; dextrose 27.5 (g/l); peptone 8.75 (g/l); and $AgNO_3$ 4.26 mM;	7–23 nm, −17.5 mV, spherical, well dispersed	Lotfy et al. 2021
Duddingtonia flagans	temperature and pH, culture medium	60°C; pH 10; filtrate supplemented with chitin	30-409 nm, −28.6 mV, spherical	Costa Silva et al. 2017
Fusarium oxysporum	temperature and culture media	28°C and modified medium for nitrate reductase	24 nm, spherical	Hamedi et al. 2017
Fusarium sp. 4F1, *Trichoderma* sp. TrS	pH and concentration of $AgNO_3$, time incubation	pH 9; 72 hrs of incubation period and 2 mM concentration of $AgNO_3$	spherical, monodispersed and stable	Pal and Hossain 2020
Penicillium oxalicum GRS-1	temperature, quantity of biomassa, concentration of $AgNO_3$ and pH	60°C; 25 g of biomass; 1.5 mM $AgNO_3$, pH 7-8	10–40 nm, spherical	Rose et al. 2019
Rhizopus stolonifer	temperature and concentration of $AgNO_3$	40°C; 10 mM $AgNO_3$	2.86 nm, spherical	AbdelRahim et al. 2017
Sclerotinia sclerotiorum MTCC 8785	culture media, quantity of biomass, concentration of $AgNO_3$, pH and temperature	potato dextrose broth; 10 g of biomass, 1 mM $AgNO_3$; pH 11; 80°C	10–15 nm, spherical	Saxena et al. 2016
Trichoderma longibrachiatum	temperature, quantity of biomass and agitation	28°C; 10 g of biomass without agitation	24.43 nm, −19.7 mV, spherical	Elamawi et al. 2018

136 *Mycosynthesis of Nanomaterials: Perspectives and Challenges*

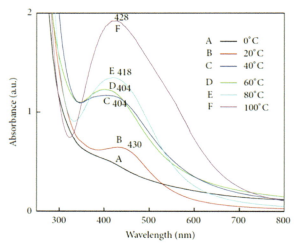

Figure 1. Effect of temperature on the size of SNPs from *Fusarium oxysporum*: UV-Vis spectral analysis shows reduction of SNPs. A: 0°C, B: 20°C, C: 40°C, D: 60°C, E: 80°C, and F: 100°C. Spectrum C and spectrum D have the blue shift which indicates the small size of synthesized SNPs. Spectrum F asymmetric spectra indicate the aggregation of particles at high temperatures (adapted and modified from Birla et al. 2013, an open-access article).

example, Hamedi et al. (2017) analyzed the efficiency of AgNPs synthesis at various temperatures, namely 23, 28, and 33°C using the cell-free filtrates of *Fusarium oxysporum* and found that AgNPs were most efficiently synthesized at 28°C. Similar observations were done by Elamawi and coauthors (2018). They recorded that AgNPs synthesis using *Trichoderma longibrachiatum* was effective at 28°C, while inhibited at 23 or 33°C. In turn, other scientists, such as Rose et al. (2019) and Costa Silva et al. (2017), reported a favorable effect of temperature increase on the production of silver nanoparticles. Rose and coauthors (2019) studied AgNPs synthesis using a fungal strain of *Penicillium oxalicum* GRS-1 and observed that the nanoparticles biosynthesized at 60°C showed the highest and narrower absorbance peak when analyzed by UV-Vis spectroscopy and were in the range of 10–40 nm. In contrast, those synthesized at 20, 30, 40 and 50°C showed broader absorbance peaks, which indicated the existence of nanoparticles with different sizes ranging from 10 to 100 nm. Similarly, Costa Silva et al. (2017), reported that optimum synthesis of AgNPs using *Duddingtonia flagans* filtrate was found to be at 60°C, while synthesis was inhibited at 30°C. AbdelRahim and coauthors (2017), synthesized AgNPs at 20, 40, and 60°C using *Rhizopus stolonifer* aqueous mycelial extract and found that obtained NPs were in the mean sizes of 25.89, 2.86, and 48.43 nm, respectively. These results clearly proved that temperature is a key factor in the manipulation of nanoparticle size. In contrast, Shahzad et al. (2019) demonstrated that AgNPs size increased from 332.8 nm to 1073.45 nm together with temperature increase from 25°C to 55°C, respectively when Dynamic light scattering (DLS) analyses were performed. The synthesis of nanoparticles at higher temperatures is probably carried out by transferring electrons from free amino acids present in fungal extracts to silver ions (Guilger-Casagrande and Lima 2019). However, very high temperatures lead to denaturation of the proteins that cover the surface of nanoparticles. This denaturation alters the nucleation of Ag^+ ions, resulting in nanoparticle aggregation and size increase (Birla et al. 2013; Guilger-Casagrande and Lima 2019). Moreover, Phanjom and Ahmed (2017) demonstrated that the higher the temperature of the reaction mixture, the lowest the reaction time. The increase in temperature from 30 to 50, 70 and 90°C reduced reaction time from 6 h to 1 h, 45 min and 20 min, respectively. However, the synthesis of AgNPs at 10°C was completely inhibited.

Effect of pH

The pH value contributes greatly to increasing or decreasing the number of H^+ ions in the reaction mixture (Manosalva et al. 2019). It is well known that a lower pH value leads to an increase in the concentration of H^+ (Siddiqi et al. 2018). The pH does not directly affect the conformational changes of proteins and enzymes, but the change in the concentration of H^+ ions can shape and alter the electronegative properties of the substrate that can affect their binding with the enzymatic active site (Wei et al. 2015; Siddiqi et al. 2018). Therefore, pH is an essential factor affecting AgNPs production by biological systems. The effect of pH of the reaction solution on the efficiency of the synthesis of nanoparticles was evaluated by many authors (Xue et al. 2016; Shahzad et al. 2019; Pal and Hossain 2020; Lotfy et al. 2021). Much more effective synthesis of metal nanoparticles at alkaline pH is associated with greater competition between protons and metal ions for the formation of bonds with negatively charged regions (Sintubin et al. 2009). For example, Saxena et al. (2016) observed that the maximum production of AgNPs was attained at pH 11 when compared with lower pH values, namely 3, 5, 7, and 9. Recently, Wang et al. (2021) investigated the effect of pH on the synthesis of AgNPs using *Aspergillus sydowii* and noticed that the absorbance was increased when pH increased from 5 to 8. The optimal pH for synthesis was found to be 8.0. Similarly, the synthesis of nanoparticles employing *Penicillium oxalicum* GRS-1 was most efficient at pH 7 and 8, as reported by Rose et al. (2019). They also reported that there was no nanoparticle formation in the acidic range of pH. The absorbance peaks recorded in UV-Vis spectroscopy analyses of reaction mixtures became more symmetrical with an increase in pH values which can be explained by the presence of OH− groups that play a crucial role in stabilizing of AgNPs. Undoubtedly, the prevention of aggregation and maintaining the stability and small size of AgNPs is possible thanks to the availability of such groups (Rose et al. 2019). Recently, Lotfy et al. (2021), showed the optimum synthesis of AgNPs using filtrate of *A. terreus* BA6 at pH 6.25, but the production of smaller AgNPs is more favorable at higher pH while Elsayed and coauthors (2018) observed that AgNPs obtained from *Aspergillus niger* NRC1731 presented greater monodispersion and stability at pH 7.

Effect of concentrations of $AgNO_3$

Similarly, as with other conditions, the concentration of precursor is one of the most important factors affecting the synthesis of silver nanoparticles (Figure 2) (Birla et al. 2013; Phanjom and Ahmed 2017). Silver nitrate is predominantly used as a salt for the synthesis of AgNPs in biological systems (Guilger-Casagrande and Lima 2019; Srikar et al. 2016). Phanjom and Ahmed (2017) used different concentrations of an aqueous solution of AgNO$_3$ ranging from 1 to 10 mM for synthesis of AgNPs using fungal cell filtrate of *Aspergillus oryzae*. Authors suggested that the metal precursor concentrations at 1–8 mM resulted in a smaller nanoparticle size (in the range of 3–28 nm) and improved dispersion. While concentrations of AgNO$_3$ at 9–10 mM have an influence on large size of nanoparticles in the range of 14–105 nm. Other authors, analyzed the effect of different AgNO$_3$ concentrations (0.5, 1, 1.5, 2, and 2.5 mM) on silver nanoparticles production and showed that the optimal substrate concentration was found to be 1.5 mM (Rose et al. 2019; Wang et al. 2021). Similarly, Elsayed et al. (2018) used different concentrations of silver nitrate (0.5–2.5 mM) in the reaction mixture with filtrate of *Aspergillus niger* NRC1731 and showed that 1.82 mM silver nitrate is the optimal concentration for the AgNPs biosynthesis. Pal and Hossain (2020) reported that the most efficient biosynthesis of AgNPs from *Fusarium* 4F1 and *Trichoderma* TRS was achieved when 2 mM concentration of AgNO$_3$ was used, while the lower concentration of silver salt (1 mM) affected the decrease in the production of AgNPs. In other studies, the use of 10 mM (final concentration in the reaction mixture) of AgNO$_3$ for the synthesis of AgNPs resulted in the formation of smaller nanoparticles (2.86 ± 0.3 nm) than the use of 1 mM and 100 mM of metal ions (54.67 ± 4.1 and 14.23 ± 1.3 nm, respectively) that additionally exhibited irregular shape (AbdelRahim et al. 2017). It should be also emphasized that higher concentrations of AgNO$_3$ used for synthesis may lead to greater toxicity against microbial cells used for synthesis (Balakumaran et al. 2015).

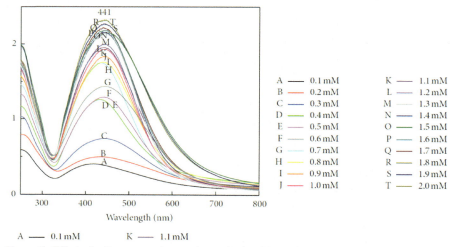

Figure 2. Effect of salt concentration on the synthesis of SNPs from *Fusarium oxysporum*. UV-Vis extinction spectra show maximum surface plasmon absorbance in 2 mM (T) concentration of SNPs. Synthesis of SNPs is proportional to the salt concentration (treatment of 1.5 mM AgNO$_3$ gives a maximum synthesis of SNPs as compared with 2mM AgNO$_3$) (adapted from Birla et al. 2013, an open-access article).

Effect of the culture medium and quantity of biomass

Fungi for the synthesis of nanoparticles require an appropriate culture medium and culture conditions for their growth. Moreover, a culture medium containing specific substrates may induce fungi for the synthesis of enzymes involved in the synthesis of AgNPs by reduction of silver ions (Guilger-Casagrande and Lima 2019). Lotfy and coauthors (2021) for efficient synthesis of silver nanoparticles by *Aspergillus terreus* BA6 used dextrose (27.5 g/L) and peptone (8.75 g/L) in the culture medium. The synthesis carried out at another ratio of the dextrose and peptone (20:10 g/L and 12.5:12.5 g/L, respectively) results in a decrease in biosynthesis efficiency. Saxena et al. (2016) analyzed a set of culture media like Potato dextrose broth (PDB), Sabouraud's dextrose (SDB), Protease production media (PP), Czapek Dox (CZAPEK), Richard medium (RM), and Glucose Yeast Extract Peptone (GYP) to grow fungal mycelia for enhanced extracellular synthesis of AgNPs from *Sclerotinia sclerotiorum*. Fungal biomass grown in PDB has shown enhanced AgNPs synthesis followed by SDB, RM, CZAPEK, PP, and GYP. Interestingly, in the case of synthesis of AgNPs from *Duddingyonia flagans*, the fungal biomass was transferred to pure water and to water containing insect carapaces as a source of chitin (substrate for fungal enzymes). The obtained fungal filtrate from biomass supplemented with chitin contained around three times more protein and showed higher nanoparticle production when compared with non-supplemented fungal filtrate (Costa Silva et al. 2017). Moreover, a significant effect of the amount of fungal biomass on the synthesis of nanoparticles was revealed in studies by Saxena et al. (2016) and Elamawi et al. (2018). Rose et al. (2019) reported that with an increase in biomass amount of

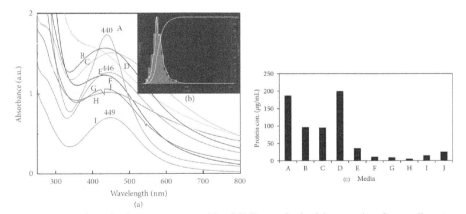

Figure 3. UV-Vis extinction spectroscopy (a) of SNPs synthesized by growing fungus *Fusarium oxysporum* on different culture media—A: MGYP broth, B: PDB, C: lipase production medium, D: protease production medium, E: sucrose peptone yeast broth, F: gluten glucose medium, G: Richard's medium, H: Czapek dox medium, I: glucose peptone yeast broth, and J: Sabouraud broth. Fungus grown in MGYP medium shows maximum synthesis with symmetry in the spectrum. Analysis of sizes distribution (b) of SNPs prepared from the fungus that grew on MGYP medium by NTA. Histogram of protein concentration (c) in different media: highest protein concentration is found in MGYP (A) and protease production medium (D) (adapted from Birla et al. 2013, an open-access article).

Penicillium oxalicum GRS-1 from 5 to 25 g, more symmetrical, sharp, and narrower absorbance peaks were obtained, indicating the small-sized and uniform distribution of Ag nanoparticles. In other studies, the greater production, smaller size, and better dispersion of the AgNPs was observed when a lower amount (7 g) of wet biomass of *Aspergillus fumigatus* BTCB10 was used for synthesis when compared to the use of 10 g of the biomass (Shahzad et al. 2019). The effect of different growth media on protein and AgNPs synthesis by *Fusarium oxysporum* is shown in Figure 3.

Conclusion and future perspectives

Fungi are considered as efficient producers of metal nanoparticles from the metal precursors due to their ability to secrete a broad range of reducing and stabilizing components (Narayanan et al. 2011). In addition, the physico-chemical parameters such as temperature, pH, precursor concentration, biomass amount, media for fungal growth, and culture conditions can be used to optimize the biosynthesis process (Manosalva et al. 2019). Optimization of physico-chemical conditions plays a crucial role not only in synthesis efficiency but also in nanoparticle properties such as morphology, bioactivity, and biocompatibility. In turn, these affect AgNPs suitability for use in a variety of therapies (de Oliveira et al. 2019). Silver nanoparticles due to their outstanding properties affect all spheres of human life, which has led nanotechnology to be widely used in various industries and biomedical applications (Seetharaman et al. 2021). Therefore, the synthesis of silver nanoparticles by using fungi may be carried out on large scale due to the low cost and simplicity of the process. Nevertheless, the toxicity concerns are a major challenge in the synthesis and safe application of AgNPs (Zhao et al. 2018).

Acknowledgement

J.T.W. acknowledges grant No. 2022/45/N/NZ9/01483 from National Science Centre, Poland, and also to NCU (No. 38/2021). MW thanks to NSC (No. 2016/23N/N/NZ9/00247) for financial assistance.

References

Abd El-Aziz, A.R.M., Al-Othman, M.R., Mahmoud, M.A., and Metwaly, H.A. 2015. Biosynthesis of silver nanoparticles using *Fusarium solani* and its impact on grain borne fungi. Dig. J. Nanomater. Biostructures 10: 655–662.

AbdelRahim, K., Mahmoud, S.Y., Ali, A.M., Almaary, K.S., Mustafa, A.E.Z.M., and Husseiny, S.M. 2017. Extracellular biosynthesis of silver nanoparticles using *Rhizopus stolonifer*. Saudi J. Biol. Sci. 24: 208–216.

Adebayo, E.A., Azeez, M.A., Alao, M.B., Oke, A.M., and Aina, D.A. 2021. Fungi as veritable tool in current advances in nanobiotechnology. Heliyon 7: e08480.

Ahmad, A., Mukherjee, P., Senapati, S., Mandal, D., Khan, M.I., Kumar, R., and Sastry, M. 2003. Extracellular biosynthesis of silver nanoparticles using the fungus *Fusarium oxysporum*. Colloids Surf. B: Biointerfaces 28: 313–318.

Alghuthaymi, M.A., Almoammar, H., Rai, M., Said-Galiev, E., and Abd-Elsalam, K.A. 2015. Myconanoparticles: synthesis and their role in phytopathogens management. Biotechnol. Biotechnol. Equip. 29: 221–236.

Mycogenic Synthesis of Silver Nanoparticles and its Optimization 141

Allend, S.O., Garcia, M.O., da Cunha, K.F., de Albernaz, D.T.F., da Silva, M.E., Ishikame, R.Y., and Hartwig, D.D. 2021. Biogenic silver nanoparticle (Bio-AgNP) has an antibacterial effect against carbapenem-resistant *Acinetobacter baumannii* with synergism and additivity when combined with polymyxin B. J. Appl. Microbiol. 132: 1036–1047.

Ameen, F., Al-Homaidan, A.A., Al-Sabri, A., Almansob, A., and AlNAdhari, S. 2021. Anti-oxidant, anti-fungal and cytotoxic effects of silver nanoparticles synthesized using marine fungus *Cladosporium halotolerans*. Appl. Nanosci. 1–9.

Ammar, H.A., Abd El Aty, A.A., and El Awdan, S.A. 2021. Extracellular myco-synthesis of nano-silver using the fermentable yeasts *Pichia kudriavzevii* HA-NY2 and *Saccharomyces uvarum* HA-NY3, and their effective biomedical applications. Bioprocess Biosyst. Eng. 44: 841–854.

Anil Kumar, S., Abyaneh, M.K., Gosavi, S.W., Kulkarni, S.K., Pasricha, R., Ahmad, A., and Khan, M.I. 2007. Nitrate reductase-mediated synthesis of silver nanoparticles from AgNO$_3$. Biotechnol. Lett. 29: 439–445.

Apte, M., Girme, G., Bankar, A., RaviKumar, A., and Zinjarde, S. 2013a. 3, 4-dihydroxy-L-phenylalanine-derived melanin from *Yarrowia lipolytica* mediates the synthesis of silver and gold nanostructures. J. Nanobiotechnology 11: 1–9.

Apte, M., Sambre, D., Gaikawad, S., Joshi, S., Bankar, A., Kumar, A.R., and Zinjarde, S. 2013b. Psychrotrophic yeast *Yarrowia lipolytica* NCYC 789 mediates the synthesis of antimicrobial silver nanoparticles via cell-associated melanin. AMB Express 3: 1–8.

Ashrafi, S.J., Rastegar, M.F., Ashrafi, M., Yazdian, F., Pourrahim, R., and Suresh, A.K. 2013. Influence of external factors on the production and morphology of biogenic silver nanocrystallites. J. Nanosci. Nanotechnol. 13: 2295–2301.

Azmath, P., Baker, S., Rakshith, D., and Satish, S. 2016. Mycosynthesis of silver nanoparticles bearing antibacterial activity. Saudi Pharm. J. 24: 140–146.

Balaji, D.S., Basavaraja, S., Deshpande, R., Mahesh, D.B., Prabhakar, B.K., and Venkataraman, A. 2009. Extracellular biosynthesis of functionalized silver nanoparticles by strains of *Cladosporium cladosporioides* fungus. Colloid Surf. B: biointerfaces 68: 88–92.

Balakumaran, M.D., Ramachandran, R., and Kalaichelvan, P.T. 2015. Exploitation of endophytic fungus, *Guignardia mangiferae* for extracellular synthesis of silver nanoparticles and their *in vitro* biological activities. Microbiol. Res. 178: 9–17.

Barabadi, H., Mohammadzadeh, A., Vahidi, H., Rashedi, M., Saravanan, M., Talank, N., and Alizadeh, A. 2021. *Penicillium chrysogenum*-derived silver nanoparticles: exploration of their antibacterial and biofilm inhibitory activity against the standard and pathogenic *Acinetobacter baumannii* compared to tetracycline. J. Clust. Sci. 1–14.

Basavaraja, S., Balaji, S.D., Lagashetty, A., Rajasab, A.H., and Venkataraman, A. 2008. Extracellular biosynthesis of silver nanoparticles using the *fungus Fusarium semitectum*. Mater. Res. Bull. 43: 1164–1170.

Bhainsa, K.C., and D'souza, S.F. 2006. Extracellular biosynthesis of silver nanoparticles using the fungus *Aspergillus fumigatus*. Colloid Surf. B: biointerfaces 47: 160–164.

Birla, S.S., Tiwari, V.V., Gade, A.K., Ingle, A.P., Yadav, A.P., and Rai, M.K. 2009. Fabrication of silver nanoparticles by *Phoma glomerata* and its combined effect against *Escherichia coli, Pseudomonas aeruginosa* and *Staphylococcus aureus*. Lett. Appl. Microbiol. 48: 173–179.

Birla, S.S., Gaikwad, S.C., Gade, A.K., and Rai, M.K. 2013. Rapid synthesis of silver nanoparticles from *Fusarium oxysporum* by optimizing physicocultural conditions. Sci World J. 2013: 796018.

Blackwell, M. 2011. The Fungi: 1, 2, 3… 5.1 million species? Am. J. Bot. 98: 426–438.

Boroumand Moghaddam, A., Namvar, F., Moniri, M., Azizi, S., and Mohamad, R. 2015. Nanoparticles biosynthesized by fungi and yeast: a review of their preparation, properties, and medical applications. Molecules 20: 16540–16565.

Costa Silva, L.P., Oliveira, J.P., Keijok, W.J., Silva, A.R., Aguiar, A.R., Guimarães, M.C.C., Ferraz, C.M. Araújo, J.V., Tobias, F.L., and Braga, F.R. 2017. Extracellular biosynthesis of silver nanoparticles using the cell-free filtrate of nematophagus fungus *Duddingtonia flagans*. Int. J. Nanomed. 12: 6373–6381.

Das, C.A., Kumar, V.G., Dhas, T.S., Karthick, V., Govindaraju, K., Joselin, J.M., and Baalamurugan, J. 2020. Antibacterial activity of silver nanoparticles (biosynthesis): A short review on recent advances. Biocatal. Agric. Biotechnol. 27: 101593.

142 *Mycosynthesis of Nanomaterials: Perspectives and Challenges*

Desai, R., Mankad, V., Gupta, S.K., and Jha, P.K. 2012. Size distribution of silver nanoparticles: UV-visible spectroscopic assessment. Nanosci. Nanotechnol. Lett 4: 30–34.

Dhabalia, D., Ukkund, S.J., Syed, U.T., Uddin, W., and Kabir, M.A. 2020. Antifungal activity of biosynthesized silver nanoparticles from *Candida albicans* on the strain lacking the CNP41 gene. Mater. Res. Express 7: 125401.

Durán, N., Marcato, P.D., Alves, O.L., De Souza, G.I., and Esposito, E. 2005. Mechanistic aspects of biosynthesis of silver nanoparticles by several *Fusarium oxysporum* strains. J. Nanobiotechnology 3: 1–7.

Elamawi, R.M., Al-Harbi, R.E., and Hendi, A.A. 2018. Biosynthesis and characterization of silver nanoparticles using *Trichoderma longibrachiatum* and their effect on phytopathogenic fungi. Egypt. J. Biol. Pest. Control. 28: 1–11.

El-Bendary, M.A., Moharam, M.E., Hamed, S.R., Abo El-Ola, S.M., Khalil, S.K., Mounier, M.M., and Allam, M.A. 2021. Mycosynthesis of silver nanoparticles using *Aspergillus caespitosus*: Characterization, antimicrobial activities, cytotoxicity, and their performance as an antimicrobial agent for textile materials. Appl. Organomet. Chem. 35: e6338.

Elgorban, A.M., Al-Rahmah, A.N., Sayed, S.R., Hirad, A., Mostafa, A.A.F. and Bahkali, A.H. 2016. Antimicrobial activity and green synthesis of silver nanoparticles using *Trichoderma viride*. Biotechnol, Biotechnol. Equip. 30: 299–304.

Elsayed, M.A., Othman, A.M., Hassan, M.M., and Elshafei, A.M. 2018. Optimization of silver nanoparticles biosynthesis mediated by *Aspergillus niger* NRC1731 through application of statistical methods: enhancement and characterization. 3 Biotech 8: 1–10.

Elshafei, A.M., Othman, A.M., Elsayed, M.A., Al-Balakocy, N.G., and Hassan, M.M. 2021. Green synthesis of silver nanoparticles using *Aspergillus oryzae* NRRL447 exogenous proteins: Optimization via central composite design, characterization and biological applications. Environ. Nanotechnol. Monit. Manag. 16: 100553.

Feroze, N., Arshad, B., Younas, M., Afridi, M.I., Saqib, S., and Ayaz, A. 2019. Fungal mediated synthesis of silver nanoparticles and evaluation of antibacterial activity. Microsc. Res. Tech. 83: 72–80.

Gade, A.K., Bonde, P.P., Ingle, A.P., Marcato, P.D., Duran, N., and Rai, M.K. 2008. Exploitation of *Aspergillus niger* for synthesis of silver nanoparticles. J. Biobased. Mater. 2: 243–247.

Gade, A., Gaikwad, S., Duran, N., and Rai, M. 2014. Green synthesis of silver nanoparticles by *Phoma glomerata*. Micron 59: 52–59.

Govindappa, M., Manasa, D.J., Vridhi Vinaykiya, Bhoomika, V., Suryanshi Dutta, Ritu Pawar, and Vinay, B. 2022. Raghavendra Screening of Antibacterial and Antioxidant Activity of Biogenically Synthesized Silver Nanoparticles from *Alternaria alternata*, Endophytic Fungus of *Dendrophthoe falcata*-a Parasitic Plant. BioNanoSci.

Guilger-Casagrande, M., and Lima, R.D. 2019. Synthesis of silver nanoparticles mediated by fungi: a review. Front. Bioeng. Biotechnol. 7: 287.

Haji Basheerudeen, M.A., Mushtaq, S.A., Soundhararajan, R., Nachimuthu, S.K., and Srinivasan, H. 2021. Marine endophytic fungi mediated silver nanoparticles and their application in plant growth promotion in *Vigna radiata* L. Int. J. Nano. Dimens. 12: 1–10.

Hamedi, S., Ghaseminezhad, M., Shokrollahzadeh, S., and Shojaosadati, S.A. 2017. Controlled biosynthesis of silver nanoparticles using nitrate reductase enzyme induction of filamentous fungus and their antibacterial evaluation. Artif. Cells Nanomed. Biotechnol. 45: 1588–1596.

Hamzah, H.M., Salah, R.F., and Maroof, M.N. 2018. *Fusarium mangiferae* as new cell factories for producing silver nanoparticles. J. Microbiol Biotechnol. 28: 1654–1663.

Hikmet, R.A., and Hussein, N.N. 2021. Mycosynthesis of Silver Nanoparticles by Candida albicans Yeast and its Biological Applications. Arch. Razi Inst. 76: 857–869.

Hulikere, M.M., and Joshi, C.G. 2019. Characterization, antioxidant and antimicrobial activity of silver nanoparticles synthesized using marine endophytic fungus-*Cladosporium cladosporioides*. Process Biochem. 82: 199–204.

Ilahi, N., Haleem, A., Iqbal, S., Fatima, N., Sajjad, W., Sideeq, A., and Ahmed, S. 2021. Biosynthesis of silver nanoparticles using endophytic *Fusarium oxysporum* strain NFW16 and their *in vitro* antibacterial potential. *Microscopy research and technique*. https://doi.org/10.1002/jemt.24018.

Ingle, A., Gade, A., Pierrat, S., Sonnichsen, C., and Rai, M. 2008. Mycosynthesis of silver nanoparticles using the fungus *Fusarium acuminatum* and its activity against some human pathogenic bacteria. Curr. Nanosci. 4: 141–144.

Ishida, K., Cipriano, T.F., Rocha, G.M., Weissmüller, G., Gomes, F., Miranda, K., and Rozental, S. 2014. Silver nanoparticle production by the fungus *Fusarium oxysporum*: nanoparticle characterisation and analysis of antifungal activity against pathogenic yeasts. Memorias do Instituto Oswaldo Cruz 109: 220–228.

Jalal, M., Ansari, M.A., Alzohairy, M.A., Ali, S.G., Khan, H.M., Almatroudi, A., and Raees, K. 2018. Biosynthesis of silver nanoparticles from oropharyngeal *Candida glabrata* isolates and their antimicrobial activity against clinical strains of bacteria and fungi. Nanomaterials 8: 586.

Kathiresan, K., Manivannan, S., Nabeel, M.A., and Dhivya, B. 2009. Studies on silver nanoparticles synthesized by a marine fungus, *Penicillium fellutanum* isolated from coastal mangrove sediment. Colloids Surf. B. 71: 133–137.

Konappa, N., Udayashankar, A.C., Dhamodaran, N., Krishnamurthy, S., Jagannath, S., Uzma, F., and Jogaiah, S. 2021. Ameliorated antibacterial and antioxidant properties by *Trichoderma harzianum* mediated green synthesis of silver nanoparticles. Biomolecules 11: 535.

Korbekandi, H., Ashari, Z., Iravani, S., and Abbasi, S. 2013. Optimization of biological synthesis of silver nanoparticles using *Fusarium oxysporum*. Iran. J. Pharm. Res. 12: 289.

Korbekandi, H., Mohseni, S., Mardani Jouneghani, R., Pourhossein, M., and Iravani, S. 2016. Biosynthesis of silver nanoparticles using *Saccharomyces cerevisiae*. Artif. Cells. Nanomed. Biotechnol. 44: 235–239.

Korcan, S.E., Kahraman, T., Acikbas, Y., Liman, R., Ciğerci, İ.H., Konuk, M., and Ocak, İ. 2021. Cyto–genotoxicity, antibacterial, and antibiofilm properties of green synthesized silver nanoparticles using *Penicillium toxicarium*. Microsc. Res. Tech. 84: 2530–2543.

Kowshik, M., Ashtaputre, S., Kharrazi, S., Vogel, W., Urban, J., Kulkarni, S.K., and Paknikar, K.M. 2003. Extracellular synthesis of silver nanoparticles by a silver-tolerant yeast strain MKY3. Nanotechnology 14: 95.

Kumar, D., Karthik, L., Kumar, G., and Roa, K.B. 2011. Biosynthesis of silver nanoparticles from marine yeast and their antimicrobial activity against multidrug resistant pathogens. Pharmacologyonline 3: 1100–1111.

Kumar, R.R., Priyadharsani, K.P., and Thamaraiselvi, K. 2012. Mycogenic synthesis of silver nanoparticles by the Japanese environmental isolate *Aspergillus tamarii*. J. Nanopart. Res. 14: 1–7.

Lawrance, N., Benaltraja, V., Ramasamy, A., Devaraj, B., Subramani, T., and Parameswaran, R. 2021. Synthesis and characterization of silver nanoparticles using *Acremonium borodinense* and their anti-bacterial and hemolytic activity. Biocatal. Agric. Biotechnol. 102222.

Li, G., He, D., Qian, Y., Guan, B., Gao, S., Cui, Y., and Wang, L. 2012. Fungus-mediated green synthesis of silver nanoparticles using *Aspergillus terreus*. Int. J. Mol. Sci. 13: 466–476.

Lotfy, W.A., Alkersh, B.M., Sabry, S.A., and Ghozlan, H.A. 2021. Biosynthesis of silver nanoparticles by *Aspergillus terreus*: Characterization, optimization, and biological activities. Front. Bioeng. Biotechnol. 9: 633468.

Madakka, M., Jayaraju, N., and Rajesh, N. 2018. Mycosynthesis of silver nanoparticles and their characterization. MethodsX 5: 20–29.

Mandal, D., Bolander, M.E., Mukhopadhyay, D., Sarkar, G., and Mukherjee, P. 2006. The use of microorganisms for the formation of metal nanoparticles and their application. Appl. Microbiol. Biotechnol. 69: 485–492.

Manosalva, N., Tortella, G., Cristina Diez, M., Schalchli, H., Seabra, A.B., Durán, N., and Rubilar, O. 2019. Green synthesis of silver nanoparticles: effect of synthesis reaction parameters on antimicrobial activity. World J. Microbiol. Biotechnol. 35: 1–9.

Mishra, A., Tripathy, S.K., and Yun, S.I. 2011. Bio-synthesis of gold and silver nanoparticles from *Candida guilliermondii* and their antimicrobial effect against pathogenic bacteria. J. Nanosci. Nanotechnol. 11: 243–248.

Mistry, H., Thakor, R., Patil, C., Trivedi, J., and Bariya, H. 2021. Biogenically proficient synthesis and characterization of silver nanoparticles employing marine procured fungi *Aspergillus brunneoviolaceus* along with their antibacterial and antioxidative potency. Biotechnol. Lett. 43: 307–316.

144 *Mycosynthesis of Nanomaterials: Perspectives and Challenges*

Mukherjee, P., Ahmad, A., Mandal, D., Senapati, S., Sainkar, S.R., Khan, M.I., and Sastry, M. 2001. Fungus-mediated synthesis of silver nanoparticles and their immobilization in the mycelial matrix: a novel biological approach to nanoparticle synthesis. Nano lett. 1: 515–519.

Mukherjee, P., Roy, M., Mandal, B.P., Dey, G.K., Mukherjee, P.K., Ghatak, J., and Kale, S.P. 2008. Green synthesis of highly stabilized nanocrystalline silver particles by a non-pathogenic and agriculturally important fungus *T. asperellum*. Nanotechnology 19: 075103.

Narayanan, K.B., and Sakthivel, N. 2010. Biological synthesis of metal nanoparticles by microbes. Adv. Colloid Interface Sci. 156: 1–13.

Narayanan, K.B., and Sakthivel, N. 2011. Green synthesis of biogenic metal nanoparticles by terrestrial and aquatic phototrophic and heterotrophic eukaryotes and biocompatible agents. Adv. Colloid Interface Sci. 169: 59–79.

Naveen, K.V., Sathiyaseelan, A., Mariadoss, A.V.A., Xiaowen, H., Saravanakumar, K., and Wang, M.H. 2021. Fabrication of mycogenic silver nanoparticles using endophytic fungal extract and their characterization, antibacterial and cytotoxic activities. Inorg. Chem. Commun.128: 108575.

Nayak, R.R., Pradhan, N., Behera, D., Pradhan, K.M., Mishra, S., Sukla, L.B., and Mishra, B.K. 2011. Green synthesis of silver nanoparticle by *Penicillium purpurogenum* NPMF: the process and optimization. J. Nanopart. Res. 13: 3129–3137.

Nelson, P.E., Dignani, M.C., and Anaissie, E.J. 1994. Taxonomy, biology, and clinical aspects of *Fusarium* species. Clin. Microbiol. Rev. 7: 479–504.

Niknejad, F., Nabili, M., Ghazvini, R.D., and Moazeni, M. 2015. Green synthesis of silver nanoparticles: advantages of the yeast *Saccharomyces cerevisiae* model. Curr. Med. Mycol. 1: 17.

Othman, A.M., Elsayed, M.A., Elshafei, A.M., and Hassan, M.M. 2017. Application of response surface methodology to optimize the extracellular fungal mediated nanosilver green synthesis. J. Genet. Eng. Biotechnol. 15: 497–504.

Pal, S., and Hossain, K.S. 2020. Optimization of mycobiosynthesis of silver nanoparticles by using *Fusarium* 4F1 and *Trichoderma* TRS isolates. Bangladesh J. Bot. 49: 343–348.

Pandiarajan, G. Govindaraj, R., Makesh Kumar, B., and Ganesan, V. 2011. Biosynthesis of silver nanoparticles using silver nitrate through biotransformation. J. Ecobiotechnol. 2: 13.

Phanjom, P., and Ahmed, G. 2017. Effect of different physicochemical conditions on the synthesis of silver nanoparticles using fungal cell filtrate of *Aspergillus oryzae* (MTCC No. 1846) and their antibacterial effects. Adv. Nat. Sci. Nanosci. Nanotechnol. 8: 1–13.

Qu, M., Yao, W., Cui, X., Xia, R., Qin, L., and Liu, X. 2021. Biosynthesis of silver nanoparticles (AgNPs) employing *Trichoderma* strains to control empty-gut disease of oak silkworm (*Antheraea pernyi*). Mater. Today Commun. 28: 102619.

Rahimi, G., Alizadeh, F., and Khodavandi, A. 2016. Mycosynthesis of silver nanoparticles from *Candida albicans* and its antibacterial activity against *Escherichia coli* and *Staphylococcus aureus*. Trop. J. Pharm. Res. 15: 371–375.

Rai, M., Yadav, A., and Gade, A. 2009. Silver nanoparticles as a new generation of antimicrobials. Biotechnol. Adv. 27: 76–83.

Rai, M., Bonde, S., Golinska, P., Trzcińska-Wencel, J., Gade, A., Abd-Elsalam, K., and Ingle, A. 2021. *Fusarium* as a novel fungus for the synthesis of nanoparticles: mechanism and applications. J. Fungi 7: 139.

Rauwel, P., Küünal, S., Ferdov, S., and Rauwel, E. 2015. A review on the green synthesis of silver nanoparticles and their morphologies studied via TEM. Adv. Mater. Sci. Eng. 2015. https://doi.org/10.1155/2015/682749.

Rose, G.K., Soni, R., Rishi, P., and Soni, S.K. 2019. Optimization of the biological synthesis of silver nanoparticles using *Penicillium oxalicum* GRS-1 and their antimicrobial effects against common food-borne pathogens. Green Process. Synth. 8: 144–156.

Roy, S., Shankar, S., and Rhim, J.W. 2019. Melanin-mediated synthesis of silver nanoparticle and its use for the preparation of carrageenan-based antibacterial films. Food Hydrocoll. 88: 237–246.

Salaheldin, T.A., Husseiny, S.M., Al-Enizi, A.M., Elzatahry, A., and Cowley, A.H. 2016. Evaluation of the cytotoxic behavior of fungal extracellular synthesized Ag nanoparticles using confocal laser scanning microscope. Int. J. Mol. Sci. 17: 329.

Saxena, J., Sharma, P.K., Sharma, M.M., and Singh, A. 2016. Process optimization for green synthesis of silver nanoparticles by *Sclerotinia sclerotiorum* MTCC 8785 and evaluation of its antibacterial properties. Springerplus 5: 861.

Seetharaman, P.K., Chandrasekaran, R., Periakaruppan, R., Gnanasekar, S., Sivaperumal, S., Abd-Elsalam, K.A., and Kuca, K. 2021. Functional attributes of myco-synthesized silver nanoparticles from endophytic fungi: A new implication in biomedical applications. Biology 10: 473.

Shahzad, A., Saeed, H., Iqtedar, M., Hussain, S.Z., Kaleem, A., and Abdullah, R. 2019. Size-controlled production of silver nanoparticles by *Aspergillus fumigatus* BTCB10: likely antibacterial and cytotoxic effects. J. Nanomater. 2019: 5168698

Shaligram, N.S., Bule, M., Bhambure, R., Singhal, R.S., Singh, S.K., Szakacs, G., and Pandey, A. 2009. Biosynthesis of silver nanoparticles using aqueous extract from the compactin producing fungal strain. Process Biochem. 44: 939–943.

Shu, M., He, F., Li, Z., Zhu, X., Ma, Y., Zhou, Z., and Zeng, M. 2020. Biosynthesis and antibacterial activity of silver nanoparticles using yeast extract as reducing and capping agents. Nanoscale Res. Lett. 15: 1–9.

Shukla, G., Gaurav, S.S., Singh, A., and Rani, P. 2021. Synthesis of mycogenic silver nanoparticles by *Fusarium pallidoroseum* and evaluation of its larvicidal effect against white grubs (*Holotrichia* sp.). Mater. Today 49: 3517–3527.

Siddiqi, K.S., Husen, A., and Rao, R.A. 2018. A review on biosynthesis of silver nanoparticles and their biocidal properties. J. Nanobiotechnology 16: 1–28.

Singh, P., Kim, Y.J., Zhang, D., and Yang, D.C. 2016. Biological synthesis of nanoparticles from plants and microorganisms. Trends Biotechnol. 34: 588–599.

Sintubin, L., Windt, W.D., Dick, J., Mast, J., Ha, D.V.D., and Verstraete, W. 2009. Lactic acid bacteria as reducing and capping agent for the fast and efficient production of silver nanoparticles. Appl. Microbiol. Biotechnol. 84: 741–749.

Skalickova, S., Baron, M., and Sochor, J. 2017. Nanoparticles biosynthesized by yeast: a review of their application. Kvasny Prumysl 63: 290–292.

Smitha, S.L., Nissamudeen, K.M., Philip, D., and Gopchandran, K.G. 2008. Studies on surface plasmon resonance and photoluminescence of silver nanoparticles. Spectrochim. Acta A Mol. Biomol. Spectrosc. 71: 186–190.

Srikar, S.K., Giri, D.D., Pal, D.B., Mishra, P.K., and Upadhyay, S.N. 2016. Green synthesis of silver nanoparticles: a review. Green Sustain. Chem. 6: 34–56.

Vanaja, M., Gnanajobitha, G., Paulkumar, K., Rajeshkumar, S., Malarkodi, C., and Annadurai, G. 2013. Phytosynthesis of silver nanoparticles by *Cissus quadrangularis*: influence of physicochemical factors. J. Nanostructure Chem. 3: 1–8.

Verma, V.C., Kharwar, R.N., and Gange, A.C. 2010. Biosynthesis of antimicrobial silver nanoparticles by the endophytic fungus *Aspergillus clavatus*. Nanomedicine 5: 33–40.

Vigneshwaran, N., Kathe, A.A., Varadarajan, P.V., Nachane, R.P., and Balasubramanya, R.H. 2006. Biomimetics of silver nanoparticles by white rot fungus, *Phaenerochaete chrysosporium*. Colloids Surf. B 53: 55–59.

Waghmare, S.R., Mulla, M.N., Marathe, S.R., and Sonawane, K.D. 2015. Ecofriendly production of silver nanoparticles using *Candida utilis* and its mechanistic action against pathogenic microorganisms. 3 Biotech 5: 33–38.

Wang, D., Xue, B., Wang, L., Zhang, Y., Liu, L., and Zhou, Y. 2021. Fungus-mediated green synthesis of nano-silver using *Aspergillus sydowii* and its antifungal/antiproliferative activities. Sci. Rep. 11: 1–9.

Wei, L., Lu, J., Xu, H., Patel, A., Chen, Z.S., and Chen, G. 2015. Silver nanoparticles: synthesis, properties, and therapeutic applications. Drug Discov. Today 20: 595–601.

Xue, B., He, D., Gao, S., Wang, D., Yokoyama, K., and Wang, L. 2016. Biosynthesis of silver nanoparticles by the fungus *Arthroderma fulvum* and its antifungal activity against genera of *Candida, Aspergillus* and *Fusarium*. Int. J. Nanomed. 11: 1899–1906.

Yari, T., Vaghari, H., Adibpour, M., Jafarizadeh-Malmiri, H., and Berenjian, A. 2022. Potential application of *Aspergillus terreus*, as a biofactory, in extracellular fabrication of silver nanoparticles. Fuel 308: 122007.

Zhao, X., Zhou, L., Riaz Rajoka, M.S., Yan, L., Jiang, C., Shao, D., and Jin, M. 2018. Fungal silver nanoparticles: synthesis, application and challenges. Crit. Rev. Biotechnol. 38: 817–835.

Zielonka, A., and Klimek-Ochab, M. 2017. Fungal synthesis of size-defined nanoparticles. Advances in Natural Sciences: Nanoscience and Nanotechnology 8: 043001.

Chapter 8

Biosynthesis of Gold Nanoparticles by Fungi
Progress, Challenges and Applications

Adriano Brandelli and *Flávio Fonseca Veras*

Introduction

Gold nanoparticles (AuNPs) regularly occur as colloidal dispersions in a solvent, usually water. Colloidal gold nanoparticles have been employed since ancient times due to the vibrant colors produced by their interaction with visible light. Although the mechanisms of action are not completely understood, gold colloids have been also mentioned as curative agents for many diseases over the centuries, and claimed for their therapeutic potential until recent times (Brown et al. 2007; Norn et al. 2011). AuNPs own important properties such as high electron density and strong optical absorption. These nanoparticles are available in the range from 1 to more than 120 nm and their plasmon band visible absorption can be observed above 3 nm. The properties of colloidal AuNPs are strongly reliant on their size and shape, thus determining their specific applications as well. More recently, these exceptional optical-electronics properties have been deeply investigated and employed in high technology applications (Hu et al. 2020).

The array of applications for AuNPs is rapidly increasing, including microelectronics (chips, sensors and probes), biomedical applications (imaging, diagnosis and therapy) and catalysis (Daniel and Astruc 2004). AuNPs are being used as conductors, resistors, and other components of electronic chips, in a diversity of sensors (Jakobs et al. 2008; Zhao et al. 2014), and as suitable probes for imaging applications (He et al. 2008). Biomedical applications of AuNPs are also increasing, as drug delivery agents, for the diagnosis of infectious agents and

---

Laboratório de Bioquímica e Microbiologia Aplicada, Instituto de Ciência e Tecnologia de Alimentos, Universidade Federal do Rio Grande do Sul, 91501-970 Porto Alegre, Brazil.
Corresponding author: abrand@ufrgs.br

heart diseases, and also in cancer diagnosis and therapy (Boisselier and Astruc 2009; Dykman and Khlebtsov 2011; Jain et al. 2012). Furthermore, the potential of AuNPs for applications in food packaging and detection of food hazards has been recently suggested (Liu et al. 2018; Paidari and Ibrahim 2021).

Despite many studies describe the manufacture and applications of AuNPs, additional possibilities may be attained by surface modification, functionalization and bioconjugation (Sperling and Parak 2010; Yoo et al. 2011). In particular, decoration with biocompatible polymers and proteins may be very useful to control colloidal stability, assembly or delivery of AuNPs to specific targets (Rai et al. 2015). The selection of appropriate ligand molecules depends on the composition of the nanoparticle core, the particle size, the solvent and the proposed application. The binding of the ligand molecules to the nanoparticle surface is mediated by some attractive interactions, either chemisorption, electrostatic attraction or hydrophobic interaction, often provided by a head group of the ligand molecule. Thiol groups have been recognized by their high affinity to noble metal surfaces, in particular to gold (about 200 kJ mol^{-1}; Sperling and Park 2010). Thus, thiol-containing molecules are usually strongly bound to AuNPs surface through covalent-like bonds.

Considering the relevance of AuNPs as an exciting material with diverse applications in optical, electronics, foods and medicine, this chapter presents a compilation of significant research using fungi as agents for biological synthesis.

Fabrication of gold nanoparticles

The preparation of AuNPs can be attained by either chemical, physical, or biological methods. The chemical reduction of metal ions is the most common and easy method for the preparation of metallic nanoparticles, including AuNPs. The chemical synthesis of AuNPs can be conducted by a great number of protocols, most of the time starting from commercial chloroauric acid (HAuCl$_4$). The method based on citrate reduction of Au^{3+} to Au^0 in water was introduced by Turkevitch, an approach that is still used nowadays to later replace the citrate ligand of these AuNPs by suitable ligands of biological interest (Kimling et al. 2006). As the use of citrate in the synthesis of AuNPs can lead to nanoparticle agglomeration, several methods have been developed to address this issue using different solutions including Tween 20, thiolic acid, and different surfactants. The application of thiolic groups has been described as a strategy to produce non-aggregate and re-dispersible solutions of AuNPs (Brust et al. 1994). Recent modifications of the Turkevitch method have allowed better size distribution and size control within the 9-120 nm range (Kimling et al. 2006).

Currently, the chemical conversion of the metal ions into metal nanoparticles can be performed using wet chemical synthesis, photochemical process, by using liquid crystal, polymer templates, or solution-based methodologies (Chouhan 2018; Lee and Jun 2019). Usually, the following three main components are employed in the chemical synthesis process of AuNPs: (a) metal precursors (often inorganic salts), (b) reducing agents (e.g., NaBH$_4$, amines, formaldehyde), and (c) capping/stabilizing agents (e.g., citrate, 2-mercapto ethanol, polyols). A number of synthesis methods reported the precise control of nanoparticle size (Singh et al. 2016), particularly for the synthesis of silver and gold NPs. An accurate control of the nanoparticle size can be achieved by adjusting the pH, temperature, concentration of salt and reducing agent (Agnihotri et al. 2014). Moreover, the different reducing power of different reductants may have a significant effect in determining the final shape of nanoparticles (Chouhan 2018).

Besides chemical methods used for the production of AuNPs from bulk gold, preparation of AuNPs using physical methods based on UV radiation reduction, laser ablation and aerosol technology have been described (Slepicka et al. 2020). However, methodologies based on chemical reduction have gained much more attention because they are less expensive and based on simple instrumentation. More recently, different reducing agents have been used in green synthesis of AuNPs. Biological synthesis using plant extracts or microorganisms has gained importance because it is associated with green synthesis and, consequently, less toxic processes with minimum release of hazardous materials into the environment are expected. In addition, the chemical methods for the synthesis of gold nanoparticles lead to the formation of some toxic chemicals adsorbed on the surface that may have undesirable effects for medical applications. Consequently, the need to develop environmentally safe nanoparticles has gained traction.

The synthesis of metal nanoparticles by green methods is usually carried out using an aqueous solution containing a salt of the metal and the plant extract or microorganism (Singh et al. 2016). In general, these methods produce heterogeneous nanoparticles with low yield. The optimization of parameters is performed by the control of temperature, pH, time, mixing rate, and aeration. After optimization, stable and homogeneous nanoparticles with good arrangement can be obtained. By manipulating the process parameters, it is possible to control shape and morphology of the nanoparticles (Figure 1). For example, the use of pear extract for biosynthesis of AuNPs resulted in triangular and hexagonal nanoparticles at alkaline pH, but AuNPs were not formed under acidic pH conditions (Ghodake et al. 2010). Different properties can be achieved by changing the size, shape and morphology of nanoparticles, and therefore much attention has been devoted to synthesize AuNPs of shapes other than spherical, including nanocubes, nanorods, triangular and hexagonal nanostructures.

Biological synthesis of gold nanoparticles by fungi

As the green synthesis of metal nanoparticles have gained importance, several biological sources have been investigated for their potential to produce AuNPs. Many fungal species have demonstrated promising characteristics as biological materials for the synthesis of AuNPs (Rai et al. 2009). The biosynthesis mechanism can be both extracellular and intracellular depending on the microbial location of AuNPs production. The intracellular mechanism can be conducted by reducing sugars, oxidoreductase enzymes and other proteins involved in the energy metabolism of fungal cells, while extracellular formation of AuNPs follows adsorption of $AuCl_4^-$ ions on cell wall enzymes mediated by electrostatic interactions. Current evidences suggest that NADPH-dependent oxidoreductases either on the cell surface or in the cytoplasm are the key enzymes in AuNPs biosynthesis, similar to that

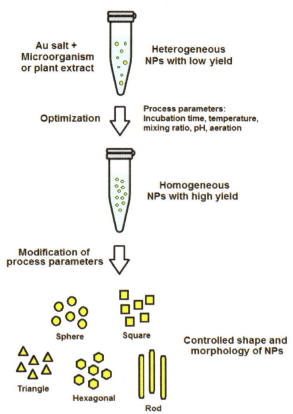

Figure 1. Influence of different process parameters during biological synthesis of gold nanoparticles.

described for AgNPs (Mikhailova 2021). Examples on yeast, filamentous fungi and macroscopic fungi used for this purpose are presented in the sequence.

Yeasts

Yeasts are unicellular organisms classified as members of the fungus kingdom. Some species have the ability to develop multicellular features by forming filaments of connected budding cells known as false hyphae. Yeasts are widely distributed in the environment and often isolated from carbohydrate-rich materials, using mostly sugars as carbon source. The yeast species *Saccharomyces cerevisiae* is famous for its relevance in converting carbohydrates into ethanol and CO_2 by fermentation, with great relevance in food and beverage industry. Besides, yeasts have many biotechnological applications including production of metabolites, enzymes, recombinant proteins and biofuels (Mattanovich et al. 2014).

The nanobiotechnological application of yeast cells and extracts in producing AuNPs has been demonstrated in many studies. Two different strains of *S. cerevisiae*, namely APP22 and CCFY-100, were studied for bioaccumulation of gold in the form of $HAuCl_4$. Transmission electron microscopy (TEM) analysis of thin sections of the cells revealed that Au^{3+} was reduced in situ to Au^0 and AuNPs were formed inside the yeast cells (Sen et al. 2011). The reduction of gold ions in the cytoplasm matrix was achieved by a very low dose of γ-energy. A kinetics study of formation and entry of AuNPs showed that they gradually moved from the cell wall to cytoplasm to nucleus and finally accumulated in the nucleolus as a function of time. TEM image of budding yeast shows that gold nanoparticles are not transferred to the new generation yeast cells. In another study, AuNPs with about 13 nm particle size were biosynthesized using an aqueous extract of Baker's yeast (*S. cerevisiae*) under visible light. The composition of the aqueous yeast extract was investigated and trimethylsilyl derivatives of butan-2,3-diol, indole-3-acetic acid, glucose and undecanoic acid were identified for the first time. This yeast extract performed as a reducing and capping agent for AuNPs (Attia et al. 2016).

The biological synthesis of AuNPs using the yeast cells of *Magnusiomyces ingens* LH-F1 has been described as well. The UV-vis spectral analysis revealed that 2.2 mg/mL biomass and 1.0 mM $HAuCl_4$ were preferable for AuNPs synthesis by this yeast strain. AuNPs of different shapes, including sphere, triangle and hexagon were observed in both scanning electron microscopy (SEM) and TEM images. Further analyses of Fourier transform infrared spectroscopy (FTIR) and sodium dodecyl sulfate-polyacrylamide gel electrophoresis (SDS-PAGE) indicated that some biomolecules were absorbed on the surface of AuNPs, which are possibly involved in the formation of nanoparticles. The synthesized AuNPs showed excellent catalytic activities for the reduction of nitrophenols to aminophenols in the presence $NaBH_4$ (Zhang et al. 2016). Whole cells and cell free extracts of *Candida parapsilosis* ATCC 7330 were also investigated for the synthesis of AuNPs (Krishnan et al. 2016). The systematic study revealed that cell free extracts were more efficient than the whole cells in the synthesis of directly usable AuNPs. Cell free extracts yielded AuNPs of particle diameter ranging from 50 to 200 nm and powder X-ray diffraction analysis confirmed the crystalline nature of the nanoparticles. The nanofabrication was influenced by total protein concentration and not only by the reductase enzymes as originally supposed. Data from FTIR spectroscopy and thermal gravimetric analysis indicate that the biosynthesized AuNPs were capped by peptides/proteins. A stable dispersion of AuNPs in pH 12 solutions was confirmed by dispersion experiments, by electron microscopic analysis and validated using a surface plasmon resonance assay. The dispersed nanoparticles were effective for the reduction of 4-nitrophenol using $NaBH_4$ as a reductant, thus confirming the formation of functional AuNPs.

Similarly, the biosynthesis of metal nanoparticles (NPs) by an extremophilic yeast strain isolated from acid mine drainage in Portugal was investigated using three distinct strategies: growing of yeast strain in presence of metal ions, using the yeast biomass, or using the supernatant obtained after 24-hour incubation of yeast biomass in water for the metal nanoparticles synthesis. An effective route for nanoparticles synthesis was obtained using the yeast biomass. The synthesis of Ag and Au nanoparticles could take place intracellularly and/or extracellularly, and the cell wall seems to plays an important role for the biosynthesis of both metal nanoparticles. When the washed yeast cells were in contact with Ag or Au solutions,

AgNPs with an average diameter of 20 nm were produced, whereas diameter ranged from 20 to 100 nm for AuNPs, as determined by TEM and confirmed by energy-dispersive X-ray spectra. The nanoparticles were well dispersed, suggesting that they are capped by stabilizing agents, most probably by proteins. The supernatant-based protocol provided evidence that yeast proteins were released to the medium, which could be responsible for the formation and stabilization of the nanoparticles, although the participation of the cell wall seems essential for AuNPs synthesis (Mourato et al. 2011).

Despite both yeast cells and yeast cell extracts can be effective for the synthesis of metal nanoparticles, synthesis with culture supernatants is often the preferred method since it facilitates the purification protocol for the recovery of the nanoparticles. Extraction of nanoparticles formed within the cells or cell-wall is laborious, time-consuming and not cost effective. In this regard, the study with whole cells of *C. parapsilosis* ATCC 7330 showed that gold nanoparticles with mean particle diameter of 27 nm were biosynthesized inside the whole cells by TEM analysis. Under optimized reaction conditions, maximum gold bioaccumulation with the 24 h culture age reached a cellular uptake of about 1000 gold atoms at the single cell level, but it was difficult to extract the AuNPs from the whole cells (Krishnan et al. 2016). Therefore, yeast-derived compounds like the pyomelanin derived from *Yarrowia lipolytica* appears as alternative biological agents for the green synthesis of AuNPs (Tahar et al. 2019). A factorial design was developed to investigate the influence of pH, temperature, gold salt and pyomelanin concentration on the size distribution of the AuNPs. The size of AuNPs was significantly affected by the different process parameters, but the mathematical model allowed to predict the experimental parameters that yield AuNPs with specific size. The nanoparticles synthesized with median size value of 104 nm presented nanocrystalline structure, mostly polygonal or spherical. They showed a zeta potential value of -28.96 mV and a moderate polydispersity index of 0.267, suggesting a high colloidal stability.

Moreover, biosynthesis of gold nanostructures is still a challenge because of the expensive source and technical difficulties for manageable construction of morphology and size. Biosynthesis of monodispersed nanoparticles, along with determination of potential responsible biomolecules, are major bottlenecks in the field of nanobiotechnology research. Some authors describe that the morphologies and sizes of gold nanostructures can be controlled during biological synthesis using yeast extracts by varying the pH value of the medium. Gold nanoplates with side length from 300 to 1300 nm and height from 15 to 18 nm can be obtained in acid condition by increasing pH values. On contrast, only nanoflowers and nanoparticles were observed under neutral or alkaline conditions (Yang et al. 2017). Organic molecules contained in the extract, such as succinic acid, lactic acid, malic acid, and glutathione, which are generated in yeast metabolism, play important role in the reduction of gold ions. As biological synthesis of gold nanostructures has gained increasing attention because of its green and sustainable approach, one-pot biosynthesis methods using inexpensive yeast extract as a green precursor merits additional scale up studies.

Filamentous fungi

There may be as many as five million species of fungi in the biosphere, and the vast majority are those named filamentous fungi or "moulds", because they are composed of very fine threads called hyphae. The hyphae keep growing and interlacing until they form a filamentous network named mycelium. Filamentous fungi are essential for the functioning of natural ecosystems, along with bacteria, they are the main agents responsible for decomposing dead organic matter, making its chemical components available to the next generation of organisms (Aleklett and Boddy 2021). Filamentous fungi are important pathogens of crop plants, and in some cases cause serious human diseases, particularly in immunocompromised individuals (Gnat et al. 2021; Shao et al. 2021). They have also been employed for biotechnological uses, including the production of antibiotics such as penicillin, other medicines, citric acid, enzymes and foods such as soy sauce and cheese (Sugiharto 2019; El Enshasy 2022). More recently, the biotechnological importance of filamentous fungi for synthesis of AuNPs has been revealed.

The scalability and cost efficiency of fungal growth even on industrial scale make extracellular or intracellular extracts of fungi excellent candidates for the synthesis of metal nanoparticles. Several methods and techniques that use fungi-derived fractions for synthesis of AuNPs have been described, but the drawbacks and limitations of these techniques are still poorly understood. Moreover, the identification of fungal components that play key roles in the synthesis is challenging. The capability of 29 thermophilic fungi for the synthesis of AuNPs was studied through three different approaches using either the extracellular fraction, the autolysate of the fungi or the intracellular fraction. Nanoparticles with different sizes (ranging between 6 nm and 40 nm) and size distributions (with standard deviations ranging between 30% and 70%) were observed depending on the fungi and experimental conditions. Using an ultracentrifugal filtration technique, it was possible to determine that the size of reducing agents is less than 3 kDa and the size of molecules that can efficiently stabilize nanoparticles is greater than 3 kDa (Molnár et al. 2018).

A rapid eco-friendly cell surface-based synthesis of stable gold nanoprisms using *Penicillium citrinum* MTCC9999 biomass was proposed by Goswami and Ghosh (2013). The fungal biomass was incubated at room temperature with $HAuCl_4$ yielding dispersed AuNPs solution showing absorption maxima at 540 nm due to surface plasma resonance of nanoparticles. Typical TEM images showed that most AuNPs were prism shaped with a mean diameter ranging from 20 to 40 nm, which were correlated with the data obtained from dynamic light scattering (DLS) experiments. Average zeta potential of AuNPs was –20 mV, suggesting that some biomolecules capped the nanoparticles imparting a net negative charge over it. The participation of protein molecules in stabilization was showed by FTIR analysis. In another study, the green synthesis of AuNPs using *P. citrinum* isolated from brown algae and its antioxidant activity were investigated (Manjunath et al. 2017). The culture filtrate of *P. citrinum* was mixed with $AuCl_4$ resulting in a color change from yellow to reddish violet. The reduction of gold metal ions was confirmed from the UV-vis spectrum. The mean particle size of AuNPs was determined as 60–80 nm by field emission scanning electron microscopy and DLS. AuNPs showed significant antioxidant potential as evaluated by scavenging of the free radical 1,1-diphenyl-2-picrylhydrazyl, and the activity was comparable to the standard ascorbic acid. The culture supernatant of the filamentous fungi *Penicillium chrysogenum*, isolated from a copper mine, was also useful for biosynthesis of AuNPs by reducing gold ions in aqueous solution (Sheikhloo and Salouti 2011). A peak at 532 nm was observed in the UV–vis spectrum, which is very specific characteristic for AuNPs. The XRD spectrum confirmed the presence of crystalline gold nanoparticles, and TEM images exhibited the intracellular formation of AuNPs in spherical, triangle and rod shapes with sizes ranging from 5 to 100 nm.

Several studies describe the use of *Aspergillus* spp. for the biological synthesis of AuNPs. An eco-friendly protocol for size-controlled synthesis of AuNPs, using the fungus *Aspergillus terreus* IF0 was described by Priyadarshini et al. (2014). The addition of $HAuCl_4$ to the aqueous fungal extract resulted in the instantaneous formation of AuNPs. Particle diameter and dispersity of nanoparticles were controlled by varying the pH of the fungal extract, resulting in synthesized particles with an average size in the range of 10–19 nm at pH 10. Fractions of the fungal extract were analyzed by FTIR analysis revealing that biomolecules larger than 12 kDa and having –CH, –NH, and –SH functional groups were responsible for bioreduction and stabilization of AuNPs. The biosynthesis of stable AuNPs using culture filtrate of *Aspergillus flavus* was also investigated (El-Bendary et al. 2018). The optimum conditions for the synthesis were 1 mM $HAuCl_4$ in aqueous solution containing 10% *A. flavus* filtrate, 0.1% Tween 20, incubation period of 30 min at 30°C under static and illumination conditions. Myco-synthesized nanoparticles were fully characterized, showing a pinkish violet color with a maximum absorbance at 530–540 nm in the UV-vis spectrum. The TEM analysis revealed spherical, hexagonal, squared and anisotropic crystalline AuNPs with size ranging from 10 to 50 nm. The average particle size was confirmed by DLS as 39.5 nm, and the FTIR analysis showed the presence of carbonyl and amide functional groups suggesting that peptides are involved in the synthesis and stability of AuNPs.

The endophytic fungus *Aspergillus clavatus*, isolated from surface sterilized stem tissues of *Azadirachta indica* A. Juss., when incubated with an aqueous solution of $AuCl_4$ ions produces a diverse mixture of intracellular AuNPs, especially nanotriangles in the size range from 20 to 35 nm. The reaction process was simple and convenient to handle and was monitored using UV–vis spectroscopy. The morphology and crystalline nature of the nanotriangles were determined from TEM, atomic force spectroscopy (AFM) and X-ray diffraction (XRD) spectroscopy (Verma et al. 2011). The AuNPs synthesized by *Aspergillus fischeri* were fairly monodispersed and uniform in shape as confirmed by the relatively tall and dominant UV-vis spectroscopy peak at 540 nm (Banerjee and Vittal 2017). TEM images showed that round shaped AuNPs with size ranges between 20 and 50 nm were produced by *A. fischeri*. The crystalline powder formed peaks at 2θ angles specific for AuNPs and FTIR analysis confirms presence of functional groups that help in stabilization of gold particulates.

The fungus *Aspergillus niger* has been also described for the synthesis of AuNPs (Soni and Prakash 2012). The fungal liquid after exposure to 10^{-3} M aqueous solution of $HAuCl_4$ for 72 h clearly became yellow indicating the synthesis of AuNPs. A

154 *Mycosynthesis of Nanomaterials: Perspectives and Challenges*

typical and properly broad absorption band centered at 530 nm was observed after exposure to Au^+ ions, and the presence of the broad resonance indicates an aggregated structure of the AuNPs in the fungal liquid. TEM images of AuNPs synthesized by *A. niger* showed AuNPs of different sizes (10–30 nm) and shapes. In a recent study, the xylanase enzymes from fungal strains *A. niger* L3 and *Trichoderma longibrachiatum* L2 were evaluated for the biofabrication of AuNPs (Elegbede et al. 2020). The synthesis of AuNPs was observed from the color change from light yellow to purple and surface plasmon resonance at 545 and 560 nm for L3-AuNPs and L2-AuNPs, respectively. The sizes of the nanoparticles ranged from about 5 to 124 nm, showing mostly spherical (L3-AuNPs) and flower-shaped (L2-AuNPs) morphologies. FTIR spectroscopy showed that protein molecules are involved in the capping and stabilization of these AuNPs.

Filamentous fungi from the genus *Trichoderma* have been also investigated for the synthesis of AuNPs. The fungal biomass of *Trichoderma koningii* after incubation for 72 h with deionized water was separated by filtration, and the cell-free filtrate was incubated with different concentrations of $HAuCl_4$ (Maliszewska et al. 2009). After monitoring the changes in the UV-vis spectra during nanoparticles production, the gold nanostructures synthesized in the fungal filtrate were formed in several different shapes, ranging from polydisperse small spheres to large polygons (triangles and hexagons). *T. konigii* releases reducing agents into the solution which can be responsible for the formation of AuNPs, although the cellular mechanism leading to the biosynthesis of AuNPs is not completely understood. Some authors indicated that glucose oxidase or NADH-dependent reductase enzymes are important factors in the biosynthesis of AuNPs, but using the denatured protein in the cell-free *T. konigii* filtrate showed an intense peak at around 530 nm appearing within 24 h (Maliszewska et al. 2009). The ability of the heat-denatured protein to synthesize gold nanocrystals suggest that the native three-dimensional structure of protein is not essential in the process. The increased rate of production of Au^0 by the denatured protein indicated a rapid reduction of Au^{3+} to Au^0 due to the exposed functional groups. The thiol group (–SH), present in the side chain of cysteine, is one of the functional groups that could be important in reduction of Au^{3+} most likely by forming Au–S bonds. A biological method for synthesis of AuNPs using *Trichoderma hamatum* SU136 aqueous mycelial extract was reported (Abdel-Kareem and Zohri 2018). The culture filtrate of the fungus was exposed to different concentrations of gold chloride, and Au^{3+} ions were reduced to Au^0 in all cases, leading to the formation of stable AuNPs. The presence of a surface plasmon band around 530 nm indicates the synthesis of AuNPs, which ranged from 5 to 30 nm. TEM analysis showed the spherical, pentagonal and hexagonal morphologies of AuNPs synthesized by *T. hamatum* SU136. The association of proteins with nanoparticles was determined by FTIR spectroscopy. After parameters optimization, the smallest size of AuNPs was attained with 0.5 mmol L^{-1} $AuCl_3$, pH 7 at 38°C. The synthesis of AuNPs by cell-free extracts of *Trichoderma viride* and *Hypocrea lixii* was investigated by Mishra et al. (2014). The biosynthesis of AuNPs was very rapid and took 10 min at 30°C using *T. viride*, while similar result was obtained by *H. lixii* at 100°C. The presence of biomolecules in cell-free extracts of both fungi was associated with the synthesis and stabilization of the gold particles. The procedure was very quick and environment friendly, not requiring subsequent processing.

A biogenic route for the synthesis of AuNPs was developed using the extract of a novel strain, *Talaromyces flavus* (Priyadarshini et al. 2014b). Production of AuNPs by the fungal extract in the presence of $HAuCl_4$ was confirmed by the results obtained from UV-vis spectroscopy, energy dispersive spectroscopy (EDS) and DLS analysis. The time-dependent kinetic study revealed that the bio-reduction process follows an autocatalytic reaction. Crystalline, irregular, and mostly flower-shaped gold nanoparticles were obtained. FTIR analysis indicates the involvement of $-NH_2$, $-SH$, and $-CO$ as the probable functional groups in the bio-reduction and stabilization process. Floral-shaped nanoparticles are known to exhibit a high surface area and enhanced plasmon resonance, but the synthetic procedures undertaken for their synthesis involves multiple steps and the use of hazardous chemicals. Compared to the conventional methods, a time-resolved, green, and economically viable method for floral-shaped nanoparticle synthesis was developed using *T. flavus*, indicating that the single step biogenic synthesis of floral AuNPs can be a major advance in catalysis as well as biomedical applications

Macroscopic fungi

Macroscopic filamentous fungi also grow by producing a mycelium below ground. They differ from moulds because they produce visible fruiting bodies, usually known as mushrooms or toadstools, which hold the spores. The fruiting body is composed of tightly packed hyphae which divide to produce the diverse parts of the fungal structure. Macroscopic fungi have intracellular metal uptake ability and maximum wall binding capacity, and therefore, they have high metal tolerance and bioaccumulation ability (Elkes et al. 2010; Kapahi and Sachdeva 2017). Similar to yeasts and filamentous fungi, two methods have been mainly described in the literature to synthesize AuNPs from macrofungi: the intracellular method, which refers to synthesis inside fungal cells by transportation of gold ions in the presence of enzymes; and the extracellular method, which refers to the treatment of aqueous filtrate containing fungal biomolecules with a metal precursor. Metal nanoparticles derived from macrofungi, including various mushroom species, such as *Agaricus bisporus*, *Pleurotus* spp., *Lentinus* spp., and *Ganoderma* spp. are well known to possess important applications due to their several biological properties (Bhardwaj et al. 2020).

The biosynthesis of AuNPs using the edible mushroom *Agaricus bisporus* extract as both reducing and stabilizing agents through microwave irradiation method was described by Eskandari-Nojehdeh et al. (2016). The effects of the microwave exposure time and the amount of $HAuCl_4$ solution on the mean particle size, concentration, and polydispersity index (PDI) of the synthesized AuNPs were investigated. Spherical and well-dispersed AuNPs with the minimum mean particle size and PDI, and maximum concentration and zeta potential of about 33.5 nm, 0.855, 148.8 ppm, and +17.2 mV, respectively, were obtained using 2.62 mL of 1 mM $HAuCl_4$ solution and 0.2 mL of mushroom extract for 55 s. Moreover, mycelial growth of the soil basidiomycetes *A. bisporus* and *Agaricus arvensis* in submerged

156   *Mycosynthesis of Nanomaterials: Perspectives and Challenges*

and solid media was examined in the presence of different concentrations of $HAuCl_4$, $AgNO_3$, Na_2SeO_3, Na_2SiO_3, and GeO_2 (Loshchinina et al. 2018). Fungal mycelial extracts and cell-free culture filtrates were able to reduce Au, Ag, Se, Si, and Ge ions, forming Au^0, Ag^0, Se^0, Si^0/SiO_2 and Ge^0/GeO_2 nanoparticles. The mycogenic nanoparticles showed different physical characteristics depending on the species of *Agaricus* and the type of extract. Specifically, Au nanospheres produced with cell-free culture filtrates of *A. bisporus* presented a mean diameter ranging from 2 to 5 nm, while those obtained from *A. arvensis* were 2–10 nm. Nanoparticles obtained by extracts of mycelia were several times larger and highly heterogeneous.

The aqueous extract of isolated mushroom *Coprinus comatus* was used to produce AuNPs by exposure to UV radiation (Naeem et al. 2021). The color change of the mixture from light yellow to purple was observed after 25 min, and the maximum absorbance reached at 530 nm in the UV-visible spectrum evidenced the formation of AuNPs. FTIR spectra revealed the presence of functional groups related to peptides, proteins, flavonoids, monosaccharides, and phenolic compounds, which are possibly associated with reduction of gold ions and capping of AuNPs. XRD analysis showed that AuNPs have a face-centered cubic crystal. The UV irradiation at different times led to an increase in the intensity of absorbance and sizes of AuNPs from 17.39 nm before the irradiation and switched to 58.16, 59.13, and 47.35 nm after 1, 2, and 3 h, respectively, but their sizes remained within the nanoscale range (less than 100 nm). An innovative approach combined the use of *Pleurotus sajor-caju*, followed by microwave treatment of silver and gold nanoparticles that ensures the capping of the active ingredient and further enhances the effects of nanomedicine (Chaturvedi et al. 2020). The nanoparticles prepared by the rapid microwave-assisted green synthesis method was compared using UV-vis spectroscopy, FTIR spectroscopy, XRD, TEM, and energy dispersive spectroscopy (EDS). The comparative study revealed that both assemblies display face-centred cubic structures and are nanocrystalline in nature. The advantage of the approach was that the sizes of gold and silver were identical in range with a similar distribution pattern.

The utilization of specific molecules obtained from macroscopic fungi has been described in the synthesis of AuNPs as well. A glucan isolated from the edible mushroom *Pleurotus florida*, cultivar Assam Florida, was used for the green synthesis of AuNPs by reducing HAuCl4 (Sen et al. 2013). Glucan acts as both reducing and stabilizing agent and the resulting AuNPs were fully characterized by UV-vis spectroscopy, TEM, XRD, SEM, and FTIR analysis. The results indicated that the concentration of $HAuCl_4$ influenced the size distribution of AuNPs. The resulting AuNPs-glucan bioconjugates function as an efficient heterogeneous catalyst in the reduction of 4-nitrophenol to 4-aminophenol, in the presence of sodium borohydride.

Applications of gold nanoparticles produced by fungi

Gold nanoparticles have a large array of applications in many fields, and their utilization in nanomedicine has gained relevance for labeling and imaging, development of biosensors, clinical diagnosis of cancer, diabetes, hepatitis, HIV, tuberculosis and potential markers for Alzheimer's disease, and different therapies

(Boisselier and Astruc 2009; Hu et al. 2020). In the following sections, some examples of prospective applications for AuNPs synthesized by fungi are discussed.

Imaging and diagnostics

Although TEM using the high atomic weight of Au is the most routine technique, several imaging methods involve the surface plasmon band. Because of their unique plasmonic properties, AuNPs are significantly transforming the field of optical imaging, allowing direct visualization of nanostructures inside the biological systems with high resolution. Presenting a typical band at 420 nm, large AuNPs can be imaged using an optical microscope in phase contrast or differential interference contrast mode. Detection with an optical microscope only involves scattered light in dark-field microcopy. Small AuNPs only absorb light, triggering heating of the milieu that can be detected by photoacoustic imaging using heat-induced liquid expansion or by photothermal imaging that record local variations of the refractive index by differential interference contrast (DIC) microscopy (Xiao and Yeung 2014; Wu et al. 2019). The classic application called immunostaining involves antibody-conjugated AuNPs that bind antigens of fixed, permeabilized cells thereby providing visualization by contrast using TEM. On the other hand, diluted AuNPs labeled with antibodies can label the outer cell surface without fixing and permeabilization, so that the inter-particle distance is larger than the optical resolution limit, which leads to single-particle imaging of cell movement (Boisselier and Astruc 2009).

Gold nanoparticles synthesized by microorganisms has emerged as a more eco-friendly, simpler and reproducible alternative to the chemical synthesis, allowing the generation of rare forms such as nanotriangles and prisms. Gold nanotriangles are of special interest since they possess distinct plasmonic features in the visible and infrared regions, which equipped them with unique physical and optical properties exploitable in vital applications such as optics, electronics, catalysis and biomedicine. The proposed mechanistic synthesis of nanotriangles by *Aspergillus* might serve as a set of design rule for the fabrication of anisotropic nanostructures with preferred architecture and can be amenable for the large scale commercial production and technical applications (Verma et al. 2011). Besides, it was found that gold nanoplates synthesized by yeast extract exhibited plasmonic property with prominent dipole infrared resonance in near-infrared region, indicating their potential in surface plasmon-enhanced applications, such as bioimaging and photothermal therapy (Yang et al. 2017).

AuNPs present several advantages in diagnostics as compared with quantum dots and organic dyes, including (a) greatly reduced or no toxicity, (b) much better contrast agents for imaging (comparing with organic dyes that suffer from rapid photobleaching), (c) surface-enhanced and distance- and refractive index-dependent spectroscopic properties (Boisselier and Astruc 2009; Singh et al. 2018). Three major approaches can be distinguished for diagnostics based on AuNPs: (a) inter-AuNP distance dependent colorimetric sensing for specific DNA hybridization for the detection of specific nucleic acid sequences in biological samples, (b) surface-functionalized AuNPs providing highly selective nanoprobes, and (c) electrochemical-based methods for signal enhancement. In this regard, the utilization of AuNPs synthesized by fungi has been described in diagnosis methods for several diseases (Chauhan et al. 2011; Sojinrin et al. 2017; Sibuyi et al. 2021). With the advent of Covid-19 pandemic, AuNPs have also proven as excellent tools for rapid and selective detection of SARS-CoV-2 and respective IgG antibodies, thus being effective nanosensors for colorimetric serological assays for diagnosis, epidemiological studies, or vaccine efficacy assessment (Pramanik et al. 2021; Lew et al. 2021).

Anticancer activity

Biologically synthesized metallic nanoparticles have been extensively researched for their function in cancer therapy (Sarkar and Kotteeswaran 2018; Chaturvedi et al. 2020; Zhang et al. 2020, Shunmugam et al. 2021). In particular, AuNPs have been considered as prospective anticancer agents due to their potential to combat various cancerous cells (Kajani et al. 2016; Ismail et al. 2018; Wang et al. 2019). The effect mechanism of these metallic NPs on tumor cells has been elucidated, including evidence on the ability to form reactive oxygen species (ROS), changes in the mitochondrial membrane permeability as well as dysfunction of this organelle, caspase activation processes and physicochemical interactions with intracellular compounds. Such events can cause cell cycle interruption or apoptosis (programmed cell death) (Sathish-Kumar et al. 2015; Espinosa-Diez et al. 2015; Rajeshkumar 2016; Liu et al. 2019; Ke et al. 2019). Besides, the gold nanoparticle effects depend on factors related to shape, size, concentration and chemical composition of the nanoparticle's surface or even the capping agent used (Mikhailova 2021). Interestingly, AuNPs are still proving to be favorable for use in association with other cancer treatment, as sensitizers in radiation therapy, photothermal agents, anticancer drug enhancer or drug delivery agents. Thus, they potentiate the damage caused by radiation, phototherapy and chemotherapy (Jelveh and Chithrani 2011; Mikhailova 2021). A brief schematic representation of some anticancer pathways mediated by AuNPs is shown in Figure 2.

In this context, gold nanoparticles biosynthesized by fungi have also show potential for applications in cancer therapy. For example, the efficacy of AuNPs produced using an aqueous extract of *S. cerevisiae* against Ehrlich ascites carcinoma cancer cells proved them a good anticancer agent, especially after exposure to visible light (Attia et al. 2016). In this case, the carcinoma cell percentage killed by AuNPs increased from 24.6% to 86.5% when photothermal treatment was applied. The light absorbing property of AuNPs results in the conversion of photoenergy to heat energy and ROS production, causing irreversible damage to the tumor cells. The AuNPs synthesized using a biomass extract from *Fusarium solani* also showed the anticancer potential against cervical cancer (He-La) and human breast cancer (MCF-7) cells (Clarance et al. 2020). Microscopic observations in this study revealed low cell density, condensed chromatins and nuclear fragmentations. Chaturvedi et al. (2020) were successful during anticancer evaluation of AuNPs produced from the *P. sajor-caju* aqueous filtrate. The authors observed these particles were capable to decrease the viability of human colon cancer HCT-116 cell line, and they also evidenced the ROS production and DNA fragmentation tests, although silver nanoparticles obtained

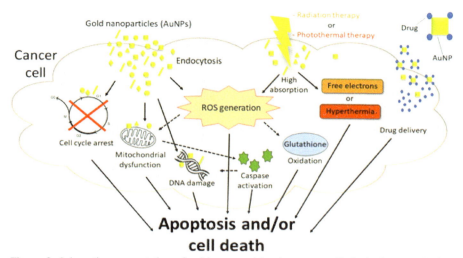

Figure 2. Schematic representation of gold nanoparticles in cancer cell. Dashed arrows indicate consequential effects of ROS production.

from the fungal extract presented better results. Other examples on the anticancer application of AuNPs synthesized by fungi are shown in Table 1.

Another important aspect when considering AuNPs as a good anticancer agent concerns their selective toxicity as minimal damage to normal cells has been reported (Hammami et al. 2021). Such an advantage has also been verified in those produced by fungi. AuNPs from the *A. flavus* culture filtrate demonstrated toxic activity against breast and lung carcinoma human cell lines and no toxic effects was verified on normal human cells, which may offer promising possibilities for the cancer treatment (El-Bendary et al. 2018). Banerjee and Vittal (2017) confirmed lower toxicity in normal cells (mouse fibroblasts) after AuNPs treatment at 1 µg/mL concentration. In contrast, these nanoparticles were highly toxic to HT-29 (colon carcinoma) cells, causing nucleus condensation and cell damage.

Antimicrobial activity (antibacterial, antifungal, antiviral)

The antimicrobial properties of metallic NPs are widely known and play an important role because of the emergence of drug-resistant microorganisms or to the pathogens for which there is no treatment or vaccine. Notably, AuNPs are becoming very interesting due to the increase of their spectrum action against viruses, Gram-positive and Gram-negative bacteria, and fungi, even though more limited in comparison with silver NPs. This points to them as a very promising trend for the microorganism control (Mikhailova 2021; Hammami et al. 2021; Qamar and Ahmad 2021). Even though there are still open questions about the real mechanisms of antimicrobial activity of gold nanoparticles, some mechanisms are similar to those reported for cancer cell (such ROS generation, DNA damage and cell heating by absorbing light). Furthermore, AuNPs can also cause increase in cell membrane and wall permeability, leakage of cell material, protein denaturation, interruption of electron transport,

Table 1. Anticancer activity of AuNPs synthesized by fungi as a green chemistry approach.

Fungal species	Size (nm)	Morphology	Carcinoma cell lines	References
Yeast				
Yarrowia lipolytica	104	Polygonal or spherical	U2OS cells (human osteosarcoma)	Tahar et al. 2019
Filamentous fungi				
Aspergillus flavus	12	Nearly spherical	HepG2 (hepatocellular carcinoma), A549 (lung carcinoma and MCF7 (breast cancer) cells	Abu-Tahon et al. 2020
Aspergillus foetidus	30–50	Spherical	A549 cells (human lung adenocarcinoma)	Roy et al. 2016
Cladosporium sp.	5–10	Spherical	MCF-7 cells	Munawer et al. 2020
Fusarium oxysporum	10–40	Spherical	ZR-75-1 (human Caucasian breast carcinoma), Daudi (human burkitt's lymphoma) and PBMC (peripheral blood mononuclear cell)	Ahmad Siddiqui et al. 2016
Fusarium solani	40–45	Needle and flower like structures with spindle shape	HeLa (cervical cancer) and MCF-7 cells	Clarance et al. 2020
Humicola spp.	18–24	Spherical	NIH3T3 (mouse fibroblast) and MDA-MB-231 (human breast adenocarcinoma) cells	Syed et al. 2012
Penicillium brevicompactum	10–120	Spherical, triangular and hexagonal	C2C12 (mouse muscle myoblast) cells	Mishra et al. 2011b
Trichoderma koningii	10–14	Spherical	LoVo (human colon adenocarcinoma) and LoVo/DX (multidrug resistance human colon adenocarcinoma sub-line) cells	Maliszewska 2013
Macroscopic fungi				
Pleurotus florida	12–50	Irregular, spherical, triangular	A-549 (Human lung carcinoma), K-562 (Human chronic myelogenous leukemia bone marrow), MDA-MB (Human adenocarcinoma mammary gland) and HeLa cells	Bhat et al. 2013
Tricholoma crassum	5–25	Rhomboid, spherical, hexagons, cuboids, triangular (isosceles and near-equilateral triangles)	Sarcoma 180 cells	Basu et al. 2018
Pleurotus sajor-caju	4–22	Spherical	HCT-116 cells (colon cancer)	Chaturvedi et al. 2020
Inonotus obliquus	23	Spherical, triangular, hexagonal, and rod shaped	MCF-1 (human breast cancer) and NCI-N87 (human stomach cancer) cells	Lee et al. 2015

and alterations in the gene expression (Lee et al. 2020; Mikhailova 2021). Another mechanism involved is specific to each microorganism, such as interaction between the particles and the ATP-driven enzyme which reduces the H^+ efflux of the *Candida* species (Ahmad et al. 2013). Or as already described in the antiviral activity in which AuNPs can interact with essential molecules (both in the host cells and in the viral genome) for the virus replication, consequently blocking the infection (Baram-Pinto et al. 2010; Mikhailova 2021).

Currently, studies on mycosynthesis of AuNPs focusing on antimicrobial action have been carried out (Table 2). Most of them report on the antibacterial activity evaluation of biosynthesized AuNPs from filamentous fungi. *Trichoderma hamatum*-synthesized AuNPs showed antimicrobial activities against the pathogenic bacteria *Bacillus subtilis*, *Staphylococcus aureus*, *Pseudomonas aeruginosa* and *Serratia* sp. (Abdel-Kareem and Zohri 2018). Similarly, AuNPs from *T. viride* were capable to inhibit the growth of *Pseudomonas syringae*, *Escherichia coli* and *Shigella sonnei* (Mishra et al. (2014). Naimi-Shamel et al. (2019) reported the antimicrobial activity of AuNPs from *Fusarium oxysporum* culture supernatant against *S. aureus*, *Bacillus cereus*, *E. coli* and *P. aeruginosa*, including multi-drug resistant strains. The authors also found an increase in the antibacterial effect of AuNPs after their conjugation with the antibiotic tetracycline, which were considered good drug-carrying agents. In general, the studies have suggested some of the aforementioned mechanisms as the likely reason behind the antibacterial activity of these particles or even the possibility of many mechanisms occurring at the same time. However, the antimicrobial effect of green synthesized AuNPs can be unpredictable since several factors such as particles size and shape, surface modification, concentration, surface charge, ligand type, and the culture medium and the source of synthesis also significantly influence this activity. Gholami-Shabani et al. (2021) produced silver, gold, copper and zinc oxide nanoparticles using biogenic synthesis from *Aspergillus kambarensis* and confirmed their antimicrobial effect, except for AuNPs. The authors suggested that the presence of fungal biomolecules may have covered the surface of the bacteria and fungi tested, preventing the action of these nanoparticles.

Unfortunately, data on the mycosynthesis of gold nanoparticles with antifungal effects are still very poor, but it must be taken into account that the fungal toxicity is generally lower than in bacteria. Production of pigments, organic acids and other compounds by target fungi may contribute to its tolerance to AuNPs (Pócsi 2011; Hanus and Harris 2013). Among the successful studies in antifungal activity of mycosynthesized AuNPs, some are particularly noteworthy. Rónavári and co-workers (2018) investigated the use of the culture supernatant of the yeast *Phaffia rhodozyma* for the synthesis of silver and gold NPs and evaluated their antifungal activity against pathogenic yeasts and some dermatophytes. They observed that the AgNPs exhibited a good antifungal property against all tested fungi. Conversely, AuNPs inhibited the growth only *Cryptococcus neoformans* strain, being the only report on the antifungal effect of AuNPs biofabricated by yeasts. In similar approach, Krishnamoorthi et al. (2021) produced gold NPs using an aqueous extract of *A. bisporus* in order to evaluate their antimicrobial potential against human pathogens, including *Candida albicans* and *Aspergillus fumigatus*. In addition to confirming *in vitro* the antimicrobial activity of the particles, the authors verified that AuNPs

Table 2. Antimicrobial properties of myco-synthesized gold nanoparticles against various microorganisms.

Fungal species	Size (nm)	Morphology	Target microorganism	References
Yeast				
Candida albicans	4–10	Spherical and nonspherical	*Staphylococcus aureus* and *Escherichia coli*	Ahmad et al. 2013
Candida guilliermondii	50–70	Spherical	*E. coli, S. aureus, Bacillus cereus, Klebsiella pneumoniae* and *Salmonella* Typhimurium	Mishra et al. 2011a
Hansenula anomala	5–50	Different shapes	*B. cereus* and *Pseudomonas putida*	Sathish Kumar et al. 2011
Phaffia rhodozyma	4–7	Spherical	*Cryptococcus neoformans*	Rónavári et al. 2018
Filamentous fungi				
Aspergillus terreus	10–19	Spherical, elongated, triangular, and rod shaped	*E. coli*	Priyadarshini et al. 2014
Bipolaris tetramera	58–261	Spherical, triangular and hexagonal	*Bacillus subtilis, B. cereus, S. aureus, E. coli* and *Enterobacter aerogenes*	Fatima et al. 2015
Cladosporium cladosporioides	30–60	-	*S. aureus, Staphylococcus epidermidis, B. subtilis, E. coli* and *Aspergillus niger*	Hulikere et al. 2017
Fusarium oxysporum	22–30	Spherical and hexagonal	*S. aureus, B. cereus, E. coli* and *Pseudomonas aeruginosa,* including multi-drug resistant strains	Naimi-Shamel et al. 2019
F. oxysporum f. sp. *cubense*	22	-	*Pseudomonas* sp.	Thakker et al. 2012
Trichoderma hamatum	5–30	Spherical, pentagonal and hexagonal	*B. subtilis, S. aureus, P. aeruginosa* and *Serratia* sp.	Abdel-Kareem and Zohri 2018
Trichoderma viride	59	Spherical	*Pseudomonas syringae, E. coli* and *Shigella sonnei*	Mishra et al. 2014
Trichoderma harzianum	32–44	Spherical	*E. coli*	Tripathi et al. 2018

Biosynthesis of Gold Nanoparticles by Fungi 163

Macroscopic fungi				
Agaricus bisporus	25	Spherical	*Aspergillus flavus*	Eskandari-Nojedehi et al. 2018
A. bisporus	10–50	Spherical	*Streptococcus pyogenes, S. aureus, K. pneumoniae, Shigella flexneri, C. albicans* and *Aspergillus fumigatus*	Krishnamoorthi et al. 2021
Inonotus obliquus	23	Spherical, triangle, hexagonal, and rod shaped	*B. subtilis, S. aureus* and *E. coli*	Lee et al. 2015
Lignosus rhinocerotis	49–125	Spherical, irregular, triangular, pentagonal, and rod shape	*S. aureus, Bacillus sp., P. aeruginosa* and *E. coli*	Katas et al. 2019
Tricholoma crassum	5–25	Rhomboid, spherical, hexagons, cuboids, triangular (isosceles and near-equilateral triangles)	*E. coli, Agrobacterium tumefaciens, Magnaporthe oryzae* and *Alternaria solani*	Basu et al. 2018

had a greater effect on *C. albicans* growth (28 mm inhibition halo in the agar well diffusion assay) and the lowest concentration necessary to inhibit the *A. fumigatus* growth when compared to the others target microorganisms.

Despite the great potential of AuNPs as an antiviral agent, there are no reports on the applications of mycogenic AuNPs for defense against viruses. The ability of AuNPs to inactivate viruses as well as inhibit viral entry and replication have attracted interest in combating viral pathogens by these nanoparticles (Vijayakumar and Ganesan 2012; Rafiei et al. 2016; Meléndez-Villanueva et al. 2019; Babaei et al. 2021). Furthermore, the use of AuNPs as carriers of antigens/protein for the development of nano-based vaccines have been shown to be promising. Sekimukai et al. (2020) have achieved positive preclinical outcomes when AuNPs were evaluated as an antigen carrier (SARS-CoV spike protein) inducing a strong antigen-specific IgG response against the coronavirus infection. Pramanik et al. (2021) proved the blocking viral replication and SARS-CoV-2 spread in HEK293T cells by anti-spike antibody attached gold nanoparticles.

Other applications

It is worth noting that AuNPs may have other very useful properties. Fungi can be used for biological synthesis of valuable gold nanostructures for microelectronics applications (Kashyap et al. 2013). The efficient synthesis of gold nanowires can be achieved through biological methods using *Aspergillus* or *Neurospora* species. In addition, quantum dots are spherical fluorescent nanocrystals composed of semiconductor materials that connect the gap between individual atoms and solid bulk semiconductors. Gold quantum dots have unique optical and electronic properties that can be valuable for photoluminescence, fluorescence and microelectronics applications. These gold nanostructures are very useful for miniaturization of electronic devices and their manufacture in large scale would be very relevant in the field of microelectronics.

Moreover, AuNPs obtained by biosynthesis with plant extracts can exert anti-inflammatory and antidiabetic effects (Mikhailova 2021). The antioxidant potential of AuNPs has been described as well (Kinoshita et al. 2021). *In vitro* assessment of the free radical scavenging activity of AuNPs produced using plant extracts or microorganisms revealed this antioxidant potential. Most studies have been developed with plant extracts, and the antioxidant activity of plant extract influences in the formation rate and size of resulting AuNPs (Stozhko et al. 2019). In several cases, different biomolecules coated on the surface of AuNPs could be related with the improved antioxidant activity (Boomi et al. 2020; Oueslati et al. 2020). Studies on these emergent biological activities of AuNPs are essentially related with the use of plant extracts for the biosynthesis, and therefore, the potential of fungi to generate gold nanostructures with improved or even unknown activities must be explored.

Conclusions

The green synthesis of nanomaterials through biological methods has been gaining importance as an environmental friendly approach to obtain useful metallic nanoparticles. The utilization of fungi as the biological agents has proven to be an effective way to obtain gold nanoparticles, since several species can produce gold-based nanomaterials. The modulation of biosynthetic conditions can result in gold nanostructures with different sizes, morphologies and physico-chemical characteristics, which enable them to be used in many different applications. These include anticancer and antimicrobial therapies, imaging and diagnosis. Despite the optimization of process parameters offering significant advances in gold nanostructures biosynthesis, further research is necessary to solve technical difficulties for the controllable assembly of morphology and size, obtaining monodispersed nanoparticles, scaling-up processes, and determination of the biomolecules effectively responsible for biosynthesis. These are still major challenges in the field of myconanotechnology.

References

Abdel-Kareem, M.M., and Zohri, A.A. 2018. Extracellular mycosynthesis of gold nanoparticles using *Trichoderma hamatum*: optimization, characterization and antimicrobial activity. Lett. Appl. Microbiol. 67: 465–475.

Abu-Tahon, M.A., Ghareib, M., and Abdallah, W.E. 2020. Environmentally benign rapid biosynthesis of extracellular gold nanoparticles using *Aspergillus flavus* and their cytotoxic and catalytic activities. Process Biochem. 95: 1–11.

Agnihotri, S., Mukherji, S., and Mukherji, S. 2014. Size-controlled silver nanoparticles synthesized over the range 5–100 nm using the same protocol and their antibacterial efficacy. RSC Adv. 4: 3974–3983.

Ahmad Siddiqui, E., Ahmad, A., Julius, A., Syed, A., Khan, S., Kharat, M., Pai, K., Kadoo, N., and Gupta, V. 2016. Biosynthesis of anti-proliferative gold nanoparticles using endophytic *Fusarium oxysporum* strain isolated from neem (*A. indica*) leaves. Curr. Top. Med. Chem. 16: 2036–2042.

Ahmad T., Wania, I.A., Lone I.H., Ganguly, A., Manzoorb, N., Ahmad, A., Ahmedc, J., and Al-Shihri, A.S. 2013. Antifungal activity of gold nanoparticles prepared by solvothermal method. Mater Res. Bull. 48: 12–20.

Aleklett, K., and Boddy, L. 2021. Fungal behaviour: a new frontier in behavioural ecology. Trends Ecol. Evolut. 36: 787–796.

Attia, Y.A., Farag, Y.E., Mohamed, Y.M.A., Hussien, A.T., and Youssef, T. 2016. Photo-extracellular synthesis of gold nanoparticles using Baker's yeast and their anticancer evaluation against Ehrlich ascites carcinoma cells. New J. Chem. 40: 9395–9402.

Babaei, A., Mousavi, S.M., Ghasemi, M., Pirbonyeh, N., Soleimani, M., and Moattari, A. 2021. Gold nanoparticles show potential in vitro antiviral and anticancer activity. Life Sci. 284: 119652.

Banerjee, K., and Vittal, R.R. 2017. *Aspergillus fischeri* mediated biosynthesis of gold nanoparticles and their beneficially comparative effect on normal and cancer cell lines. Pharm. Nanotechnol. 5: 220–229.

Baram-Pinto, D., Shukla, N., Gedanken, A., and Sarid, R. 2010. Inhibition of HSV-1 attachment, entry, and cell-to- cell spread by functionalized multivalent gold nanoparticles. Small 6: 1044–1050.

Basu, A., Ray, S., Chowdhury, S., Sarkar, A., Mandal, D.P., Bhattacharjee, S., and Kundu, S. 2018. Evaluating the antimicrobial, apoptotic, and cancer cell gene delivery properties of protein-capped gold nanoparticles synthesized from the edible mycorrhizal fungus *Tricholoma crissum*. Nanoscale Res. Lett. 13: 154.

Bhardwaj, K., Sharma, A., Tejwan, N., Bhardwaj, S., Bhardwaj, P., Nepovimova, E., Shami, A., Kalia, A., Kumar, A., Abd-Elsalam, K.A., and Kuca, K. 2020. *Pleurotus* macrofungi-assisted nanoparticle synthesis and its potential applications: a review. J. Fungi (Basel). 6: 351.

Bhat, R., Sharanabasava, V.G., Deshpande, R., Shetti, U., Sanjeev, G., and Venkataraman, A. 2013. Photo-bio-synthesis of irregular shaped functionalized gold nanoparticles using edible mushroom *Pleurotus florida* and its anticancer evaluation. J. Photochem. Photobiol. B. 125: 63–69.

166 *Mycosynthesis of Nanomaterials: Perspectives and Challenges*

Boisselier, E., and Astruc, D. 2009. Gold nanoparticles in nanomedicine: preparations, imaging, diagnostics, therapies and toxicity. Chem. Soc. Rev. 38: 1759–1782.

Boomi, P., Ganesan, R., Prabu Poorani, G., Jegatheeswaran, S., Balakumar, C., Gurumallesh Prabu, H., Anand, K., Marimuthu Prabhu, N., Jeyakanthan, J., and Saravanan, M. 2020. Phyto-engineered gold nanoparticles (AuNPs) with potential antibacterial, antioxidant, and wound healing activities under *in vitro* and *in vivo* conditions. Int. J. Nanomedicine. 15: 7553–7568.

Brown, C.L., Bushell, G., Whitehouse, M.W., Agrawal, D.S., Tupe, S.G., Paknikar, K.M., and Tiekink, E.R.T. 2007. Nanogold-pharmaceutics: (i) The use of colloidal gold to treat experimentally-induced arthritis in rat models; (ii) Characterization of the gold in *Swarna bhasma*, a microparticulate used in traditional Indian medicine. Gold Bull. 40: 245–250.

Brust, M., Walker, M., Bethell, D., Schiffrin, D.J., and Whyman, R. 1994. Synthesis of thiol-derivatised gold nanoparticles in a two-phase liquid-liquid system. J. Chem. Soc. Chem. Commun. 7: 801–802.

Chaturvedi, V.K., Yadav, N., Rai, N.K., Ellah, N.H.A., Bohara, R.A., Rehan, I.F., Marraiki, N., Batiha, G.E-S., Hetta, H.F., and Singh, M.P. 2020. *Pleurotus sajor-caju*-mediated synthesis of silver and gold nanoparticles active against colon cancer cell lines: A new era of herbonanoceutics. Molecules. 25: 3091.

Chauhan, A., Zubair, S., Tufail, S., Sherwani, A., Sajid, M., Raman, S.C., Azam, A., and Owais, M. 2011. Fungus-mediated biological synthesis of gold nanoparticles: potential in detection of liver cancer. Int. J. Nanomed. 6: 2305–2319.

Chouhan, N. 2018. Silver nanoparticles: synthesis, characterization and applications. pp. 21–57. *In*: Maaz, K. (ed.). Silver Nanoparticles - Fabrication, Characterization and Applications. London: IntechOpen.

Clarance, P., Luvankar, B., Sales, J., Khusro, A., Agastian, P., Tack, J.C., Al-Khulaifi, M.M., Al-Shwaiman, H.A., Elgorban A.M., Syed, A., and Kim, H.J. 2020. Green synthesis and characterization of gold nanoparticles using endophytic fungi *Fusarium solani* and its *in-vitro* anticancer and biomedical applications. Saudi J. Biol. Sci. 27: 706–712.

Daniel, M.C., and Astruc, D. 2004. Gold nanoparticles: Assembly, supramolecular chemistry, quantum-size-related properties, and applications toward biology catalysis, and nanotechnology. Chem. Rev. 104: 293–346.

Dykman, L.A., and Khlebtsov, N.G. 2011. Gold nanoparticles in biology and medicine: Recent advances and prospects. Acta Nat. 3: 34–55.

El-Bendary, M.A., Moharam, M.E., Hamed, S.R., Khalil, S.K.H., Elkomy, G.M., and Mounier, M.M. 2018. Myco-synthesis of gold nanoparticles using *Aspergillus flavus*: Characterization, optimization and cytotoxic activity. Curr. Trends Microbiol. 12: 67–79.

El Enshasy, H.A. 2022. Fungal morphology: a challenge in bioprocess engineering industries for product development. Curr. Op. Chem. Eng. 1: 100729.

Elegbede, J.A., Lateef, A., Azeez, M.A., Asafa, T.B., Yekeen, T.A., Oladipo, I.C., Aina, D.A., Beukes, L.S., and Gueguim-Kana, E.B. 2020. Biofabrication of gold nanoparticles using xylanases through valorization of corncob by *Aspergillus niger* and *Trichoderma longibrachiatum*: Antimicrobial, antioxidant, anticoagulant and thrombolytic activities. Waste Biomass Valor. 11: 781–791.

Elkes, C.C., Busuioc, G., and Ionita, G. 2010. The bioaccumulation of some heavy metals in the fruiting body of wild growing mushrooms. Notulae Botanicae Horti Agrobotanici Cluj-Napoca 38: 147–151.

Eskandari-Nojehdehi, M., Jafarizadeh-Malmiri, H., and Rahbar-Shahrouzi, J. 2016. Optimization of processing parameters in green synthesis of gold nanoparticles using microwave and edible mushroom (*Agaricus bisporus*) extract and evaluation of their antibacterial activity. Nanotechnol. Rev. 5: 537and548.

Eskandari-Nojedheri, M., Jafarizadeh-Malmiri, H., and Rahbar-Shahrouzi, J. 2018. Hydrothermal green synthesis of gold nanoparticles using mushroom (*Agaricus bisporus*) extract: physico-chemical characteristics and antifungal activity studies. Green Process Synth. 7: 38and47.

Espinosa-Diez, C., Miguel, V., Mennerich, D., Kietzmann, T., Sánchez-Pérez, P., Cadenas, S., and Lamas, S. 2015. Antioxidant responses and cellular adjustments to oxidative stress. Redox Biol. 6: 183–197.

Fatima, F., Bajpai, P., Pathak, N., Singh, S., Priya, S., and Verma, S.R. 2015. Antimicrobial and immunomodulatory efficacy of extracellularly synthesized silver and gold nanoparticles by a novel phosphate solubilizing fungus *Bipolaris tetramera*. BMC Microbiol. 15: 52.

Gnat, S., Łagowski, D., Nowakiewicz A., Dylazg M. 2021. A global view on fungal infections in humans and animals: infections caused by dimorphic fungi and dermatophytoses. J. Appl. Microbiol. 131: 2688–2704.

Ghodake, G.S., Deshpande, N.G., Lee, Y.P., and Jin, E.S. 2010. Pear fruit extract-assisted room-temperature biosynthesis of gold nanoplates. Colloids Surf B. 75: 584–589.

Gholami-Shabani, M., Sotoodehnejadnematalahi, F., Shams-Ghahfarokhi, M., Eslamifar, A., and Razzaghi-Abyaneh, M. 2022. Physicochemical properties, anticancer and antimicrobial activities of metallic nanoparticles green synthesized by *Aspergillus kambarensis*. IET Nanobiotechnol 16: 1–13.

Goswami, A., and Ghosh, S. 2013. Biological synthesis of colloidal gold nanoprisms using *Penicillium citrinum* MTCC9999. J. Biomater. Nanobiotechnol. 4: 20–27.

Hammami, I., Alabdallah, N.M., Al Jomaa, A., and Kamoun, M. 2021. Gold nanoparticles: Synthesis properties and applications. J. King Saud Univ. Sci. 33: 101560.

Hanus, M.J., and Harris, A.T. 2013. Nanotechnology innovations for the construction industry. Prog. Mater Sci. 58: 1056–1102.

He, H., Xie, C., and Ren, J. 2008. Nonbleaching fluorescence of gold nanoparticles and its applications in cancer cell imaging. Anal Chem. 80: 5951–5957.

Hu, X., Zhang, Y., Ding, T., Liu, J., and Zhao, H. 2020. Multifunctional gold nanoparticles: A novel nanomaterial for various medical applications and biological activities. Front Bioeng. Biotechnol. 8: 990.

Hulikere, M.M., Joshi, C.G., Danagoudar, A., Poyya, J., Kudva, A.K., and Dhananjaya, B.L. 2017. Biogenic synthesis of gold nanoparticles by marine endophytic fungus *Cladosporium cladosporioides* isolated from seaweed and evaluation of their antioxidant and antimicrobial properties. Process Biochem. 63: 137–144.

Ismail, E.H., Saqer, A.M.A., Assirey, E., Naqvi, A., and Okasha, R.M. 2018. Successful green synthesis of gold nanoparticles using a *Corchorus olitorius* extract and their antiproliferative effect in cancer cells. Int. J. Mol. Sci. 19: 2612.

Jain, S., Hirst, D.G., and O'Sullivan, J.M. 2012. Gold nanoparticles as novel agents for cancer therapy. Br. J. Radiol. 85: 101–113.

Jakobs, R.T.M., van Herrikhuyzen, J., Gielen, J.C., Christianen, P.C.M., Meskers, S.C.J., and Schenning, A.P.H.J. 2008. Self-assembly of amphiphilic gold nanoparticles decorated with a mixed shell of oligo(*p*-phenylene vinylene)s and ethyleneoxide ligands. J. Mater Chem. 18: 3438–3441.

Jelveh, S., and Chithrani, D.B. 2011. Gold nanostructures as a platform for combinational therapy in future cancer therapeutics. Cancers. 3: 1081–1110.

Kajani, A.A., Bordbar, A.-K., Esfahani, S.H.Z., and Razmjou, A. 2016. Gold nanoparticles as potent anticancer agent: Green synthesis, characterization, and *in vitro* study. RSC Adv. 6: 63973–63983.

Kapahi, M., and Sachdeva, S. 2017. Mycoremediation potential of *Pleurotus* species for heavy metals: A review. Bioresour. Bioprocess. 4: 32.

Kashyap, P.L., Kumar, S., Srivastava, A.K., and Sharma, A.K. 2013. Myconanotechnology in agriculture: a perspective. World J. Microbiol. Biotechnol. 29: 191–207.

Katas, H., Lim, C.S., Azlan, A.Y.H.N., Buang, F., and Busra, M.F.M. 2019. Antibacterial activity of biosynthesized gold nanoparticles using biomolecules from *Lignosus rhinocerotis* and chitosan. Saudi Pharm J. 27: 283–292.

Ke, Y., Al Aboody, M.S., Alturaiki, W., Alsagaby, S.A., Alfaiz, F.A., Veeraraghavan, V.P., and Mickymaray, S. 2019. Photosynthesized gold nanoparticles from *Catharanthus roseus* induces caspase-mediated apoptosis in cervical cancer cells (HeLa). Artif Cells Nanomed. Biotechnol. 47: 1938–1946.

Kimling, J., Maier, M., Okenve, B., Kotaidis, V., Ballot, H., and Plech, A. 2006. Turkevich method for gold nanoparticle synthesis revisited. J. Phys. Chem. B. 110: 15700–15707.

Kinoshita, A., Lima, I., Guidelli, E.J., and Baffa Filho, O. 2021. Antioxidative activity of gold and platinum nanoparticles assessed through electron spin ressonance. Eclet Quim. 46: 68–74.

Krishnan, S., Narayan, S., and Chadha, A. 2016. Whole resting cells vs. cell free extracts of *Candida parapsilosis* ATCC 7330 for the synthesis of gold nanoparticles. AMB Expr 6: 92.

Krishnamoorthy, R., Bharathakumar, S., Malaikozhundan, B., and Mahalingam, P.U. 2021. Mycofabrication of gold nanoparticles: Optimization, characterization, stabilization and evaluation of its antimicrobial potential on selected human pathogens. Biocatal. Agric. Biotechnol. 35: 102107.

168 *Mycosynthesis of Nanomaterials: Perspectives and Challenges*

Lee, S.M., and Jun, B.-H. 2019. Silver nanoparticles: Synthesis and application for nanomedicine. Int. J. Mol. Sci. 20: 865.

Lee, K.D., Nagajyothi, P.C., Sreekanth, T.V.M., and Park, S., 2015. Eco-friendly synthesis of gold nanoparticles (AuNPs) using *Inonotus obliquus* and their antibacterial, antioxidant and cytotoxic activities. J. Ind. Eng. Chem. 26: 67–72.

Lee, K.X., Shameli, K., Yew, Y.P., Teow, S.Y., Jahangirian, H., Rafiee-Moghaddam, R., and Webster, T.J. 2020. Recent developments in the facile bio-synthesis of gold nanoparticles (AuNPs) and their biomedical applications. Int. J. Nanomed. 15: 275–300.

Lew, T.T.S., Aung, K.M.M., Ow, S.Y., Amrun, S.N., Sutarlie, L., Ng, L.F.P., and Su, X. 2021. Epitope-functionalized gold nanoparticles for rapid and selective detection of SARS-CoV-2 IgG antibodies. ACS Nano 15: 12286–12297.

Liu, G., Lu, M., Huang, X., Li, T., and Xu, D. 2018. Application of gold-nanoparticle colorimetric sensing to rapid food safety screening. Sensors 18: 4166.

Liu, R., Pei, Q., Shou, T., Zhang, W., Hu, J., and Li, W. 2019. Apoptotic effect of green synthesized gold nanoparticles from *Curcuma wenyujin* extract against human renal cell carcinoma A498 cells. Int. J. Nanomed. 4: 4091–4103.

Loshchinina, E.A., Vetchinkina, E.P., Kupryashina, M.A., Kursky, V.F., and Nikitina, V.E. 2018. Nanoparticles synthesis by *Agaricus* soil basidiomycetes. J. Biosci. Bioeng. 126: 44–52.

Manjunath, H.M., Joshi, C.G., and Raju, N.G. 2017. Biofabrication of gold nanoparticles using marine endophytic fungus - *Penicillium citrinum*. IET Nanobiotechnol. 11: 40–44.

Mattanovich, D., Sauer, M., and Gasser, B. 2014. Yeast biotechnology: Teaching the old dog new tricks. Microb. Cell Fact. 13: 34.

Maliszewska, I. 2013. Microbial mediated synthesis of gold nanoparticles: Preparation, characterization and cytotoxicity studies. Dig. J. Nanomater Biostruct. 8: 1123–1131.

Maliszewska, I., Aniszkiewicz, Ł., and Sadowski, Z. 2009. Biological synthesis of gold nanostructures using the extract of *Trichoderma koningii*. Acta Phys. Polonica A. 116: S163–S165.

Meléndez-Villanueva, M.A., Morán-Santibañez, K., Martínez-Sanmiguel, J.J., Rangel-López, R., Garza-Navarro, M.A., Rodríguez-Padilla, C., Zarate-Triviño, D.G., and Trejo-Ávila L.M. 2019. Virucidal activity of gold nanoparticles synthesized by green chemistry using garlic extract. Viruses 11: 1111.

Mikhailova, E.O. 2021. Gold nanoparticles: biosynthesis and potential of biomedical application. J. Funct. Biomater. 12: 70.

Mishra, A., Kumari, M., Pandey, S., Chaudhry, V., Gupta, K.C., and Nautiyal, C.S. 2014. Biocatalytic and antimicrobial activities of gold nanoparticles synthesized by *Trichoderma* sp. Bioresour. Technol. 166: 235–242.

Mishra, A., Tripathy, S.K., and Yun, S.-I. 2011a. Bio-s of gold and silver nanoparticles from *Candida guilliermondii* and their antimicrobial effect against pathogenic bacteria. J. Nanosci. Nanotechnol. 11: 243–248.

Mishra, A., Tripathy, S.K., Wahab, R., Jeong, S.-H., Hwang, I., Yang, Y.-B., Kim, Y.-S., Shin, H.-S., and Yun, S.-I. 2011b. Microbial synthesis of gold nanoparticles using the fungus *Penicillium brevicompactum* and their cytotoxic effects against mouse Mayo Blast Cancer C2C12 cells. Appl. Microbiol. Biotechnol. 92: 617–630.

Molnár, Z., Bódai, V., Szakacs, G., Erdély, B., Fogarassy, Z., Sáfrán, G., Varga, T., Kónya, Z., Tóth-Szeles, E., Szűcs, and R., Lagzi, I. 2018. Green synthesis of gold nanoparticles by thermophilic filamentous fungi. Sci. Rep. 8: 3943.

Mourato, A., Gadanho, M., Lino, A.R., and Tenreiro, R. 2011. Biosynthesis of crystalline silver and gold nanoparticles by extremophilic yeasts. Bioinorg. Chem. Appl. 2011: 546074.

Munawer, U., Raghavendra, V.B., Ningaraju, S., Krishna, K.L., Ghosh, A.R., Melappa, G., and Pugazhendhi, A. 2020. Biofabrication of gold nanoparticles mediated by the endophytic *Cladosporium* species: Photodegradation, *in vitro* anticancer activity and *in vivo* antitumor studies. Int. J. Pharm. 588: 119729.

Naeem, G.A., Jaloot, A.S., Owaid, M.N., and Muslim, R.F. 2021. Green synthesis of gold nanoparticles from *Coprinus comatus*, Agaricaceae, and the effect of ultraviolet irradiation on their characteristics. Walailak J. Sci. Technol. 18: 9396.

Biosynthesis of Gold Nanoparticles by Fungi 169

Naimi-Shamel, N., Pourali, P., and Dolatabadi, S. 2019. Green synthesis of gold nanoparticles using Fusarium oxysporum and antibacterial activity of its tetracycline conjugant. J. Mycol. Méd. 29: 7–13.

Norn, S., Permin, H., Kruse, P.R., and Kruse, E. 2011. Guld i medicinens tjeneste-Med traek af dansk indsats ved tuberkulose og reumatoid artrit [History of gold-with danish contribution to tuberculosis and rheumatoid arthritis]. Dan Medicinhist Arbog. 39: 59–80.

Oueslati, M.H., Tahar, L.B., and Harrath, A.H. 2020. Catalytic, antioxidant and anticancer activities of gold nanoparticles synthesized by kaempferol glucoside from *Lotus leguminosae*. Arab. J. Chem. 13: 3112–3122.

Paidari, S., and Ibrahim, S.A. 2021. Potential application of gold nanoparticles in food packaging: A mini review. Gold Bull. 54: 31–36.

Pócsi, I. 2011. Toxic metal/metalloid tolerance in fungi—a biotechnology-oriented approach. pp. 31–58. *In*: Bánfalvi, G. (ed.). *Cellular Effects of Heavy Metals*. Springer Science+Business Media B.V., Dordrecht.

Pramanik, A., Gao, Y., Pativandla, S., Mitra, D., McCandless, M.G., Fassero, L.A., Gates, K., Tandon, R., and Ray, P.C. 2021. The rapid diagnosis and effective inhibition of coronavirus using spike antibody attached gold nanoparticles. Nanoscale Adv. 3: 1588–1596.

Priyadarshini, E., Pradhan, N., Sukla, L.B., and Panda, P.K. 2014. Controlled synthesis of gold nanoparticles using *Aspergillus terreus* IF0 and its antibacterial potential against Gram negative pathogenic bacteria. J. Nanotechnol. 2014: 653198.

Priyadarshini, E., Pradhan, N., Sukla, L.B., Panda, P.K., and Mishra, B.K. 2014b. Biogenic synthesis of floral-shaped gold nanoparticles using a novel strain, *Talaromyces flavus*. Ann. Microbiol. 64: 1055–1063.

Qamar, S.U.R., and Ahmad, J.N. 2021. Nanoparticles: Mechanism of biosynthesis using plant extracts, bacteria, fungi, and their applications. J. Mol. Liq. 334: 116040.

Rai, M., Yadav, A., Bridge, P., and Gade, A. 2009. Myconanotechnology: A new and emerging science. pp. 258–267. *In*: Rai, M., and Bridge, P.D. (eds.) *Applied Mycology*. CAB International, Oxfordshire UK.

Rai, M., Ingle, A.P., Gupta, I., and Brandelli, A. 2015. Bioactivity of noble metal nanoparticles decorated with biopolymers and their application in drug delivery. Int. J. Pharm. 496: 159–172.

Rajeshkumar, S. 2016. Anticancer activity of eco-friendly gold nanoparticles against lung and liver cancer cells. J. Genet. Eng. Biotechnol. 14: 195–202.

Rafiei, S., Rezatofighi, S.E., Roayaei Ardakani, M., and Rastegarzadeh, S. 2016. Gold nanoparticles impair foot-and-mouth disease virus replication. IEEE Trans. Nanobiosci. 15: 34–40.

Rónavári, A., Igaz, N., Gopisetty, M. K., Szerencsés, B., Kovács, D., Papp, C., Vágvölgyi, C., Boros, I. M., Kónya, Z., Kiricsi, M., and Pfeiffer, I. 2018. Biosynthesized silver and gold nanoparticles are potent antimycotics against opportunistic pathogenic yeasts and dermatophytes. Int. J. Nanomed. 13: 695–703.

Roy, S., Das, T.K., Maiti, G.P., and Basu, U. 2016. Microbial biosynthesis of nontoxic gold nanoparticles. Mater Sci. Eng. B. 203: 41–51.

Sarkar, S., and Kotteeswaran, V. 2018. Green synthesis of silver nanoparticles from aqueous leaf extract of pomegranate (*Punica granatum*) and their anticancer activity on human cervical cancer cells. Adv. Nat. Sci: Nanosci. Nanotechnol. 9: 025014.

Sathish Kumar, K., Amutha, R., Arumugam, P., and Berchmans, S. 2011. Synthesis of gold nanoparticles: an ecofriendly approach using *Hansenula anomala*. ACS Appl. Mater Interfaces. 3: 1418–1425.

Sathish Kumar, M., Pavagadhi, S., Mahadevan, A., and Balasubramanian, R. 2015. Biosynthesis of gold nanoparticles and related cytotoxicity evaluation using A549cells. Ecotoxicol. Environ. Saf. 114: 232–240.

Sen, K., Sinha, P., and Lahiri, S. 2011. Time dependent formation of gold nanoparticles in yeast cells: A comparative study. Biochem. Eng. J. 55: 1–6.

Sen, I.K., Maity, K., and Islam, S.S. 2013. Green synthesis of gold nanoparticles using a glucan of an edible mushroom and study of catalytic activity. Carbohydr. Polym. 91: 518–528.

Sekimukai, H., Iwata-Yoshikawa, N., Fukushi, S., Tani, H., Kataoka, M., Suzuki, T., Hasegawa, H., Niikura, K., Arai, K., and Noriyo, N. 2020. Gold nanoparticle-adjuvanted S protein induces a strong antigen-specific IgG response against severe acute respiratory syndrome-related coronavirus

170 Mycosynthesis of Nanomaterials: Perspectives and Challenges

infection, but fails to induce protective antibodies and limit eosinophilic infiltration in lungs. Microbiol. Immunol. 64: 33–51.

Shao, D., Smith, D.L., Kabbage, M., and Roth M.G. 2021. Effectors of plant necrotrophic fungi. Front Plant Sci. 12: 687713.

Sheikhloo, Z., and Salouti, M. 2011. Intracellular biosynthesis of gold nanoparticles by the fungus *Penicillium chrysogenum*. Int. J. Nanosci. Nanotechnol. 7: 102–105.

Shunmugam, R., Balusamy, S.R., Kumar, V., Menon, S., Lakshmi, T., and Perumalsamy, H. 2021. Biosynthesis of gold nanoparticles using marine microbe (*Vibrio alginolyticus*) and its anticancer and antioxidant analysis. J. King Saudi. Univ. Sci. 33: 101260.

Sibuyi, N.R.S., Moabelo, K.L., Fadaka, A.O., Meyer, S., Onani, M.O., Madiehe, A.M., and Meyer, M. 2021. Multifunctional gold nanoparticles for improved diagnostic and therapeutic applications: A review. Nanoscale Res. Lett. 16: 174.

Singh, P., Kim, Y.-J., Zhang, D., and Yang, D.-C. 2016. Biological synthesis of nanoparticles from plants and microorganisms. Trends Biotechnol. 34: 588–599.

Singh, P., Pandit, S., Mokkapati, V.R.S.S., Garg, A., Ravikumar, V., and Mijakovic, I. 2018. Gold nanoparticles in diagnostics and therapeutics for human cancer. Int. J. Mol. Sci. 19: 1979.

Slepicka, P., Kasálková, N.S., Siegel, J., Kolská, Z., and Švorčík, V. 2020. Methods of gold and silver nanoparticles preparation. Materials. 13: 1.

Sojinrin, T., Conde, J., Liu, K., Curtin, J., Byrne, H., Cui, D., and Tian, F. 2017. Plasmonic gold nanoparticle for detection of fungi and human cutaneous fungal infections. Anal. Bioanal. Chem. 409: 4647–4658.

Soni, N., and Prakash, S. 2012. Synthesis of gold nanoparticles by the fungus *Aspergillus niger* and its efficacy against mosquito larvae. Rep. Parasitol. 2: 1–7.

Sperling, R.A., and Park, W.J. 2010. Surface modification, functionalization and bioconjugation of colloidal inorganic nanoparticles. Phil. Trans. Royal. Soc. A. 368: 1333–1383.

Stozhko, N.Y., Bukharinova, M.A., Khamzina, E.I., Tarasov, A.V., Vidrevich, M.B., and Brainina, K.Z. 2019. The effect of the antioxidant activity of plant extracts on the properties of gold nanoparticles. Nanomaterials. 9: 1655.

Sugiharto, S. 2019. A review of filamentous fungi in broiler production. Ann. Agric. Sci. 64: 1–8.

Syed, A., Raja, R., Kundu, G.C., Gambhir, S., and Ahmad, A. 2012. Extracellular biosynthesis of monodispersed gold nanoparticles, their characterization, cytotoxicity assay, biodistribution and conjugation with the anticancer drug doxorubicin. J. Nanomed. Nanotechnol. 44: 123–131.

Tahar, I.B., Fickers, P., Dziedzic, A., Płoch, D., Skóra, B., and Kus-Liskiewicz, M. 2019. Green pyomelanin-mediated synthesis of gold nanoparticles: modelling and design, physico-chemical and biological characteristics. Microb. Cell Fact. 18: 210.

Thakker, J.N., Dalwadi, P., and Dhandhukia, P.C. 2012. Biosynthesis of gold nanoparticles using *Fusarium oxysporum* f. sp. *cubense* JT1, a plant pathogenic fungus. ISRN Biotechnol. 2013: 515091.

Tripathi, R.M., Shrivastav, B.R., and Shrivastav, A. 2018. Antibacterial and catalytic activity of biogenic gold nanoparticles synthesised by *Trichoderma harzianum*. IET Nanobiotechnol. 12: 509–513.

Verma, V.C., Singh, S.K., Solanki, R., and Prakash, S. 2011. Biofabrication of anisotropic gold nanotriangles using extract of endophytic *Aspergillus clavatus* as a dual functional reductant and stabilizer. Nanoscale Res. Lett. 6: 16.

Vijayakumar, S., and Ganesan, S. 2012. Gold nanoparticles as an HIV entry inhibitor. Curr. HIV Res. 10: 643–646.

Wang, L., Xu, J., Yan, Y., Liu, H., Karunakaran, T., and Li, F. 2019. Green synthesis of gold nanoparticles from *Scutellaria barbata* and its anticancer activity in pancreatic cancer cell (PANC-1). Artif Cells Nanomed. Biotechnol. 47: 1617–1627.

Wu, Y., Ali, M.R.K., Chen, K., Fang, N., and El-Sayed, M.A. 2019. Gold nanoparticles in biological optical imaging. Nanoday. 24: 120–140.

Xiao, L., and Yeung, E.S. 2014. Optical imaging of individual plasmonic nanoparticles in biological samples. Annu. Rev. Anal. Chem. 7: 89–111.

Yang, Z., Li, Z., Lu, X., He, F., Zhu, X., Ma, Y., He, R., Gao, F., Ni, W., and Yi, Y. 2017. Controllable biosynthesis and properties of gold nanoplates using yeast extract. Nano-Micro Lett. 9: 5.

Yoo, M., Bang, J., Paek, K., and Kim, B.J. 2011. A strategy to decorate the surface of NPs and control their locations within block copolymer templates. *In*: Reddy, B. (ed.). Advances in Diverse Industrial Applications of Nanocomposites. Intech Open, London, UK.

Zhang, D., Ma, X., Gu, Y., Huang, H., and Zhang, G. 2020. Green synthesis of metallic nanoparticles and their potential applications to treat cancer. Front Chem. 8: 799.

Zhang, X., Qu, Y., Shen, W., Wang, J., Li, H., Zhang, Z., Li, S., and Zhou, J. 2016. Biogenic synthesis of gold nanoparticles by yeast *Magnusiomyces ingens* LH-F1 for catalytic reduction of nitrophenols. Colloid Surfaces A. 497: 280–285.

Zhao, C., Zhong, G., Kim, D.E., Liu, J., and Liu, X. 2014. A portable lab-on-a-chip system for gold-nanoparticle-based colorimetric detection of metal ions in water. Biomicrofluidics. 8: 052107.

CHAPTER 9

Biosynthesis of Zinc Oxide Nanoparticles and Major Applications

Shrutika Chaudhary,[1] Saurabh Shivalkar[2] and Amaresh Kumar Sahoo[2,]*

Introduction

In the past few decades, zinc oxide nanoparticles (ZnO NPs) have shown their potential for different purposes ranging from biomedical applications to optoelectronics. Hence, the large-scale production of the ZnO NPs is one of the most demanding areas of material sciences, which has recently been explored by various research groups globally. Moreover, recent developments in the chemistry and material sciences provide the flexibility to synthesize user-defined NPs' by various chemical and physical methods. However, there are still certain drawbacks related to the chemical synthesis and applications of ZnO NPs like stability in biological fluid and toxicity. Optimization of the dose and physicochemical characteristics of NPs play very crucial roles in their toxicity. Thus, several improvements are essential to remove the shortcomings of the as-synthesized NP during the fabrication and optimization of physicochemical characteristics of the NPs. To address these concurrent issues, extensive work has been reported on the synthesis of ZnO NPs using biosynthesis routes. Likewise, plant-based methods involving roots, leaves, fruit, stem, flower and microorganism-based methods including bacteria and fungi cells have been employed as 'nano-factory' for the synthesis of NPs. For example, *M. oliefera* leaf extracts were used to synthesize nanoparticles to reduce large-scale water contamination (Gautam et al. 2020). Both bacteria and fungi-based

---

[1] Department of Biotechnology, Delhi Technological University, Delhi.
[2] Department of Applied Sciences, Indian Institute of Information technology, Allahabad.
* Corresponding author: asahoo@iiita.ac.in

nanoparticles are synthesized but fungi have proven to be more beneficial as they can tolerate changes in physiological conditions. Herein, we have mainly focused on the biosynthesis of ZnO NPs by using fungi as a source organism. Fungus cells provide faster growth of biomass that contains huge amounts of metabolites and proteins as compared to other microorganisms including bacteria. Studies also showed that metal ions tolerance is high in the case of the fungi as compared to bacteria, which helps in the easy synthesis of NPs. Further, it provides the scope of amenable processing of biomass and collection of cell-free solution that is suitable for economical large-scale production of NPs. Fungi follow two strategies for the synthesis of nanoparticles—(a) intracellular synthesis and (b) extracellular synthesis. Among these, extracellular synthesis is easier and simple as compared to the intracellular one. Moreover, the extracellular method can easily be applied for large-scale production and also possess economic viability (Agarwal et al. 2017). Fungi produce many primary and secondary metabolites that are found to have several health benefits for humans. They also can produce nanoparticles that can be used for antimicrobial activity, antioxidant activity, and anticancer activity (Abdelhakim et al. 2020). Additionally, fungal-based nanoparticles are not only established as potent bactericidal agents but also employed as effective anti-cancer agents as demonstrated by various recent studies.

A myriad of different types of metal and metal oxide nanoparticles has been synthesized due to huge progress in the material sciences with specific size and surface properties that showed immense prospect in various biomedical applications. In this regard, the low level of toxicity of ZnO NPs has proven to be a much better option for biomedical applications compared with other nanoparticles. Moreover, ZnO NPs possess strong antimicrobial properties against both Gram-positive and Gram-negative bacteria even at a very low concentration, attracting the interest of researchers over these properties. It should be mentioned here that zinc is an important trace element for humans that has multiple biological functions. It's also considered safe for the environment and other animals. Thus, the US Food and Drug Administration (FDA 21CFR182.8991) has marked zinc as "Generally Recognized as Safe" (GRAS). Moreover, it can protect the skin from the harmful UV-light radiation of the sunlight. Thus, now-a-days several cosmetic products and commercial items contain ZnO NPs as an important constituent. Apart from these wide ranges of biomedical usages, it has huge applications in optoelectronics, development of energy storage devices and nano-sensors. The ZnO NPs also act as a photocatalyst, where it could capable of degrading toxic environmental pollutants (Rajan et al. 2016). Therefore, large-scale, biocompatible and economical synthesis of ZnO NPs is 'on-demand' for a long time. Biosynthesis of the ZnO NPs found to be very effective due to the lower level of toxicity as compared to chemical synthesis as well as economical. Various biosynthesis routes have been proposed and demonstrated to be suitable for the large-scale production of ZnO NPs. However, in this chapter, we are primarily focusing on the fungal-mediated synthesis of ZnO NPs and their potential application in different fields with their limitations and future prospects.

Biosynthesis of zinc oxide nanoparticles

The most common and easy way to do the synthesis of the ZnO NPs is the chemical method that offers the scope of a higher yield of production. The monitoring of the size and shape is also becoming more convenient by the chemical route. Additionally, recent progress in this line of interest allows doing surface functionalization by specific ligand/functional groups, which provides the development of a much easier way to monitor its physicochemical properties. However, chemical methods were found to be very critical owing to their generation of toxic by-products, use of costly reagents, and multiple steps involved in the purification of the NPs. In this regard, 'green-synthesis' by using biological routes become very popular and is considered as a viable alternative, which possesses lower toxicity and is economical too.

Plant extract mediated: Plant has always been the best source of food and medicines since the existence of humankind. In the past few decades, it has been used for the synthesis of nanoparticles synthesis. The most common parts used for this purpose are leaves and flowers. Though NPs can also be synthesized using stem, root, fruit or seed, it is much easier to produce NPs from leaves and flowers (Agarwal et al. 2017). The phytochemicals, proteins and vitamins present in plants aid in the stable and large-scale production of pure nanoparticles without any contamination issues (Agarwal et al. 2017). The secondary metabolites present in the plants like saponins, tannins, flavonoids, etc., perform the role of stabilizing and reducing agents (Boroumandmoghaddam et al. 2015). For the synthesis of NPs, the bioactive molecules first separate from the crude plant extract, which contains numerous molecules by using a specific solvent and/or mixture of solvents with proper polarity. Bioactive molecules are then concentrated by evaporating organic solvent(s) before use for the biosynthesis of NPs. The above technique provides the idea that solvent-mediated extract contains more than one plant bioactive molecule, which participates in the synthesis of NPs. Thus, controlling the size and characterization of the NPs by using plant extract is a big challenge. Moreover, these NPs also have issues with aggregation and precipitation. Minute variation of the reaction condition such as pH and temperature play a huge role to control the characteristics of the NPs. In an approach, leaf extracts of *Coriandrum sativum* were used for the ZnO NPs synthesis and were confirmed with the formation of a pale white precipitate (Sabir et al. 2014). Similarly, *Citrus medica L.* fruit peels were used for the ZnO NPs production. The fruit peels were washed using distilled water and supplemented with zinc acetate. For the formation of ZnO NPs, the mixture was agitated for 3 hours and a white precipitate was formed confirming the production of nanoparticles. The nanoparticles were collected after centrifugation and dried (Keerthana et al. 2021). The major advantages of using plant extracts are that the process is easy and less time taking without the involvement of costly instruments (Agarwal et al. 2017).

Bacteria mediated: Apart from plant sources, bacteria have also been used for ZnO NPs production. Bacterial cells could also be used as a reaction center for intracellular synthesis of the NPs. Different enzymes like nitrate reductase and NADH-dependent reductase are used for the reduction of metal ions to their specific nanoparticles (Boroumandmoghaddam et al. 2015). As bacteria culture media contain several secreted molecules, thus the extracellular synthesis of NPs could also be achieved by using bacterial cells. There are some advantages associated with using bacteria for NPs production like they are inexpensive, easy to grow, and easy to control as well (Pantidos and Horsfall 2014). In an approach, *Epicoccum nigrum* was used to synthesize ZnO NPs. The ZnO NPs produced had antibacterial activity and possessed cytotoxic activity against colon cancer cells (Abdel-Aziz et al. 2018). Similarly, *Trichoderma* species isolated from diseased rice leaves were used to produce ZnO NPs and possessed antibacterial effects against *Xanthomonas oryzae* (Shobha et al. 2020). In another approach, ZnO NPs were produced from *B. licheniformis* bacteria and were capable of degrading methylene blue dye via photocatalytic activity (Agarwal et al. 2017). Despite some advantages, there are many disadvantages associated with bacteria such as no control over the size of the nanoparticles, slight contamination won't produce nanoparticles, high cost for production because the media used for the growth of bacteria is costly, and many more (Agarwal et al. 2017).

Fungi as a potential resource for nanoparticles synthesis

Plants and microbes are frequently used for the synthesis of metal nanoparticles as they can reduce metals to their nano-size range particles via some enzymes. Among these resources where a plethora of plants and microbes have been used, fungi have proven to be the best resource for nanoparticle synthesis. This is so because of their wide range of stability and tolerance towards metals. The major advantage of using fungi is that they provide good biomass production and do not require additional steps to extract the filtrate (Guilger-Casagrande and Lima 2019). Moreover, they are rebellious towards variation in temperature, pressure, pH, agitation and can be applied at a large scale for synthesis (Guilger-Casagrande and Lima 2019). The conversion of metal salts to metal nanoparticles follows a cell-wall-based procedure where Zn ions are reduced to their nano range sizes when added and trapped onto the surface of fungi. It is easy to scale up when we use fungi as our resource for the synthesis of nanoparticles. Apart from this, this method is economically feasible and the presence of mycelia provides a large surface area for the synthesis of nanoparticles. Last but not the least, there is higher production of proteins in fungi as compared to the bacteria resulting in the amplification of good and high productivity of nanoparticles (Pantidos and Horsfall 2014). The merits of using fungi as the bio-factory for the production of ZnO NPs are shown in Figure 1. If we compare fungi with bacteria for nanoparticles production, bacterial production of NPs is very common but fungal-based NPs are more advantageous as their mycelia help in more interaction due to larger surface area. Fungi also produce many enzymes helping in the faster NPs production. Additionally, the fungal cell wall plays a major role by absorbing the metal ions over their surface and activating certain enzymes to reduce/oxidize them into their nanoparticles hence, proving that fungi-mediated synthesis is a better alternative for ZnO NPs synthesis (Khandel and Shahi 2018).

Accumulation of metal nanoparticles by fungi can be done physiologically or biologically by binding with the metabolites, polymers or certain polypeptides extracellularly. Metabolism-dependent mechanisms are followed to accumulate the nanoparticles synthesized from fungi (Gade et al. 2010). If we compare and analyze

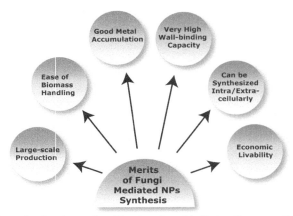

Figure 1. Schematics showing merits of fungi used as a bio-factory for nanoparticles production.

the advantages as well as disadvantages of both extracellular and intracellular synthesis of nanoparticles, the size of intracellularly synthesized NPs is smaller as compared to the extracellular ones. The reason behind this can be due to the nucleation of NPs inside the cell. Moreover, the extracellular synthesis is faster than the intracellular one and its downstream processing is easy. The major difficulty in following the intracellular synthesis is complicated downstream processing. Additionally, purifying and accumulating nanoparticles requires high precision and advanced techniques. Therefore, increasing the overall cost of production makes it a cost-inefficient method (Yadav et al. 2015).

Fungi-mediated synthesis of zinc oxide nanoparticles

Several nanoparticles have been synthesized but the potential of zinc oxide nanoparticles against pathogens and microbes has proven to be very beneficial. The process of fungal-mediated synthesis of ZnO NPs is shown in Figure 2(a). For the fungi-mediated synthesis of zinc oxide nanoparticles, the as-mentioned mechanisms and processes are generally followed:

Intracellular Synthesis: In intracellular synthesis, the metal precursor is internalized when added to the mycelial culture. The metal ions get attached over the surface of the cell wall of fungi by electrostatic interaction. They get reduced to the nano-ranged sizes by the enzymes present in the cell wall as shown in Figure 2(b). The NPs formation takes place in the cytoplasm and is accumulated in purified form by ultrasonication (MohdYusof et al. 2019). Post-synthesis different techniques like filtration, centrifugation or chemical treatment are required to extract and release NPs from the biomass. The intracellular synthesis requires proper handling of the fungal biomass as they contain the nanoparticles. Thus, worldly-wise instruments are required with precision to extract the nanoparticles from the biomass (Gade et al. 2010).

Biosynthesis of Zinc Oxide Nanoparticles and Major Applications   177

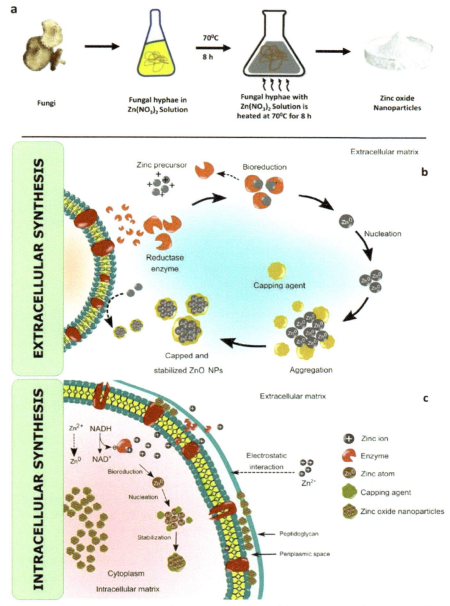

Figure 2. (a) Fungal mediated synthesis of ZnO nanoparticles; **(b)** Schematic representation of extracellular synthesis mechanism of ZnO NPs (MohdYusof et al. 2019); **(c)** Schematic representation of intracellular synthesis mechanism of ZnO NPs (MohdYusof et al. 2019).

Extracellular Synthesis: In the case of extracellular synthesis, the method becomes easier due to aqueous filtrate having fungal molecules to produce ZnO NPs as soon as a metal precursor is added. The nitrate reductase enzyme initiates the nucleation of the nanoparticles by reducing the metalions. The proteins present in the fungi aid the stability of the nanoparticles by acting as their capping agents as shown in Figure 2(c). Finally, the production of the nanoparticles is confirmed when a white precipitate is observed (MohdYusof et al. 2019). There are no requirements for using advanced techniques for the extraction of ZnO NPs; hence it is highly preferred over the intracellular synthesis method (Guilger-Casagrande and Lima 2019). Moreover, extracellular synthesis of the NPs has several advantages over intracellular synthesis of NPs like (i) Easy separation and downstream processing (ii) Cost-effective (iii) Required less energy (iv) Scope of large-scale production for industrial applications. On the other hand, the intracellular synthesis of NPs required additional steps of purification of the NPs by disrupting the microbial cells. Also, purification of the NPs from intracellular components possesses an additional hurdle.

Aspergillus sp. and *Saccharomyces cerevisiae* are the most commonly used fungi for zinc oxide nanoparticles synthesis. In an approach, Preethi et al. prepared an aqueous button mushroom extract and added it dropwise into zinc acetate dehydrate solution under constant stirring. The white precipitate was formed when NaOH was added drop by drop to the mixture prepared earlier. The nitrate reductase enzyme helps in the reduction and nucleation of NPs. The protein secreted by the fungi acted as a capping agent to stabilize the structure of NPs. Lastly, the white precipitate formed post-synthesis indicated the successful formation of ZnO NPs. Monitoring and characterization of already synthesized ZnO NPs are done using UV-Vis spectroscopy. After that, centrifugation was done at 5000 rpm for 10 minutes to collect the pellet accumulated from the precipitate. Later, the pellet was washed thrice using distilled water to remove impurities followed by washing with ethanol to remove minor impurities. Then the precipitate was dried for more characterization in the hot air oven at a 60°C temperature (Preethi et al. 2019). Recently in 2019 Gao et al. prepared ZnO NPs by growing *Aspergillus niger* in an enriched medium and added a loop of it to the liquid media. The medium was incubated for long 96 hours on a rotator shaker with physiological conditions like 28°C temperature at a speed of 200 rpm. Filtration was done post-completion of incubation to separate the biomass and collect the cell filtrate (Gao et al. 2019). Similarly, fungal culture was prepared using *Aspergillus terreus* by growing aerobically it in an orbital shaker at 160 rpm for 4 days at 32°C. The white precipitate was formed when $ZnSO_4$ salt was added to the culture and was incubated again at the same temperature at 150 rpm for 2 days. This process of white precipitate formation is called a transformation process. The aggregate was then filtered, centrifuged and lyophilized (Khamis et al. 2017). In other praiseworthy efforts, K. Gupta and TS Chundawat used *Fusarium oxysporum* as the source to produce zinc oxide nanoparticles (ZnO NPs). The main reason behind the selection *of Fusarium oxysporum* was that it had high tolerance towards zinc metal and also had the potential to synthesize ZnO NPs extracellularly. Later, the characterization was done using SEM, FTIR, XRD, DTA and UV-Vis spectroscopy, etc., for the structural and functional analysis of the nanoparticles with their quality, stability and purity (Gupta and Chundawat 2020). Similarly, Kadam et al. in 2019 demonstrated the synthesis of ZnO NPs using endophytic fungi *Cochliobolus geniculatus*. These fungi had tolerance towards zinc and were isolated from the leaf of *Nothapodytes foetida*. The identification of the fungi was done using the rDNA internal transcribed spacer region sequence. Characterization of the ZnO NPs produced from *Cochliobolus geniculatus* proved that the NPs were well distributed, polydisperse, and deprived of crystallinity and agglomeration (Kadam et al. 2019). In a similar way, *A. fumigatus* JCF was isolated from vegetable waste. Further characterization and sequencing of the fungi were done and then cultured on an agar plate for 96 hours and later refrigerated at 4°C. The inoculation and incubation of the fungi were done following centrifugation at 3500 rpm for 20 minutes. The supernatant collected was mixed with $ZnSO_4$ salt leading to the formation of white precipitate indicating the successful production of ZnO NPs (Rajan et al. 2016).

Sol-gel Process: Sol-gel process is generally considered simple, economic and reproducible for different types of metal oxide NPs. In this process, hydrolyzation, condensation and polymerization techniques are used for converting suspension ('sol') into a gel or solid structure. Herein, effective removal of the organic matrix using thermal decomposition leads to the formation of metal oxide NPs. The advantage of the sol-gel process is that the physiological and morphological properties can easily be controlled. Recently, this method has been immensely used in the application of several biological derivatives and extracts as a chelating agent. From a sustainable point of view, this method is widely considered due to its green protocol and utilization of biological feedstock as a precursor material. Therefore, a renewable source such as fungal biomass can be used as a precursor material for the synthesis of ZnO NPs. In research, *Periconium* sp. was chosen for the biosynthesis of ZnO NPs as this is an endophytic fungus, most commonly found within the leaf, shrub, root, etc., of the trees and shrubs. Endophytic fungi are known to have secondary metabolites that can be used for various biomedical applications. Therefore, an aqueous extract of *Periconium* sp. was taken for the synthesis of ZnO NPs using the sol-gel process (Ganesan et al. 2020).

Factors affecting the myco-synthesis of nanoparticles

Fungal growth and production of nanoparticles are highly dependent on the conditions, especially physiological conditions provided to the nanoparticles. Interactions of fungi with the environment where they are grown influence their growth and development. Therefore, it is necessary to provide optimized conditions and culture in such a way that the maximum yield of nanoparticles can be obtained. Some parameters like temperature, pH, metal species, parent compound behaviour, and incubation period were highly effective in the yield of NPs as shown in Figure 3. Furthermore, the size, shape and dispersity of the nanoparticles can be controlled by biomass concentration of fungal strain and their colloidal interaction conditions. It was concluded that 0.1 mmol/L precursor salt concentration, 72 h of incubation at pH 5.5 and temperature 28°C resulted in a larger NP yield (Alghuthaymi et al. 2015).

Similarly, Motazedi et al. demonstrated the parameters influencing the absorption of ZnO NPs to the zinc acetate solution ratio (1:1.5, 1:3 and 1:6) with a pH range of 5–12. Further, characterization was done using UV-vis spectroscopy where absorption was observed at 376 nm (Motazedi et al. 2020). In another approach, Ganesan et al.

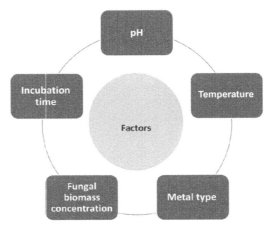

Figure 3. Major factors that affect the synthesis of ZnO NPs using fungal species (Alghuthaymi et al. 2015).

Table 1. List of fungi and their optimized condition of various factors for the synthesis of ZnO NPs.

S. No.	Fungi used for ZnO NP synthesis	Optimized conditions			References
		pH	Temperature	Incubation time	
1.	*Saccharomyces cerevisiae*	3.0	30°C	48 hours	Motazedi et al. 2020
2.	*Cochliobolus geniculatus*	7.2–7.8	28 ± 1 °C	72 hours	Kadam et al. 2019
3.	*Agarius bisporus*	7.2	37°C	24 hours	Preethi et al. 2019
4.	*Aspergillus niger*	6.2 ± 0.2	37°C	48 hours	Gao et al. 2019
5.	*Aspergillus fumigatus*	6.5	32°C	96 hours	Rajan et al. 2016
6.	*Penicillium chrysogenum*	6.0	30°C	48 hours	Mohamed et al. 2021

demonstrated the synthesis of ZnO NPs via fungal extract of *Periconium* sp. cultured in potato dextrose broth for 21 days. The culture broth was filtered and dried in a hot air oven at 60°C for 12 h (Ganesan et al. 2020). The optimized conditions for different fungi have been listed in Table 1.

Limitations of the mycosynthesis

Bio-synthesis of the NPs depends on metal ions tolerance of the microbes, which offers intracellular stress within the cells. Thus, the higher concentration of metal ions may have adverse effects on microbial growth and proliferation. Interestingly, in such stress condition microbes changes their chemical composition by up and/ or down-regulating productions of several biomolecules, which served as potential reducing and/or oxidizing agents for several metal ions. Therefore, microbial cells act as a natural 'nano-factory' suitable for the synthesis of metal and metal oxide NPs. However, not all the types of microbial cells are found to be useful for the synthesis of NPs, since every microbe produces different metabolites and adopted altered

Table 2. Merits and demerits use of various sources for the biosynthesis of ZnO NPs.

Source for ZnO NPs synthesis	Pros	Cons
Plant	- Pure nanoparticles can be obtained - Eco-friendly method - Cheap process - Less time consuming - Applicable for large scale production - Stable production - Pure nanoparticles can be obtained	- Very time taking method - Variation in temperature affects the production of NPs.
Bacteria	- Cost friendly method - Easy to grow - Easy to control - Easy to manipulate	- Screening for microbes is time taking. - Prone to contamination - No control over the size of nanoparticles. - Media cost is high.
Fungi	- High tolerance in comparison with bacteria - Better metal bioaccumulation - Highly stable - Good biomass production - Economically feasible - Mycelia provides a large surface area for NPs production	- Require additional step for extraction when synthesized intracellularly

biochemical pathways for their survivability. The selection of specific microbes, their culture conditions, and the number of metal ions are very important parameters that hugely control the rate of formation of NPs. For example, a study showed that several soil fungus strains have better zinc ions tolerance, and thus provide the scope of biosynthesis of ZnO NPs. It is worth mentioning that finding the exact biomolecules responsible for the synthesis of the NPs is very challenging and several cases did not illustrate yet. Therefore, biosynthesis has inherent limitations such as (i) Controlling the size and growth of the NPs (ii) Monitoring the surface charge of the NPs and (iii) Attachment of the specific surface ligands on the NPs. (iv) Reproducibility of the yield and rate of synthesis NPs (v) Formation of the polydisperse NPs, resulted in wide size and shape distribution of NPs. Table 2 shown below states the pros and cons of using fungi, bacteria and plants as a precursor for the synthesis of ZnO NPs.

Use of zinc oxide nanoparticles

Applications in the biomedical field

It is a well-known fact that zinc is an essential trace element in the body. Thus, regular uptakes of zinc are very important for normal body functions like metabolism and synthesis of nucleic acid and proteins, neurogenesis as well as hematopoiesis, it plays important role in several vital cellular functions. Zinc also served as a co-factors and vital component for various enzymes. Interestingly, the absorption of the zinc becomes much more convenient and faster by delivering in the form of ZnO NPs—having smaller sizes—that help the easy release of the zinc ions in physiological conditions. Apart from these normal body functions, the ZnO NPs are having several biomedical applications which are given below.

Anticancer activity of ZnO NPs: Cancer is one of the deadliest diseases and some reports suggested that it can be cured to some extent by using ZnO NPs. To justify the anticancer activity of ZnO NPs several *in vitro* and *in vivo* studies were carried out. For example, Abdelhakim et al. used ZnO NPs against three cancer cell lines *MCF-7*, *HFB-4* and *HepG-2*. The MTT assay was done to assess the activity of ZnO NPs against these three cancer cell lines. Results showed that cell viability was significantly decreased with an increased dose of ZnO NPs. The ZnO NPs were synthesized using *Alternaria tenuissima* (Abdelhakim et al. 2020). In another *in vitro* study, cytotoxicity of ZnO NPs was checked against human breast cancer cells *MCF-7* where an MTT assay was performed to analyze the viability of the cancer cells. It was observed that the viability was reduced to below 10% when 100 µg/mL concentrations of ZnO NPs were used as shown in Figure 4 (Motazedi et al. 2020). Similarly, Es-haghi et al. demonstrated the synthesis of ZnO NPs from *Aspergillus niger* which had the potential to reduce the cytotoxicity of *MCF-7* (Human Breast Cancer Cell Line) to almost 1% at 125 µg/mL concentrations of ZnO NPs within 72 h (Es-haghi et al. 2019).

Anti-diabetic activity of ZnO NPs: Recent studies demonstrated that oral admiration of the ZnO NPs in case mice showed anti-diabetic effects as it significantly increases

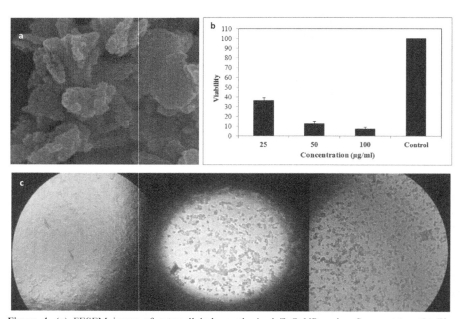

Figure 4. (a) FESEM image of extracellularly synthesized ZnO NPs using *S. cerevisiae*; **(b)** The cytotoxic effect of ZnO NPs on *MCF-7* cancer cells viability content; (b) *MCF7*-human breast cancer cell; **(c)** images *MCF7* after exposed to100µg/ml of biosynthesized ZnO NPs (Motazedi et al. 2020).

the zinc ions concertation in the pancreas, liver and adipose tissue. This also improved the glucose tolerance of the mice. Administration of the ZnO NPs increased serum insulin levels and reduced blood glucose levels. The report also revealed that ZnO NPs may decrease non-esterified fatty acids and reduce triglycerides. Zinc helps in the stabilization of insulin structure and also aids in the synthesis, storage and secretion of insulin. Studies showed that in the case of diabetic rats, ZnO NPs increase the production rate of insulin, reduce the blood glucose level and increase their glucokinase (GK) activity proving that ZnO NPs have anti-diabetic activities as well (Bedi and Kaur 2015). Therefore, systematic and controlled uptakes of ZnO NPs might address the issue of anti-diabetic activity as they have the potential to reduce the blood glucose level.

Antioxidant activity and free radical scavenger: Apart from anticancer activities, ZnO NPs possess antioxidant properties which can help in controlling the oxidative stress inside the cells and protect them from oxidative damage. ZnO NPs were synthesized using *Pleurotus djamor* where DPPH (2,2-diphenyl-1-picryl-hydrazyl-hydrate) radical scavenging activity was tested. It was found that the ZnO NPs had a significant impact and also inhibited oxidative stress, protecting cells from severe damage as shown in Figure 5 (Manimaran et al. 2021). The motivation behind this method was to scavenge DPPH in the presence of antioxidants, resulting in the decolourization of the DPPH solution. Similarly, the ZnO NPs synthesized from *Perconium* sp. exhibited 85.52% of radical scavenging activity which was near the activity of ascorbic activity (Ganesan et al. 2020).

Development of cosmetic products: The ZnO NPs have excellent ultraviolet (UV) protection power thus found to be suitable for the protection of UV radiation from sunlight. Several cosmetics, dermal and sunscreen products are available in the market where ZnO NPs used as integrated components. The ZnO NPs also have potent antimicrobial activity along with UV-light absorption ability; therefore these have been employed in personal care products. It would be mentioned here that another easy way to deliver drugs apart from the oral administration is the dermatological delivery of drugs. Though there can be some limitations because the skin is the protective covering of the human body the small size of nanoparticles and their tunable surface chemistry can favour dermatological delivery (Shivalkar et al. 2021d).

Design of sensors by ZnO NPs: Nanomaterials can potentially be used as sensors due to their inherited properties of large surface area, efficient adsorption ability, higher active sites on the surface, high conductivity and strong catalytic ability. These properties generally contribute to the underlying principles of biosensors as they can easily immobilize the biomolecules or some biomarkers to label respective biomolecules (Shivalkar and Sahoo 2021c). Exploring these potentials, researchers have utilized ZnO NPs for the modification of Glassy Carbon Electrode to develop an electrochemical sensor for the detection of amino acids, uric acid and dopamine. Herein, ZnO NPs were used as nano-catalyst in the electrochemical reaction that catalyzes the reaction by enhancing the electron transfer efficiency. During this electrocatalytic activity, a clear oxidation peak of amino acid, uric acid and dopamine

184 *Mycosynthesis of Nanomaterials: Perspectives and Challenges*

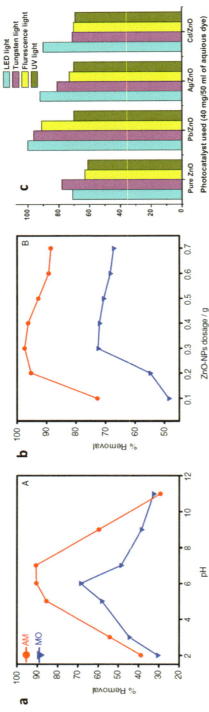

Figure 5. **(a)** Effect of pH on adsorption of dyes by ZnO-NPs after equilibrium time of 50 min [AM (red line), MO (blue line)]; **(b)** Effect of adsorbent dose on adsorption of dyes by ZnO-NPs after equilibrium time of 50 min [AM (red line), MO (blue line)] (Zafar et al. 2019), and **(c)** Photocatalytic activities of modified ZnO nanoparticles under different light sources (Nagaraju et al. 2020).

was observed. The sensitivity of this ZnO NPs-based electrochemical biosensor states that the concentration range for amino acid (50–1000 μM), uric acid (5–150 μM), and dopamine (5–150 μM) can easily be detected. However, the challenge of detecting the severely overlapped oxidation potential of amino acid, uric acid and dopamine still need to be overcome (Pan et al. 2020). Similarly, an enzyme-based glucose biosensor was developed using the composite of ZnO NPs and graphene-carbon nanotube hybrid. The glucose oxidase enzyme was immobilized on the surface of the ZnO NPs-Graphene-Carbon Nanotube composite. Well-defined redox peaks were observed in this setup signifying excellent electron transfer at the electrode surface. Therefore, it shows high electrocatalytic potential for glucose sensing in presence of oxygen. The sensitivity of this sensor ranges from 10 μM to 6.5 mM of glucose with a limit of detection of 4.5 μM (Hwa and Subramani 2014). In another example, ZnO NPs-Chitosan composite was used to develop a hybrid nanocomposite film on the indium-tin-oxide glass plate with immobilized cholesterol oxidase on its surface for the detection of cholesterol. The sensitivity of this sensor was 1.41×10^{-4} A mg dl^{-1} of cholesterol with a limit of detection of 5 mg dl^{-1} (Khan et al. 2008). Additionally, these ZnO NPs in combination with multi-walled carbon nanotubes (MWCNTs) were used for the immobilization of probe DNA strands (Majumdar et al. 2020).

Applications in the agricultural field

It has always been observed farms get easily affected by some pathogens like fungi bacteria or other microorganisms leading to the growth of a diseased plant or the death of a healthy plant. Thus, we need a source to tackle such issues in the agricultural sector where we can save our plants with a cost-friendly and easy approach. ZnO NPs were found to have antifungal and antibacterial properties, so they can be used as nano-fertilizers or nano-pesticides against harmful pests or microbes. To justify the beneficial properties of ZnO NPs, they were applied to the peanut seeds to boost the growth of the seeds at different ZnO NPs concentrations. It was observed that the plant showed vigorous growth, stem elongation and root growth with seed germination too (Sabir et al. 2014). It has also been reported that carbon dots (C-dots) in combination with different nanoparticles possess antibacterial properties specifically antibiotic-resistant recombinant bacteria (Verma et al. 2020). Some plants are deficient in zinc leading to a decrease in plant yield, zinc-deficient crops and many more. The effect of ZnO NPs was also seen on the corn (*SC704*) where normal irrigation was performed and the nano-colloids of ZnO NPs were added at the first, second and third week of corn emergence. An improvement in the leaf area, growth and dry weight of leaf was observed after the introduction of zinc oxide nano-colloids (Taheri et al. 2015). In another approach, Nanoprime and an unprimed germination assay of ZnO NPs were carried out on the ladyfinger by following the paper towel method. It was observed after a week that there was an improvement seen in the leaf shape, pod shape, pod length, root height and even the total plant height (Keerthana et al. 2021). When ZnO NPs are used as nano-fertilizers, they help in the controlled release of fertilizers and reduce the toxicity among plants and in the environment as well. Furthermore, if it is used as nano-sensors for plants, they can help in the detection of soil moisture content and their nutrient level with a rapid

186 *Mycosynthesis of Nanomaterials: Perspectives and Challenges*

response against pathogens (Bedi and Kaur 2015). The consumption of fertilizers, pesticides or herbicides can also be reduced if ZnO NPs are used as their substitute. For an instance, the growth and yield of wheat plants increased to around 25% when they were treated with ZnO NPs. Thus, we can conclude that the ZnO NPs possess several advantages when applied in the agriculture sector for their increment in the growth, development, yield and disease-resistant properties (Bedi and Kaur 2015).

In another example, for the production of bioethanol from rice straw, they are subjected to delignification using ammonia to reduce the lignin content. In the presence of micro-aerobic conditions, the sugars get converted to ethanol via fermentation. It was observed that when ZnO NPs were used, the ethanol production increased and a maximum yield of 0.0359 g/g (dry weight) ethanol was obtained when 200 mg/L of ZnO NPs were used. Earlier the production was 0.159 g per 10 g of rice straw and later it was 0.359 g per 10 g of rice straw (Gupta and Chundawat 2020).

Applications in environmental pollution remediation

Nanoparticles also can absorb or adsorb pollutants on their surface. Nanoparticles are capable of degrading toxic dyes released from industries such as methyl orange, amaranth and azo dyes (Shivalkar et al. 2021a; b). In an approach, Zafar et al., used ZnO NPs for the adsorption of dyes released from industries. They chose methyl orange (MO) and amaranth (AM) as their model dyes and analyzed them at different pH. It was observed that at pH 5, the adsorption of AM increased up to 95% at room temperature and the removal of MO also increased at a pH 6 as shown in Figure 5(a). Moreover, the amount of ZnO NPs also affects the adsorption rate because when the quantity was increased from 0.1 g to 0.3 g, the removal rate increased from 68.5% to 82.5% for MO and 76.7% to 97.4% for AM as shown in Figure 5(b) (Zafar et al. 2019).

The chemical structure of these dyes is very complex and is highly soluble in water bodies causing harm to the ecosystem (Gautam et al. 2021). Furthermore, the ZnO NPs also exhibit photocatalytic activity as they were tested against reactive blue 21 (RB21) dye. This is one of the unique properties of ZnO NPs, which act as photocatalysts and break organic pollutants to achieve complete decontamination of water (Shivalkar et al. 2021b). When the dye was exposed to UV radiation, the degradation percentage was not up to the mark. But when the photodegradation was tested in the presence of ZnO NPs, the degradation was almost 80–90% in 270 min thus, proving the ability of ZnO NPs in reducing environmental pollution (Davar et al. 2015). In other praiseworthy efforts, Nagaraju et al. demonstrated the photocatalytic degradation of chlorobenzene in the presence of ZnO NPs and modified ZnO NPs like Pb/ZnO, Ag/ZnO and Cd/ZnO. Four lights (Fluorescence light, tungsten light, UV light and LED light) were used to test the degradation properties. It was observed that the degradation rate was very high around 90% by using tungsten and 96%–100% in the presence of LED light as shown in Figure 5(c) (Nagaraju et al. 2020).

A new approach was seen, where ZnO NPs were used as an agent to alleviate the bio-deterioration of cellulose-based papers. For this, Whatman filter paper was used as a model due to its high cellulose content of around 98%. *Aspergillus niger* strain A2 isolates were grown over the surface of the filter paper to check the activity of

ZnO NPs against it. At 2 mM concentration, the ZnO NPs exhibited good preservative properties and can be used as an agent to protect papers from fungal deterioration. They were also found to have increased the tensile strength of papers and kept their cellulosic materials intact (Fouda et al. 2019). Furthermore, these NPs are capable to degrade methylene blue dye under the influence of light. Within 1 hr, the absorbance of dye was decreased to 664 nm upon 20 min exposure to sunlight in the presence of ZnO NPs. At a concentration of 200 mg of ZnO NPs, maximum degradation was obtained about 100% depicting the successful application of ZnO NPs in dye removal (Abdelhakim et al. 2020). Similarly, Bismarck brown dye was degraded by ZnO NPs within 72 hrs, where 100 µl of ZnO NPs were used and maximum degradation of colour around 89% was observed (Kalpana et al. 2018).

Other applications

The antimicrobial property of ZnO NPs was utilized to produce antibacterial fabrics and wound healing materials. Bactericidal potency of the NPs depends on (i) the Size of the NPs—the smaller the size better the bactericidal activity, (ii) the shape of the NPs—it was found that certain anisotropic NPs offer better bactericidal potency as compared to spherical NPs. (iii) Surface charge—as bacterial cell wall is negatively charged (in case of both Gram-positive and Gram-negative bacteria), hence positively charged NPs show better activity as compared to its counterparts. (iv) Surface ligands-biological activity and specificity of the NPs greatly relies on the chemical nature of the surface passivating agents of the NPs.

The cotton fabric was cured with ZnO NPs and tested against *Staphylococcus aureus* and *Escherichia coli.* The cotton fabric was placed on the agar plate and the measurement was done in mm for the zone of inhibition (Kalpana et al. 2018). These NPs provide durability and antimicrobial effects to the textiles. Moreover, they can also be used in the production of antimicrobial and durable surgical gloves, cosmetics, toothbrushes, and masks (Gade et al. 2010). ZnO is generally regarded as safe (GRAS) by the food and drug administration (FDA) and can be helpful in the growth and development of humans and animals. The antimicrobial property of ZnO NPs can be useful in food packaging applications. It was examined that ultraviolet radiation could add to the activity of ZnO NPs by increasing their antibacterial properties (Espitia et al. 2016). The other important properties of zinc include the defense mechanism of the human body, antioxidant immune system, hormone secretion, reproduction, etc. (MohdYusof et al. 2019).

Mycotoxins are common toxins produced in animal feed that can contaminate and destroy the feedstock. ZnO NPs were used to test the antifungal properties and it was found that it had antifungal activity as well (MohdYusof et al. 2019). Adding more to the advantages and applications of ZnO NPs, their antibacterial properties were utilized to create a protective film for the prevention of copper bio-corrosion. The main concept behind the corrosion is the biofilm formation over the copper surface and its oxidation to the metal oxides. Additionally, ZnO NPs have the property to prevent bio-corrosion and inhibit bacterial growth over the surface of copper. These NPs act as a protective covering over the surface of the metal. It was observed that the corrosion efficiency was 73% proving the potential of ZnO NPs as

188 *Mycosynthesis of Nanomaterials: Perspectives and Challenges*

a protective covering for bio-corrosion prevention (Preethi et al. 2019). In another example, the anti-biofilm activity of ZnO NPs was checked on the biofilm formed by *Candida tropicalis*. The results produced by XTT [2, 3-bis(2-methoxy-4-nitro-5-sulfophenyl)-5-(phenylamino) carbonyl) - 2H-tetrazolium hydroxide] assay and ATPase assay showed that ZnO NPs were able to inhibit the fluconazole-susceptible strain of *Candida tropicalis*. The anti-candidal efficiency of ZnO NPs can be used as a cleaning agent for the prevention of microbes (Jothiprakasam et al. 2017). Table 3 shown below describes the use of various fungi for the ZnO NPs synthesis and their use in different fields.

Table 3. Application of ZnO NPs synthesized by different fungi with their synthesis method and sizes.

S. No.	Fungi used for ZnO NP synthesis	Synthesis method	Size of the ZnO NPs	Application	References
1.	*Penicillium chrysogenum*	Extracellular	9–35 nm	Antibacterial, antibiofilm and antifungal activity	Mohamed et al. 2021
2.	*Penicillium chrysogenum*	Extracellular	8.74–23.07 nm	Bio-deterioration of cellulose-based materials	Fouda et al. 2019
3.	*Aspergillus fumigatus*	Extracellular	60–80 nm	Antibacterial activity	Rajan et al. 2016
4.	*Alternaria tenuissima*	Extracellular	15.45 nm	Antioxidant, antimicrobial, anticancer and photocatalytic activity	Abdelhakim et al. 2020
5.	*Periconium* sp.	Sol-gel process	40 nm	Antimicrobial and antioxidant activity	Ganesan et al. 2020
6.	*Aspergillus niger*	Extracellular	30–70 nm	Antioxidant and anticancer activity	Es-haghi et al. 2019
7.	*Pleurotus djamor*		70–80 nm	Histopathlogy, mosquito larvicidal, antibacterial, antioxidant and anticancer	Manimaran et al. 2021
8.	*Trichoderma harzianum*	Extracellular	27–40 nm width and 134–200 nm length	Antifungal activity	Consolo et al. 2020
9.	*Fusarium oxysporum*	Extracellular	18–25 nm	Bioethanol production from rice straw	Gupta and Chundawat 2020
10.	*Agarius biosporus*	Extracellular	14.48 nm	Antibiofilm, anti-corrosion and antibacterial assay	Preethi et al. 2019

Toxicity and other drawbacks of mycosynthesized zinc oxide nanoparticles

We have discussed several applications of the ZnO NPs ranging from biomedical, agriculture, and environmental remedy. Zinc is also considered safe and a trace element that has a beneficial impact on human health and the environment. However, systematic toxicity evaluation of ZnO NPs was not performed in many cases. Additionally, the issue of hemocompatibility and genotoxicity were not even checked extensively as it is a fact that the toxicity of any nanoscale material may vary with size, shape and surface ligands. Therefore, it's obvious that the use of ZnONPs-based materials needs further studies before its real-life applications. As ZnO NPs have antimicrobial properties, they can have application to food packaging also. Nevertheless, it should be made sure that there is no transfer of nanoparticles to the food products, creating contamination or toxicity. Though the toxicity mechanism of ZnO NPs is still unclear it has been noted that these NPs bind easily with the membrane of the cells, releasing zinc ions which can lead to an increase in the oxidative stress of the cells, and lipid peroxidation or DNA damage as well. Furthermore, their oral consumption can also harm the renal function of the human body; increase the blood urea nitrogen and many more. Their toxicity is also related to the dosage provided to the human body and may cause liver, kidney or heart failure. Though their low dosage does not harm any part of the human body thus; they can be used at a very low concentration rather than using higher concentrations (MohdYusof et al. 2019). It was suggested that we can use ZnO NPs by fabricating them with some polymers so that there won't be any toxicity concerns in the future. The synthesis duration of mycosynthesized nanoparticles is also time-consuming as it takes a few hours to a day for the production of nanoparticles (Hulkoti and Taranath 2014). During the intracellular synthesis of ZnO NPs, issues may arise in the extraction and purification of nanoparticles because the process becomes complicated and requires precision. It has also been found that the nanoparticles get degraded after some time due to mycosynthesis. Thus, there is a need for a thorough study regarding understanding and enhancing the stability of nanoparticles. Furthermore, their potential in cancer treatment is good but a detailed study is required considering the toxicological factors of the ZnO NPs. These are some of the drawbacks of limiting the use of ZnO NPs. As discussed, the easiest way to eradicate toxicological concerns is by fabricating them with some polymers so that it can increase the stability; reduce toxicity, and increase efficiency and long-time durability of the zinc oxide nanoparticles.

Conclusion and future perspectives

In brief, the method of biosynthesis has the potential and economical solution for the large-scale production of NPs, essentially fungi are one of the easiest and most convenient ways of synthesis of ZnO NPs. There are several advantages associated with myco-synthesis like it has huge tolerance of metal ions and physiological

190 *Mycosynthesis of Nanomaterials: Perspectives and Challenges*

variables (temperature, pressure, and pH) that helps in the fast and easy synthesis of ZnO NPs. Moreover, fungi produce many important primary and secondary metabolites which aid high-yield nanoparticles production. Extracellular synthesis is mostly followed for the production of nanoparticles as it is economical and energy-efficient in comparison with intracellular synthesis. ZnO NPs have many applications in the biomedical field including the design of new anticancer, antimicrobial as well as anti-diabetic agents. Even in the agricultural sector, it showed a great deal of scope of development of nano-pesticides, nano-herbicides, and nano-fertilizers as they help in the growth and proliferation of roots, shoot elongation, and leaf expansion. The major advantage of ZnO NPs as nano-fertilizers is that there are no requirements for a higher dose of fertilizers as it works more efficiently for a longer duration of time even at low concentrations. Therefore, the lower toxicity, cost effectiveness and scope of large-scale production of ZnO by myco-synthesis have great prospects, and proper optimization and standardization obviously will meet the criteria of several futuristic applications. However, there are a few issues that still exist in the biosynthesis of ZnO NPs like the stability and durability of the NPs. Also, the issues of systematic toxicity need to be addressed more profoundly and rigorously before extensive real-life applications.

Abbreviations

NPs—Nanoparticles; ZnO—Zinc Oxide; C-Dots—Carbon Dots; NaOH—Sodium Hydroxide; SEM—Scanning Electron Microscope; FTIR—Fourier Transform Infrared; XRD—X-Ray Diffraction; DTA—Differential Thermal Analysis; AM—Amaranth; MO—Methyl Orange; MTT—3-(4,5-dimethylthiazol-2-yl)-2,5-diphenyl tetrazolium bromide; MRI- Magnetic Resonance Imaging; GK—Glucokinase; FESEM—Field Emission Scanning Electron Microscope; XTT-2, 3-bis(2-methoxy-4-nitro-5-sulfophenyl)-5-(phenyl amino) carbonyl)—2H-tetrazolium hydroxide; DPPH—(2,2-diphenyl-1-picryl-hydrazyl-hydrate); GRAS—Generally Recognized as Safe; FDA—Food and Drug Administration.

References

Abdel-Aziz, M.M., Safwat, N.A., and Amin, B.H. 2018. Selective toxicity of mycosynthesized zinc nanoparticles toward colon cancer and its most associative pathogen (*Streptococcus gallolyticus*). Egypt. J. Exp. Biol. (Bot.) 14: 133–142.

Abdelhakim, H.K., El-Sayed, E.R., and Rashidi, F.B. 2020. Biosynthesis of zinc oxide nanoparticles with antimicrobial, anticancer, antioxidant and photocatalytic activities by the endophytic *Alternaria tenuissima*. J. Appl. Microbiol. 128: 1634–1646.

Agarwal, H., Venkat Kumar, S., and Rajeshkumar, S. 2017. A review on green synthesis of zinc oxide nanoparticles – An eco-friendly approach. Resource-Efficient Technologies 3: 406–413.

Alghuthaymi, M., Almoammar, H., Rai, M., Said-Galiev, E., and Abd-Elsalam, K. 2015. Myconanoparticles: Synthesis and their role in phytopathogens management. Biotechnol. Biotechnol. Equip. 29: 1–16.

Bedi, P.S., and Kaur, A. 2015. An overview on uses of zinc oxide nanoparticles. World J. Pharm. Pharm. Sci. 04: 1177–1196.

Boroumandmoghaddam, A., Namvar, F., Moniri, M., Tahir, P., Azizi, S., and Mohamad, R. 2015. Nanoparticles biosynthesized by fungi and yeast: A review of their preparation, properties, and medical applications. Molecules (Basel, Switzerland). 20: 16540–16565.

Consolo, V.F., Torres-Nicolini, A., and Alvarez, V.A. 2020. Mycosinthetized Ag, CuO and ZnO nanoparticles from a promising *Trichoderma harzianum* strain and their antifungal potential against important phytopathogens. Sci. Rep. 10: 20499.

Davar, F., Majedi, A., and Mirzaei, A. 2015. Green synthesis of ZnO nanoparticles and its application in the degradation of some dyes. J. Am. Ceram. Soc. 98.

Es-haghi, A., Soltani, M., Karimi, E., Namvar, F., and Homayouni, M. 2019. Evaluation of antioxidant and anticancer properties of zinc oxide nanoparticles synthesized using *Aspergillus niger* extract. Mater. Res. Express. 6.

Espitia, P.J.P., Otoni, C.G., and Soares, N.F.F. 2016. Chapter 34 - zinc oxide nanoparticles for food packaging applications. pp. 425–431. *In*: Antimicrobial Food Packaging San Diego: Academic Press.

Fouda, A., Abdel-Maksoud, G., Abdel-Rahman, M.A., Eid, A.M., Barghoth, M.G., and El-Sadany, M.A.-H. 2019. Monitoring the effect of biosynthesized nanoparticles against biodeterioration of cellulose-based materials by *Aspergillus niger*. Cellulose. 26: 6583–6597.

Gade, A., Ingle, A., Whiteley, C., and Rai, M. 2010. Mycogenic metal nanoparticles: Progress and applications. Biotechnol. Lett. 32: 593–600.

Ganesan, V., Hariram, M., Vivekanandhan, S., and Muthuramkumar, S. 2020. *Periconium* sp. (endophytic fungi) extract mediated sol-gel synthesis of ZnO nanoparticles for antimicrobial and antioxidant applications. Mater Sci. Semicond. Process. 105: 104739.

Gao, Y., ArokiaVijayaAnand, M., Ramachandran, V., Karthikkumar, V., Shalini, V., Vijayalakshmi, S., and Ernest, D. 2019. Biofabrication of zinc oxide nanoparticles from *Aspergillus niger*, their antioxidant, antimicrobial and anticancer activity. J. Clust. Sci. 30: 937–946.

Gautam, P.K., Shivalkar, S., and Banerjee, S. 2020. Synthesis of *M. oleifera* leaf extract capped magnetic nanoparticles for effective lead [Pb (II)] removal from solution: Kinetics, isotherm and reusability study. J. Mol. Liq. 305: 112811.

Gautam, P.K., Shivalkar, S., and Samanta, S.K. 2021. Environmentally benign synthesis of nanocatalysts: recent advancements and applications. pp. 1163–1181. *In*: Handbook of nanomaterials and nanocomposites for energy and environmental applications. Cham: Springer International Publishing.

Guilger-Casagrande, M., and Lima, R. de. 2019. Synthesis of silver nanoparticles mediated by fungi: A review. Front. Bioeng. Biotechnol. 7: 287.

Gupta, K., and Chundawat, T.S. 2020. Zinc oxide nanoparticles synthesized using *Fusariumoxysporum* to enhance bioethanol production from rice-straw Biomass Bioenergy 143: 105840.

Hulkoti, N.I., and Taranath, T.C. 2014. Biosynthesis of nanoparticles using microbes—A review. Colloids Surf. B. 121: 474–483.

Hwa, K.-Y., and Subramani, B. 2014. Synthesis of zinc oxide nanoparticles on graphene–carbon nanotube hybrid for glucose biosensor applications. Biosens. Bioelectron. 62: 127–133.

Jothiprakasam, V., Sambantham, M., Chinnathambi, S., and Vijayaboopathi, S. 2017. *Candida tropicalis* biofilm inhibition by ZnO nanoparticles and EDTA. Arch. Oral Biol. 73: 21–24.

Kadam, V.V., Ettiyappan, J.P., and Mohan Balakrishnan, R. 2019. Mechanistic insight into the endophytic fungus mediated synthesis of protein capped ZnO nanoparticles. J. mater. sci. eng., B. 243: 214–221.

Kalpana, V.N., Kataru, B.A.S., Sravani, N., Vigneshwari, T., Panneerselvam, A., and Devi Rajeswari, V. 2018. Biosynthesis of zinc oxide nanoparticles using culture filtrates of *Aspergillus niger*: Antimicrobial textiles and dye degradation studies. OpenNano. 3: 48–55.

Keerthana, P., Vijayakumar, S., Vidhya, E., Punitha, V.N., Nilavukkarasi, M., and Praseetha, P.K. 2021. Biogenesis of ZnO nanoparticles for revolutionizing agriculture: A step towards anti-infection and growth promotion in plants. Ind. Crops Prod. 170: 113762.

Khamis, Y., Hashim, A., Hussien, A., and Abd-Elsalam, K. 2017. Fungi as ecosynthesizers for nanoparticles and their application in agriculture. pp. 55–75. *In*: Fungal nanotechnology. Springer International Publishing.

Khan, R., Kaushik, A., Solanki, P.R., Ansari, A.A., Pandey, M.K., and Malhotra, B.D. 2008. Zinc oxide nanoparticles-chitosan composite film for cholesterol biosensor. Anal. Chim. Acta. 616: 207–213.

Khandel, P., and Shahi, S.K. 2018. Mycogenic nanoparticles and their bio-prospective applications: current status and future challenges. J. Nanostruct. Chem. 8: 369–391.

Majumdar, M., Shivalkar, S., Pal, A., Verma, M.L., Sahoo, A.K., and Roy, D.N. 2020. Chapter 15—Nanotechnology for enhanced bioactivity of bioactive compounds. pp. 433–466. *In*: Biotechnological Production of Bioactive Compounds. Elsevier.

192 *Mycosynthesis of Nanomaterials: Perspectives and Challenges*

Manimaran, K., Balasubramani, G., Ragavendran, C., Natarajan, D., and Murugesan, S. 2021. Biological applications of synthesized ZnO nanoparticles using *Pleurotus djamor* against mosquito larvicidal, histopathology, antibacterial, antioxidant and anticancer effect. J. Clust. Sci. 32: 1635–1647.

Mohamed, A.A., Abu-Elghait, M., Ahmed, N.E., and Salem, S.S. 2021. Eco-friendly mycogenic synthesis of ZnO and CuO nanoparticles for *in vitro* antibacterial, antibiofilm, and antifungal applications. Biol. Trace Elem. Res. 199: 2788–2799.

MohdYusof, H., Mohamad, R., Zaidan, U.H., and Abdul Rahman, N.A. 2019. Microbial synthesis of zinc oxide nanoparticles and their potential application as an antimicrobial agent and a feed supplement in animal industry: a review. J. Anim. Sci. Biotechnol. 10: 57.

Motazedi, R., Rahaiee, S., and Zare, M. 2020. Efficient biogenesis of ZnO nanoparticles using extracellular extract of *Saccharomyces cerevisiae* : Evaluation of photocatalytic, cytotoxic and other biological activities. Bioorganic Chemistry 101: 103998.

Nagaraju, P., Puttaiah, S.H., Wantala, K., and Shahmoradi, B. 2020. Preparation of modified ZnO nanoparticles for photocatalytic degradation of chlorobenzene. Appl. Water Sci. 10: 137.

Pan, Y., Zuo, J., Hou, Z., Huang, Y., and Huang, C. 2020. Preparation of electrochemical sensor based on zinc oxide nanoparticles for simultaneous determination of AA, DA, and UA. Front. Chem. 8: 892.

Pantidos, N., and Horsfall, L.E. 2014. Biological Synthesis of metallic nanoparticles by bacteria, fungi and plants. J. Nanomed. Nanotechnol. 5: 0–0.

Preethi, P.S., Narenkumar, J., Prakash, A.A. et al. 2019. Myco-synthesis of zinc oxide nanoparticles as potent anti-corrosion of copper in cooling towers. J. Clust. Sci. 30: 1583–1590.

Rajan, A., Cherian, E., and GurunathanDrB. 2016. Biosynthesis of zinc oxide nanoparticles using *Aspergillus fumigatus* JCF and its antibacterial activity. *In*: International Journal of Modern Science and Technology 1: p-52–57.

Sabir, S., Arshad, M., and Chaudhari, S.K. 2014. Zinc oxide nanoparticles for revolutionizing agriculture: Synthesis and applications. Sci. World J. e925494.

Shivalkar, S., Gautam, P.K., Chaudhary, S., Samanta, S.K., and Sahoo, A.K. 2021a. Recent development of autonomously driven micro/nanobots for efficient treatment of polluted water. J. Environ. Manage. 281: 111750.

Shivalkar, S., Gautam, P.K., Verma, A., Maurya, K., Sk, M.P., Samanta, S.K., and Sahoo, A.K. 2021b. Autonomous magnetic microbots for environmental remediation developed by organic waste derived carbon dots. J. Environ. Manage. 297: 113322.

Shivalkar, S., and Sahoo, A.K. 2021c. Bio-molecules sensing using physical and microfluidics devices. MEMS Applications in Biology and Healthcare. AIP Publishing LLC; pp. 6-1-6–36.

Shivalkar, S., Verma, A., Singh, V., and Sahoo, A.K. 2021d. Dermatological delivery of nanodrugs. Nanotechnology in medicine. John Wiley & Sons, Ltd; p. 259–280.

Shobha, B., Lakshmeesha, T.R., Ansari, M.A., Almatroudi, A., Alzohairy, M.A., Basavaraju, S., Alurappa, R., Niranjana, S.R., and Chowdappa, S. 2020. Mycosynthesis of ZnO nanoparticles using *Trichoderma* spp. isolated from rhizosphere soils and its synergistic antibacterial effect against Xanthomonasoryzaepv. Oryzae. J. Fungi. 6: 181.

Taheri, M., Qarache, H., Qarache, A., and Yoosefi, M. 2015. The effects of zinc-oxide nanoparticles on growth parameters of corn (SC704). STEM Fellowship Journal. 1: 17–20.

Verma, A., Shivalkar, S., Sk, M.P., Samanta, S.K., and Sahoo, A.K. 2020. Nanocomposite of Ag nanoparticles and catalytic fluorescent carbon dots for synergistic bactericidal activity through enhanced reactive oxygen species generation. Nanotechnology 31: 405704.

Yadav, A., Kon, K., Kratosova, G., Duran, N., Ingle, A.P., and Rai, M. 2015. Fungi as an efficient mycosystem for the synthesis of metal nanoparticles: progress and key aspects of research. Biotechnol. Lett. 37: 2099–2120.

Zafar, M.N., Dar, Q., Nawaz, F., Zafar, M.N., Iqbal, M., and Nazar, M.F. 2019. Effective adsorptive removal of azo dyes over spherical ZnO nanoparticles. J. Mater. Res. Technol. 8: 713–725.

CHAPTER 10

Fungi-Mediated Synthesis of Carbon-Based Nanomaterials

Pramod U. Ingle,[1] *Kunal Banode,*[2] *Suchitra Mishra,*[2]
Aniket K. Gade[1,4,*] *and Mahendra Rai*[1,3]

Introduction

The interface between nanotechnology and mycology is defined as Myconanotechnology. A relatively diverse field, it can be treated as an area of immense potential. It gets its diverse status owing to the varied range and diversity present in fungi. We already know that Mycology is the study of fungi, and we have been studying that nanotechnology is the study constituting the synthesis, design, and application of technologically relevant and important particles of nano size (Adebayo et al. 2021). Myconanotechnology or the more direct term—Fungal nanotechnology (FN) is an existing concept, but the term itself was coined recently. This word (Myconanotechnology) was primarily introduced in the year 2009 by Prof. Rai from India who cites: Myco = Fungi, Nanotechnology = material production and utilization in the 1–100 nm size range. We can describe the same as the production and subsequent usage of nanoparticles (NPs) via fungus, especially in agricultural, biomedical, and environmental commodities. Prima facie, FN delves into a number of attributes related to the synthesis of metal and metal oxide nanoparticles by the preservation of the environment and processing techniques. There are several nanomaterials that have been synthesized by fungi, to name a few; silver, gold, magnesium, palladium, zinc, and copper, Some others have also been synthesized,

---

[1] Nanobiotechnology Lab, Department of Biotechnology, Sant Gadge Baba Amravati University, Amravati, Maharashtra, India – 444602.

[2] University Department of Pharmaceutical Sciences, Rashtrasant Tukadoji Maharaj University, Nagpur-440033.

[3] Department of Microbiology, Nicolaus Copernicus University, 87-100 Torun, Poland.

[4] Department of Biological Science and Biotechnology, Institute of Chemical Technology, Matunga, Mumbai, Maharshtra, India-400019.

* Corresponding author: aniketgade@gmail.com

such as metal sulfides, selenium, titanium dioxide, and cellulose, by other key fungal species, which includes Fusarium, mushrooms, endophytic fungus, *Trichoderma* and yeast. It was possible to easily develop cost-effective synthesis and nanoparticle extraction techniques if we closely monitor and observe the actual process of nanoparticle production and the impact of various variables on metal ion reduction. As an added benefit, it would also take care of mycogenic nanoparticles, protection, control, and risk assessment. Interestingly, Fungi are known to have the ability to produce extracellular enzymes that can hydrolyze complicated macromolecules. A strong source of concern for the application of fungus as the main producer for various types of metallic NPs is the metabolic capability of its usage in bioprocesses (Rai et al. 2009; Alghuthaymi et al. 2015; Prasad 2017; Jagtap et al. 2021).

Rhizopus oryaze metabolites were used as a biocatalyst for the green synthesis of magnesium oxide (MgO-NPs) nanoparticles. In addition, biogenic selenium nanoparticles (Se-NPs) were produced by *Bacillus megaterium* and were utilized as an antifungal agent against *R. solani*, the causal organisms of damping and root rot disease in *Vicia faba*, and also for induction of plant growth. Bacillus-mediated silver nanoparticles (AgNPs) with an onion-isolated endophytic bacteria *Bacillus endophyticus* strain H3, bactriosynthesized AgNPs with a concentration of 40 g/mL, had a high rice-blast antifungal activity with an inhibition rate of 88% mycelial. Also, spore germination and *M. oryzae* appressorium have been considerably suppressed by AgNPs. The Fusarium genus, credited as one of the most common fungal species, has an important role to play and can be considered a nanofactory for the fabrication of various nanoparticles. Although, the generation of NPs using lichens has been completely unexplored until now despite the usage of lichens as natural factories for synthesizing NPs being documented. Lichens are known to possess the virtue to produce several forms of NMs, such as bimetallic alloys, nanocomposites and metal and metal oxide NPs, via a reducible activity (Ibrahim et al. 2020; Hamida et al. 2021; Hashem et al. 2021; Rai et al. 2021; Saad et al. 2021).

The synthesis of various metal nanoparticles via fungal species can be summed up as Mycofabrication. Several fungal species are using both intracellular and extracellular methods, they are eco-friendly, clean, non-toxic agents for the synthesis of metal and metal oxide nanoparticles. The mycogenic synthesis of nanoparticles, a vital element of myconanotechnology, paves the way for a revolutionary, new, and practicable multidisciplinary science with significant results; owing to the vast diversity and spectrum of fungi. Myconanoparticles have found extensive and expansive usage in clean wastewater, in the detection and control of the pathogenic agent, as nematicides, in food conservation, and for many other products. Also, the mycogenic nanoparticles generated by various fungal species may be used in some potential agricultural applications to improve crop production by increasing growth and protection against infections. In addition, this will improve the toxicity to plant ecosystems of chemical pesticides, insecticides, and herbicides. Fungal-mediated nanoparticles have shown effective inhibition against pathogens causing infectious diseases in humans, especially against that deemed multi-resistant to traditional antibacterial agents (Abd-Elsalam 2021a).

A wide range of scientific domains, like agriculture, medicines, pharmaceuticals, and electronics have provided extensive usage to fungal-mediated nanoparticles.

Resultantly, some evaluations focused on the possible applications of mycogenic nanoparticles against, post-harvest antibiotics, plant diseases, mycotoxin management, plant pests, and a few animal pathogens. But that's not all; it has been found that fungal nanomaterials exhibit promise and a high potential for precision agriculture, enhanced diagnostics, targeted smart delivery systems, and biosensors. The macrofungi-derived NPs produced by Major mushroom species such as *Lentinus* spp., *Agaricus bisporus*, *Pleurotus* spp., and *Ganoderma* spp. produce macrofungi-derived NPs which are extensively known to have strong antibacterial, antiviral, antifungal, nutritional, immune-modulatory, antioxidant, and anticancer activities. In order to develop the more dependable, cheaper, and effective product(s) against major fungal infections of plants and animals; the development of antifungal nanohybrid agents with conjugates of inorganic or organic compounds, biological components, and biopolymers was researched extensively (Abd-Elsalam 2021a).

Fungi: a rich source of biomass and macromolecules

As we very well know, fungi are eukaryotic organisms that are categorized, rightly so, as decomposers. There are over 1.5 million species of fungi thriving in different habitats on Earth; wherein we have skimmed the surface by taxonomically identifying only 70,000 species. Additionally, about 5.1 million fungal species were found when high through-put sequencing method was undertaken. Even if the fungal population in the world is reported differently by different studies, it can be assumed that fungi are actually ubiquitous and in fact, their population might actually be more than anything ever reported in the conducted studies. Secretion of important enzymes to breakdown compounds of great complexity into simpler forms and use of diverse energy sources, extracellular breakdown of substrate house the major cosmopolitan characteristics of the fungal kingdom (Blackwell 2011).

Bio-reduction processes using plants or microbes make up the green synthesis of nanoparticles. Fungi can be utilized as an effective bio-reductant to synthesize metal nanoparticles both extracellularly and intracellularly, owing to the organic acids, proteins, enzyme hydrogenase, and nitrate-dependent reductase released by them. For instance, some common fungi successfully used for the biosynthesis of nanoparticles are *Fusarium* sp., *Aspergillus* sp., *Cladosporium* sp., *Trichothecium* sp., *Trichoderma* sp., and *Penicillium* sp. What is more, they are mostly from the phyla Basidiomycetes, Ascomycetes and Phycomycetes (Bakshi et al. 2017).

Carbon-based materials

The very important members of NMs family are Carbon-based nanomaterials (CNMs) which consist of various entities, comprising unique properties in their own stead. We can distinguish the CNMs into 0D-NMs (i.e., fullerenes, carbon dots, and particulate diamonds), 1D-NMs [i.e., CNFs, CNTs, and diamond nanorods], 2D-NMs (i.e., diamond nanoplatelets, graphene and graphite sheets), and 3D-NMs [i.e., nanocrystalline diamond (NCD) films, nanostructured diamond-like carbon (DLC) films and fullerite]. It is well known that the CNMs composed entirely of sp^2 bonded graphitic carbon are in all reduced dimensionalities including fullerenes

CNTs (1D-NMs), (0D-NMs), and graphene (3D-NMs) (Barhoum et al. 2019). Figure 1, below represents the various types of carbon nanoallotropes reported.

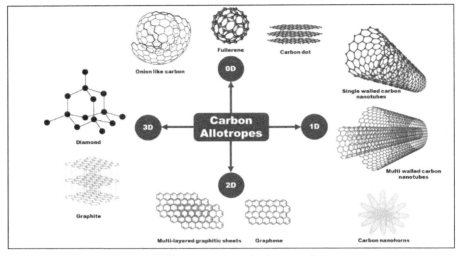

Figure 1. Different structures of carbon nanoallotropes.

Fullerenes nanostructures

Discovered in 1985, Fullerenes are other important members of the carbon family. Fullerenes can be defined as closed convexity molecules made of sp^2-hybridized carbon atoms. Structurally, each of these atoms is linked to the other three neighbouring carbon atoms morphing into pentagonal and hexagonal faces only. The fullerene family was formed with pentagonal and hexagonal faces with the atomic carbon clusters of Cn (n > 20). In unsubstituted fullerenes, carbon atoms are sp^2-hybridized and tricoordinate, resulting in the formation of a spherical conjugated unsaturated system. The immense bend of carbon atoms of a smaller structure such as C_{20} is also inappropriate for the sp^2-hybrid state of carbon whereas C_{60} is considered as the smallest stable pristine fullerene. This is due to all pentagonal faces being separated from each other by hexagonal ones (Barhoum et al. 2019).

Carbon nanotubes

CNTs are well-ordered, hollow nanostructures of carbon atoms bonded through sp2 bonds. These bonds end up giving them unparalleled mechanical strength and high thermal and electrical conductivity. Depending upon their structure, CNTs can be categorized into different types: (a) single-walled carbon nanotubes (SWCNTs); (b) double-walled carbon nanotubes (DWCNTs); and (c) multiwalled carbon nanotubes (MWCNTs). To change the surface properties and also change CNTs dispersion in the polar solvent (water), modifying the surface by adding functional groups to the CNT surface (functionalization) is considered. During the synthesis of The SWNTs, they are ended with cap-like structures at each terminal and the rings form ends by C–C bonds. The MWCNTs comprise a few layers of graphene sheets (2–10), with a thickness of more than one atom. The outer diameter of MWCNTs has a range from 2 to 100 nm and the interior diameter is approximately 1–3 nm. Their length is known to range from 0.2 to several micrometers. The degree of chirality in CNTs is a representative quantity of their conductivity and electrical properties. Apart from this, chirality is also responsible for determining the diameter of nanotubes, plus metallic or semimetallic characteristics. According to the graphitic arrangement pattern, MWNTs are of two kinds. There is the Russian-doll model wherein they are ordered in concentric layers. The second paradigm is called a parchment model, in which the graphite is rolled around itself to look like a rolled newspaper. Apart from these, there exists another type, known as double-walled nanotubes (DWNTs) comprising of a similar structure to SWNTs. Both are hugely important in varied fields and can be obtained with simple and easy techniques. There are also carbon nanohorns (CNHs) which are similar to CNTS and they are single-walled nanomaterials with makeup of graphene sheets (2–3 nm diameter) with the tip capped by a five-member ring. They resemble petals of dahlia and are often labelled as "dahlia-like" aggregates, unlike CNTs, carbon nano-horns (CNH structurally look like a horn having a cone shape. Additionally, these have several applications in drug delivery and find their usage primarily in drug delivery to cancer cells (Barhoum et al. 2019).

Graphite and graphene nanostructures

The structural observation of graphite is intriguing. It tells us that graphite consists of sp^2-hybridized carbon atoms, wherein each of these carbon atoms is bonded to three other carbon atoms, leaving one free electron in a p-π orbit; giving graphite a structure of a two-dimensional layer, Graphite possesses excellent thermal and electrical conductivity, owing to the π electrons which are delocalized within the layers. Interestingly, this kind of layered structure makes graphite extremely soft, as the layers can slide in parallel with ease. Carbon atoms can be arranged into materials with different structures with different hybridizations, such as graphite, diamond, fullerene, and CNTs. Graphene is a word combination between graphite and the suffix-ene, named in 1962 by Hanns-Peter Boehm who described single-layer carbon foils. Graphene was discovered in 1987, structurally found to be single sheets of graphite produced from graphite intercalation compounds. Graphene comprises hexagonal networks of the sp^2 carbon layer, where each two adjacent carbon atoms are strongly and covalently bonded. When carefully observed, this atomically thin sheet was found to be very stable under normal conditions, exhibiting high quality and continuity on a microscopic scale. It is very interesting that graphene is also considered as the "strongest" material (by weight) and has a negative thermal expansion coefficient (Barhoum et al. 2019).

Carbon nanofibers

CNFs are mainly comprised of graphene nanosheets, that have been distinguished from the straight wire and alternate fibers sp^2-based (one double bond, with two single

198 *Mycosynthesis of Nanomaterials: Perspectives and Challenges*

bonds), wherein it has a greater aspect proportion than 100. Put under transmission electron microscopy (TEM), it revealed that the layers of graphene planes in nearly all CNFs are generally not arranged along the axis of the fiber. The CNFs have been divided in the previous research, according to the angle of the graphene layers composing the strand, as the following:

1. Stacked of graphene layers piled at an angle between parallel and perpendicular to the fiber axis.
2. Stacked of graphene layers piled perpendicularly to the fiber axis.

Owing to the growth mechanism of CNFs, these layering preparations are possible, where they rely on the gaseous carbon feedstock and geometric sides of the metallic catalyst particle (e.g., hydrocarbon or carbon monoxide gas) that is inserted during the synthesis process. Such general classifications leave further space for additional classifications of CNFs as vapor-grown CNFs and electrospun CNFs (Barhoum et al. 2019).

Fungi-based synthesis of carbon-nanoparticles

Apart from plants, a large number of microorganisms have been proven to synthesize nanoparticles, either intra or extracellularly. In recent years, applications of nanoparticles in green energy, optics, electronics, water treatment systems, medicine and diagnostics, and even many more can be attributed to their size in the range of 1–100 nm. Recently, yeasts and fungi are getting more attraction to their ability to synthesize nanoparticles intra and extracellularly (Saxena et al. 2014). Yeast strain MKY3, *Candida glabrata, Schizosaccharomyces pombe, Rhodosporidium diobovatum, Saccharomyces cerevisiae,* and *Cryptococcus humicola* were reported to have been applied for the fabrication of nanoparticles from heavy metals (Kowshik et al. 2003; Mourato et al. 2011; Vainshtein et al. 2014; Saxena et al. 2014). Mycogenic route for nanoparticles synthesis has been well recognized because this totipotent eukaryotic microorganism has several remarkable features which have been well documented. Fungi can be used as an excellent source of various extracellular enzymes which influences nanoparticle synthesis. There are various reasons why fungi can be chosen as better nano factories over bacteria and plants. Fungi are excellent source of secretory proteins, easy to isolate and culture (Rai et al. 2009), growth conditions are manipulated easily as desired (Saha et al. 2010), and most importantly extracellular synthesis of nanomaterials (Gade et al. 2008). All these properties contribute to the carbon profile of fungal biomass making them an exceptionally important member of the synthesis of nanoparticles. Thus fungal biomass can also be used for the synthesis of carbon-based nanomaterials through various methods of fabrication and synthesis.

Carbonaceous nanomaterials derived from fungi

Fungi being called as 'nanofactories' have been proved to synthesize various nanomaterials in a facile, green, and eco-friendly way to obtain nanoscale materials with variable shapes, size, chemical composition, and controlled dispersity owing to their probable usage for human welfare (Sunkar and Nachiyar 2013; Abd-Elsalam and Hashim 2013; Abd-Elsalam 2021b). Commercially available nanoscale materials are mainly derived from micronutrients like Mn, Cu, Fe, Zn, Mo, N, B. along with these another group of nanomaterials, like carbon nano-onions and chitosan nanoparticles may be used instead of traditional crop fertilizers to improve crop quality and growth, and in the next decade, novel nanofertilizers are predicted to inspire and turn existing fertilizer processing industries (He et al. 2018; Marella et al. 2021). Polymeric carbon nanomaterials like chitosan nanoparticles synthesized by polymerizing methacrylic acid have been proven to be an effective nanocarrier for the slow release of fertilizers in *Pisum sativum* var. Master B plants (Khalifa and Hasaneen 2018). Densely porous graphene-like carbon nanomaterial was synthesized by controlled hydrothermal digestion of fungus Auricularia biomass in presence of KOH and N_2 at 800°C (Long et al. 2015). Filamentous fungi are a kind of renewable resource with fast growth, low cost, easy access, and environmental friendliness. In another report by Chen and colleagues (2019), a porous carbon material was derived from fungal hyphae of *Irpex lacteus* in presence of mixed alkali as an activator. The carbon nanotubes were fixed onto fungal hyphae to get nanocomposite by a biological method. The fungal hyphae can be used as an efficient platform for nanomaterial assembly (Zhu et al. 2018).

Methods of carbon nanomaterials synthesis from fungi

Synthetic techniques for carbon nanomaterials are categorized into two classes, 'bottom-up' and 'top-down' courses. These can be accomplished by means of chemical, electrochemical, or physical systems. Top-down strategies involve the fragmentation of carbon matter into carbon nanoparticles, and strategies comprising arc discharge, laser ablation, and electrochemical approaches. Bottom-up strategies incorporate template strategy, thermal routes, pyrolytic process, hydrothermal and aqueous methods, supported synthetic technique, reverse micelle technique, microwave-assisted strategy, and substance oxidation. The yield of CDs could be enhanced during arrangement or posttreatment. Alteration of CDs is additionally vital to obtaining favorable surface properties, which are key for solvency and applications (Wang and Hu 2014; Roy et al. 2015; Zuo et al. 2016; Singh et al. 2016). The Figure 2, below shows the schematics mechanism of microwave-assisted synthesis of carbon dots by a top-down approach.

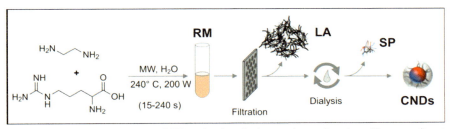

Figure 2. Scheme of the microwave (MW)-assisted synthetic procedure and work up. Three constituents resulted within the reaction mixture (RM): (i) large aggregates (LA—extracted by filtration), (ii) small particles (SP—extracted by dialysis), and (iii) final carbon nanodots (CNDs). (Figure source: Rigodanza et al. 2021—Open access).

Methods of detection and characterization

Characterization is a decisive step in the study of assets of nanomaterials to appraise their full latency in various applications. Nano-carbon-based materials have properties that are sensitive to shape, size, agglomeration state, and concentration. It is therefore perilous to characterize and quantify these factors *in situ*. Classical characterization techniques that depend on microscopic observations were often time-consuming and provide mostly qualitative results. The modern methods based on detection under specific wavelengths provide qualitative as well as quantitative properties of nanomaterials.

Spectrophotometric detection of carbon nanomaterials

Spectroscopy has been studied as an alternative tool for identifying, characterizing, and studying nanomaterials *in situ* and in a quantitative way. Spectrometric measurements are primarily applied for studying surface properties like shape, charge, and presence or absence of various active functional groups, in the dispersed state. Mycosynthesized and fungal biomass-based carbon nanomaterials are primarily detected by spectrophotometric analysis. This determines the absorption maxima of the synthesized nanomaterial. Also, it helps to predict the size and shape of the carbon nanomaterial. The position of absorption peak determines the size and shape of nanoparticles synthesized. Absorption at lower wavelengths indicates the formation of spherical particles, whilst the formation of rod-shaped nanoparticles is indicated by a shift in absorption maxima towards higher wavelengths (Rigodanza et al. 2021). The Figure 3, below shows the formation of spherical carbon nanodots with the spectral maxima of 413 nm in the dispersed form.

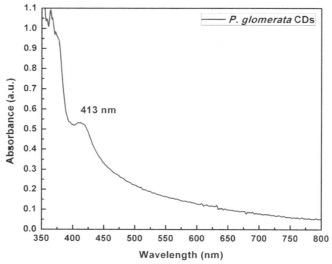

Figure 3. UV-Visible spectrometric analysis of *Phoma glomerata* biomass-based carbon dots with absorption maxima at 413 nm.

UV-Visible spectrometry is applied to quantity absorption in liquid samples. In contrast to bundled Carbon Nanotubes (CNTs), the completely dispersed CNTs are very dynamic in the 200 to 1200 nm wavelength region. The UV-Vis technique is used to detect individual CNTs by relating the intensity of absorption at a precise wavelength to the concentration of CNTs dispersed in the solution based upon the Beer-Lambert law (Grossiord et al. 2005). A stable suspension is obtained after a small interval of time with a small change in absorption maxima (Jiang et al. 2003). When nanocarbon suspension is subjected to sonication absorbance tends to increase at the beginning and becomes stable whose magnitude differs with surfactant concentration (Yu et al. 2007). The spectrometric analysis provides real-time monitoring dispersion in an aqueous medium.

Nanoparticle Tracking Analysis (NTA)

The principle of NTA is based upon Brownian motion and light scattering and determines the particle size distribution of samples in liquid suspension. The laser beam is passed through the nanoparticle suspension sample which is then scattered by nanoparticles and finally detected under a 20x magnification microscope (operated at about 30 frames per second (fps)). It tracks the particles under Brownian motion within the field of view of about 100 x 80 x 10 μm. The particle motion is captured as a video file which is further analyzed by the provided respective software.

The above Figure 4, shows the particle size distribution of carbon cots synthesized from dehydrated biomass of fungus *Phoma glomerata* by pyrolytic digestion in presence of a strong acid like sulphuric acid. The range displayed in the above figure predicts the particles with variable sizes have been synthesized. The average size shown is about 12.7 nm with a mode value of 14.2 nm and a standard deviation (SD) of 3.7 nm. This confirmed the synthesis of polydispersed carbon dots synthesis. NTA was reported to primarily apply to the characterization of studies of different carbonaceous nanomaterials, including CNT-nematic liquid crystal composite materials (Trushkevych et al. 2007; 2008). NTA also determines the oxidative potential of a panel of metallic and carbonaceous nanoparticles (Hohl et al. 2009). Recently the technique has expanded its scope in studies related to analysis of carbon nanotubes and nanocolloids as well.

Zeta potential measurement and stability analysis

Zeta potential, or ζ potential, is an enumerative abbreviation for electrokinetic potential of materials in colloidal systems. The theoretical value of zeta potential is an electric potential in the interfacial double layer of a dispersed particle against a far interfacial point in the continuous phase (Wu 2022). On the virtue of zeta potential surface charge is determined for nanocrystals which is an attribute for the physical stability of nanosuspensions (Jiang et al. 2008). Good physical stability of nanosuspensions is indicated by a large positive or negative value of zeta potential for nanocrystals which is presented by electrostatic repulsion of individual particles. A better physical colloidal stability is achieved when zeta potential value other than −30 mV to +30 mV is observed for any nano-colloid. On the other hand,

202 Mycosynthesis of Nanomaterials: Perspectives and Challenges

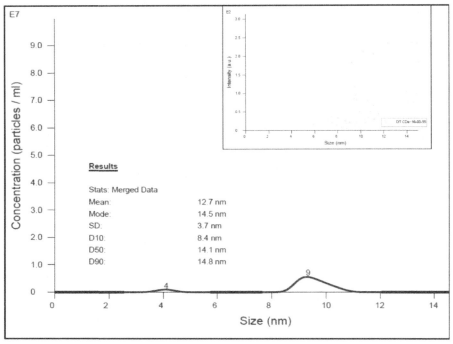

Figure 4. (a) Size distribution of *P. glomerata* CDs by Nanoparticle Tracking Analysis (version NanoSight 3.2) with an average size of 12 nm; (b) Inset figure shows the distribution of CDs over nanoscale (Author's collection).

van der Waals's attractive forces act upon particles with lower zeta potentials, tending their aggregation and precipitation of colloid i.e. physical instability (Freitas and Müller 1998; Shah et al. 2014). Other factors influencing the nanosuspensions are the presence of surfactants, material properties, and solution chemistry. The carbon

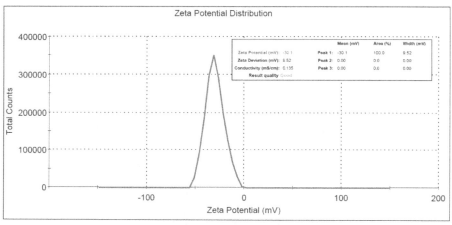

Figure 5. Zeta potential measurement of *P. glomerata* biomass CDs, average ZP = –30.1 mV (Author's collection).

dots (CDs) synthesized from fungal biomass show an average zeta potential of –30.1 mV and a standard deviation of 9.52 mV at 25°C (Figure 5, below).

Many studies have reported the synthesis of negatively charged CDs and other carbon-based nanomaterials. Kung and colleagues (2020) have reported the microwave-assisted synthesis of negatively charged CDs using citric acid and urea as precursors. Authors reported zeta potential for synthesized of –11.6 mV when subjected to dynamic light scattering (DLS) and showed to have antimicrobial activity against methicillin-resistant *Staphylococcus aureus* (MRSA) and vancomycin-intermediate *Staphylococcus aureus* (VISA). The zeta potential and stability studies are supported by Bing et al. (2016), Yang et al. (2016), and Travlou et al. (2018).

Fourier Transform Infrared (FTIR) spectrometry analysis

Fungi when cultured in *in vitro* conditions and exposed to a hypotonic atmosphere, they tend to secrete a huge range of secondary metabolites. The members of the genus *Phoma* (*P. sorghina*, *P. herbarum*, *P. exigua* var. exigua, *P. macrostoma*, *P. glomerata*, *P. macdonaldii*, etc.) have been reported to synthesize and secrete a vast range of bioactive secondary metabolites like phytotoxin and anthraquinone pigments. These are applied as mycopesticides, agrophytochemicals, and dyes (Rai et al. 2009). These secondary metabolites contribute to the carbon pool of fungal biomass and can be used for the synthesis of various carbon nanomaterials. The presence of these metabolites is detected by a technique called as FTIR, i.e., Fourier Transform Infrared (FTIR) spectrometry. FTIR thus predicts the probable percent intensity of a particular functional group and a respective biomolecule present in the fungal biomass analyzed.

The given FTIR spectrum (Figure 6) of *P. glomerata*-derived CDs showed the presence of peculiar peaks assigned to various functional groups present. The peak at 1731 cm^{-1} corresponds to the C = O group, and 1582 cm^{-1} corresponds to C = C stretch. Similarly, transmittance peaks at 1412 cm^{-1} and 1263 cm^{-1} are assigned to O = H and C = N stretch of hydroxyl and amino compounds respectively (Kung et al. 2020). The presence of these functional groups determines the presence of various biomolecular entities contributing to the formation of CDs from fungal biomass.

Fluorescence and Potoluminescence detection

Various water-soluble fluorescent carbon nanomaterials have been synthesized from natural sources like fruits, grains, milk, plants, and fungi. Jeong et al. (2014) have reported the synthesis of fluorescent carbon nanoparticles from mango fruit. The resulting product showed blue, green, and yellow fluorescence. An absorption band for most of the CDs lies in the range of around 260–323 nm, irrespective of the method of synthesis. Most interestingly the size-dependent optical absorption or photoluminescence are the classic signs of quantum confinement in case CDs. Similarly, eucalyptus twigs are employed for the synthesis of fluorescent carbon nanoparticles by a facile and green hydrothermal method followed by the differential washing technique. Multiemissive carbon nanoparticles emitting red, blue, and green fluorescence was obtained (Damera et al. 2020).

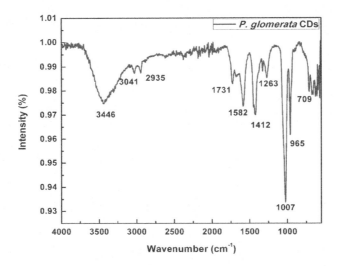

Figure 6. FTIR spectrum of CDs derived from *P. glomerata* biomass.

The Figure 7, above shows the photoelectric excitation-emission for carbon dots synthesized from *P. glomerata* biomass by hydrothermal digestion by the authors. Absorption maxima were observed at 279 nm and emission was observed at 461 nm. The emission of blue light is indicated in the inset figure when exposed to UV light.

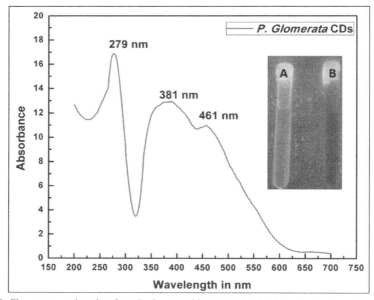

Figure 7. Fluorescent carbon dots from *P. glomerata* biomass and their photoelectric absorption-emission maxima.

Importance and application of fungal-based carbon nanomaterials (FBCNs)

The combination of mycology and nanotechnology has become a new approach called myconanotechnology. Nowadays, metals used in nanotechnology has replaced by Fungi due to its insoluble complexes like metal sulfides in colloidal form. As we studied other microorganisms, fungi are more suitable as they are biocompatible and biodegradable. Fungi are suitable due to its wide varieties, easily isolated, cultured, and maintained (Bakshi et al. 2018). Every member of the carbon-based material family displays diverse properties and has been broadly used in energy production, agriculture, bioreactors, drug delivery, tissue engineering, biosensor designing, diagnosis, cancer therapy, and imaging (Hong et al. 2015). Fungal-based carbon nanomaterials are becoming attractive nanomaterials in the field of science and technology. Some applications of fungal-based carbon nanomaterial are summarized here (Zhao et al. 2014, 2015).

In agriculture

Nanobiochar is biochar reduced to nano-scale material that is basically related to agriculture which has better physical, chemical, and surface properties including mechanical, thermal, optical, electrical, and structural diversity. It has worked in various manners like enhancement of plant growth and soil qualities, treatment of plant disease, bioreduction of contaminants and pesticides, wastewater purification, and enzyme immobilization. Nanobiochar has an excellent capacity for absorbing pollutants, nutrients, and contaminants and has mobility in the soil as compared to biochar so used as a potential waste management substitute (Chausali et al. 2021).

In physics

In 1991, Prof. Sumio Iijima was the first to observe the development of multiwall CNTs from carbon arc discharge. After some years, Prof. Iijima and Donald Bethune individually alleged single-wall CNTs formation (Monthioux and Kuznetsov 2006). In the following years, research on CNTs flourished speedily. CNTs were defined as hollow cylinders entailing graphitic sheets and classified into single-walled carbon nanotube (SWCNT) with an interlayer spacing of 3.4 Å and multi-walled carbon nanotube (MWCNT) (Odom et al. 1998). Due to its unique mechanical, electrical and structural diversity these CNT are very useful for biomedical purpose.

Biomedical applications

Atomic force microscopy

Atomic force microscopy (AFM) has great potential as a tool for structural biology, a field in which there is increasing demand to characterize larger and more complex biomolecule systems. However, the poorly characterized silicon and silicon nitride probe tips currently employed in AFM limit its biological applications (Hafner et al. 2001).

Table 1. Applications of Fungal-based carbon nanomaterial as biosensor.

Sr. No	Types of fungal based Carbon Nanomaterials	Technique	Use	References
1	SWNTs	NIR photoluminescence	Photometry	Jena et al. (2017)
2	Carbon nanotube non-woven fabrics	Conjugation of glucose oxidase	Glucose biosensors	Zhu et al. (2014)
3	SWNTs	Detecting nitric oxide and sensing epinephrine	Detecting human blood proteins in corona patients	Ulissi et al. (2014)
4	ssDNA functionalized SWNTs	NIR photoluminescence	Lipid content in the endosomal lumen of live cells	Jena et al. (2017)
5	DNA-mediated SERS property of SWNTs	T-rich de-oxy-ribonucleic acid (DNA)-mediated surface-enhanced Raman scattering	Ct DNA in human blood.	Zhou et al. (2016)
6	SWCNT	Efflux from *Escherichia coli* (bacteria) and *Pichia pastoris* (yeast)	Detection of individual proteins	Landry et al. (2017)
7	MWCNT	Field effect transistor	detecting arginase-1	Baldo et al. (2016)
8	CNT-based	Field effect transistor	DNA solution	Xuan et al. (2017)
9	Reduced graphene nanowire (RGNW)	Electrochemical oxidation signals of the discrete nucleotide bases	four bases of DNA (guanine, tyrosine, adenine and cytosine)	Akhawan et al. (2015)
10	COOH functionalized Graphene Oxide	DPV	Detected DNA	Cheng et al. (2017)
11	DNA biosensor	Graphene field effect transistors	Detection limit of 1 fM for 60-mer DNA	Ping et al. (2016)
12	Au NP-Gr-FETs	Thiol-functionalized Au NP-Gr-FETs	Detect DNA with a detection limit of 1 nM	Gao et al. (2016)
13	single-layer graphene (SLG)	FET biosensor	Detect a very low concentration of DNA (10 fM)	Zheng et al. (2015)
14	Graphene surface-modified vertically aligned silicon nanowire	sensitivity and selectivity	Detecting oligonucleotides	Kim et al. (2016)
15	Graphene surface-modified	Adsorption of DNA on GO	Detection of *Mycobacterium tuberculosis* DNA	Huang and Liu (2018)

Biosensors

Their high conductivity, high aspect ratio, sensitivity, high chemical stability, and fast electron-transfer rate make them exceptionally appropriate for biosensing applications. The elementary component of CNT-based biosensors is an immobilized biomolecule on their surface, which enhances the recognition and the process of signal transduction (Zhao et al. 2002; Lin et al. 2004; Jacobs et al. 2010; Kumar et al. 2015).

Graphene oxide is capable of dynamically cooperating with the probe for the transduction of desired response to the target molecules. This process is accomplished by Raman scattering, fluorescence, and electrochemical reaction On the basis of which GOs are generally applied in nanomaterial based biosensors (Kim et al. 2017; Suvarnaphaet and Pechprasarn 2017). Graphene nanomaterials are widely used for the discerning electrochemical sensing of single- and double-stranded DNA (Liu et al. 2012; Wang et al. 2013; Tang et al. 2015).

For drug delivery

Due to non-invasive penetration across the biological membranes fungal based CNT is very popular for drug delivery. Drug molecules are conjugated to CNT sidewalls *via* covalent or non-covalent bonding bridging the drug molecule and functionalized CNT. Henceforth, the consumption of the inner void space of CNT for loading drug delivers an ideal remoteness of the drug from the physiological environment (Eatemadi et al. 2014; Cheng et al. 2015; Kang et al. 2017).

Table 2. Applications of Fungal based carbon nanomaterial in Drug delivery.

Sr. No	Formulation	Drug	References
1	CNT	Cisplatin	Mejri et al. (2015)
2	(CaPO4)-crowned drug-loaded multiwall CT	Anticancer drug	Banerjee et al. (2015)
3	PEG functionalized MWCNTs (DOX/ES-PEG-MWCNTs)	Doxorubicin (DOX)	Mehra and Jain (2015)
4	Functionalized CNT	Increase Drug Solubility	Dhar et al. (2008); Liu et al. (2008)
5	CNT	Doxorubicin	Huang et al. (2011)
6	CNT	paclitaxel	Singh et al. (2016)
7	CNT	Docetaxel	Raza et al. (2016)
8	CNT	Oxaliplatin	Lee et al. (2016)
9	Multi-walled carbon nanotubes (MWCNTs)	Ibuprofen	Shi et al. (2015)
10	Near-infrared (NIR) light-responsive release of MWCNT	Methylene blue (MB)	Estrada et al. (2013)
11	Phospholipid monolayer coated graphene	Doxorubicin (DOX)	Liu et al. (2012)
12	Poly-N-isopropyl acrylamide (PNIPAM) grafted GO (GPNM)	Doxorubicin and indomethacin	Kundu et al. (2015)

208 *Mycosynthesis of Nanomaterials: Perspectives and Challenges*

Table 3. Application of Fungal based carbon nanomaterial in cancer therapy.

Sr. No	Formulation	Drug	Therapy	References
1	MWCNT	Candesartan (CD).	Photothermal therapy.	Su et al. (2017)
2	MWCNT	DOX-loaded TAT-chitosan	Chemo and photothermal therapy.	Kim et al. (2017)
3	SWNT-Au-PEG-FA nanomaterials improved cancer-killing efficacy	Anticancer Drug	Photothermal therapy	Wang et al. (2012)
4	Graphene Oxide	Anticancer drug	cancer therapy	Kumar et al. (2017)
5	RGO-gold nanorods	DOX	Photothermal therapy and chemotherapy	Song et al. (2014, 2015, 2016)
6	GO (GO (HPPH)-PEG-HK). The GO (HPPH)-PEG-HK	Anticancer Drug	cytotoxic CD8+ CT Imaging	Yu et al. (2017)
7	Graphene nanosheet (graphene nanosponge)		Chemotherapeutics	Su et al. (2016)
8	Carbon quantum dot (CQD)	DOX	Photothermal therapy	Zhang et al. (2015, 2017); Kim et al. (2012, 2015)

In cancer therapy

FBCNs are broadly applied in biomedical field because of their adaptable properties. These have become an attractive candidates for the transport of genes, an anticancer drugs, and proteins for chemotherapy (Aldieri et al. 2013; Lin et al. 2016; Li et al. 2017; Liu et al. 2017).

Conclusion

Over the last decade, wide spread research efforts have been conducted on FBCNs as one of the most widely accepted class of nanomaterials. Due to their characteristic optical, electrochemical, mechanical, electrical and ecofriendly properties they have been expansively used in multiple research areas. Furthermore, in arrears to their mutable size and shape, surface properties, FBCNs have drawn great attention in biomedical engineering, agriculture, environment, waste and water management. FBCNs could efficiently interact with biomolecules and responds to the light stimulus simultaneously. By which it could be used for developing biomedical applications in future. To reduce its toxicity, several chemical modification need to be done and fruitfully used in bio-applications counting drug delivery, detection of biomolecules, tissue engineering, and cancer therapy. Moreover, we similarly emphasize on some freshly found key characteristics of FBCNs and their exploitations for advanced bio-applications. Fungal system with respect to its superiority over other biological systems for the synthesis of different nanomaterials in general and carbon based nanomaterials in particular can be established.

Acknowledgements

MR thankfully acknowledges the financial support rendered by the Polish National Agency for Academic Exchange (NAWA) (Project No. PPN/ULM/2019 /1/00117/U/ 00001) to visit the Department of Microbiology, Nicolaus Copernicus University, Toruń, Poland.

References

Abd-Elsalam, K.A., and Hashim, A.F. 2013. Hidden fungi as microbial and nano-factories for anticancer agents. Fungal. Genom. Biol. 3: e115. doi:10.4172/2165-8056.1000e115.

Abd-Elsalam, K.A. 2021a. Special Issue : Fungal Nanotechnology. 4–6.

Abd-Elsalam, K.A. 2021b. Special Issue: Fungal Nanotechnology. J. Fungi. 7: 583. https://doi.org/10.3390/jof7080583.

Adebayo, E.A., Azeez, M.A., Alao, M.B., Oke, A.M., and Aina, A.D. 2021. Fungi as veritable tool in current advances in nanobiotechnology. Heliyon. 7(11): e08480. 10.1016/j.heliyon.2021.e08480.

Akhawan, A., Amenta, V., and Aschberger, K. 2015. Carbon nanotubes: potential medical applications and safety concerns. Adv. Rev. 7: 371–386. doi: 10.1002/wnan.1317.

Aldieri, E., Fenoglio, I., Cesano, F., Gazzano, E., Gulino, G., and Scarano, D. 2013. The role of iron impurities in the toxic effects exerted by short multiwalled carbon nanotubes (MWCNT) in murine alveolar macrophages. J. Toxicol. Environ. Health A. 76: 1056–1071. doi: 10.1080/15287394.2013.834855.

Alghuthaymi, M.A., Almoammar, H., Rai, M., Said-Galive, E., and Abd El-salam, K. 2015. Myconanoparticles: synthesis and their role in phytopathogens management. Biotechnol. Biotechnol. Equip. 29(2): 221–236. doi: 10.1080/13102818.2015.1008194.

Bakshi, M., Mahanty, S., and Chaudhuri, P. 2017. Fungi-Mediated Biosynthesis of Nanoparticles and Application in Metal Sequestration. doi: 10.1201/9781315153353.

Bakshi, M., Mahanty S., and Chaudhuri, P. 2018. Nanobiochar and biochar based nanocomposites: Advances and applications. 100191. https://doi.org/10.1016/j.jafr.2021.100191.

Baldo, S., Buccheri, S., Ballo, A., Camarda, M., LaMagna, A., Castagna, M.E., Romano, A., Iannazzo, D., Di Raimondo, F., Neri, G. and Scalese, S. 2016. Carbon nanotube-based sensing devices for human Arginase-1 detection. Sens. Biosensing Res. 7: 168–173. doi: 10.1016/j.sbsr.2015.11.011.

Banerjee, S.S., Todkar, K.J., Khutale, G.V., Chate, G.P., Biradar, A.V., Gawande, M.B., Zboril, R. and Khandare, J.J. 2015. Calcium phosphate nanocapsule crowned multiwalled carbon nanotubes for pH triggered intracellular anticancer drug release. J. Mater. Chem. B. 3: 3931–3939. doi: 10.1039/C5TB00534E.

Barhoum, A., Shalan, A.E., El-Hout E.I., Ali, G.A.M., Abdelbasir, S.M., Serea, E.S.A., Ibrahim, A.H., and Pal, K. 2019. A Broad Family of Carbon Nanomaterials : Classification, Properties, Synthesis, and Emerging Applications. 10.1007/978-3-319-42789-8_59-1.

Bing, W., Sun, H., Yan, Z., Ren, J., and Qu, X. 2016. Programmed bacteria death induced by carbon dots with different surface charge. Small. 12(34): 4713–4718. doi:10.1002/smll.201600294.

Blackwell, M. 2011. The Fungi: 1, 2, 3 ... 5.1 million species?', Am. J. Bot. 98: 426–438. doi: 10.3732/ajb.1000298.

Chausali, N., Saxena, J., and Prasad, R. 2021. Nanobiochar and biochar based nanocomposites: Advances and applications. J. Agri. Food Res. 5. 100191. doi:10.1016/j.jafr.2021.100191.

Cheng, D., Yang, L., Li, X., Zhou, J., Chen, Q., Yan, S., Li, N., Chu, M., Dong, Y., and Xie, Z. 2017. An electrochemical DNA sensing platform using carboxyl functionalized graphene as the electrode modified material. J. Electrochem. Soc. 164, H345–H351. doi: 10.1149/2.0951706jes.

Chen, S., Wang, Z., Xia, Y., Zhang, B., Chen, H., Chen, G., and Tang, S. 2019. Porous carbon material derived from fungal hyphae and its application for the removal of dye. RSC Advances 9(44): 25480–25487. doi: 10.1039/c9ra04648h.

Cheng, Y.J., Luo, G.F., Zhu, J.Y., Xu, X.D., Zeng, X., Cheng, D.B., Li, Y.M., Wu, Y., Zhang, X.Z., Zhuo, R.X., and He, F. 2015. Enzyme-induced and tumor-targeted drug delivery system based on multifunctional mesoporous silica nanoparticles. ACS Appl. Mater. Interfaces. 7: 9078–9087. doi: 10.1021/acsami.5b00752.

Dhar, S., Liu, Z., Thomale, J., Dai, H., and Lippard, S.J. 2008. Targeted single-wall carbon nanotube-mediated Pt (IV) prodrug delivery using folate as a homing device. J. Am. Chem. Soc. 130: 11467–111476. doi: 10.1021/ja803036e.

Damera, D.P., Manimaran, R., Krishna Venuganti, V.V., and Nag, A. 2020. Green synthesis of full-color fluorescent carbon nanoparticles from eucalyptus twigs for sensing the synthetic food colorant and bioimaging. ACS Omega. 5(31): 19905–19918. doi: 10.1021/acsomega.0c03148.

Eatemadi, A., Daraee, H., Karimkhanloo, H., Kouhi, M., Zarghami, N., Akbarzadeh, A., Abasi, M., Hanifehpour, Y., and Joo, S.W. 2014. Carbon nanotubes: properties, synthesis, purification, and medical applications. Nanoscale Res. Lett. 9: 393–405. doi: 10.1186/1556-276X-9-393.

Estrada, A.C., Daniel-da-Silva, A.L., and Trindade, T. 2013. Photothermally enhanced drug release by κ-carrageenan hydrogels reinforced with multi-walled carbon nanotubes. RSC Adv. 3(27): 10828. doi: 10.1039/c3ra40662h.

Freitas, C., and Müller, R.H. 1998. Spray-drying of solid lipid nanoparticles (SLNTM). Europ. J. Pharma. Biopharma. 46(2): 145–151. doi: 10.1016/s0939-6411(97)00172-0.

Gade, A., Bonde, P.P., Ingle, A.P., Marcato, P., Duran, N., and Rai, M.K. 2008. Exploitation of *Aspergillus niger* for synthesis of silver nanoparticles. J. Bio. Based Mater. Bioener. 2(3): 1–5.

Gao, Z., Kang, H., Naylor, C.H., Streller, F., Ducos, P., Serrano, M.D., Ping, J., Zauberman, J., Rajesh., Carpick, R.W., Wang, Y.J., Park, Y.W., Luo, Z., Ren, L., and Johnson, A.T.C. 2016. Scalable production of sensor arrays based on high-mobility hybrid graphene field effect transistors. ACS Appl. Mater. Interfaces. 8: 27546–27552. doi: 10.1021/acsami.6b0923.

Grossiord, N., Regev, O., Loos, J., Meuldijk, J., and Koning, C.E. 2005. Time-dependent study of the exfoliation process of carbon nanotubes in aqueous dispersions by using UV-visible spectroscopy. Anal. Chem. 77(16): 5135–5139.

Hafner, J.H., Cheung, C.L., Woolley, A.T., and Lieber, C.M. 2001. Structural andfunctional imaging with carbon nanotube AFM probes. Prog. Biophy. Mol. Biol. 77: 73–110.

Hamida, R.S., Ali, M.A., Abdelmeguid, N.E., Al-Zaban, M.I., Baz, L., and Bin-Meferij, M.M. 2021. Lichens—A potential source for nanoparticles fabrication: A review on nanoparticles biosynthesis and their prospective applications. J. Fungi. 7(4): 291. doi:10.3390/jof7040291.

Hashem, A.H., Abdelaziz, A.M., Askar, A.A., Fouda, H.M., Khalil, A.M.A., Abd-Elsalam, K.A., and Khaleil, M.M. 2021. *Bacillus megaterium*-mediated synthesis of selenium nanoparticles and their antifungal activity against *Rhizoctonia solani* in Faba Bean Plants. J. Fungi. 7(3): 195. doi:10.3390/jof7030195.

He, X., Deng, H., and Hwang, H. 2018. The current application of nanotechnology in food and agriculture. J. Food Drug Anal. doi:10.1016/j.jfda.2018.12.002.

Hohl, M.S.S., Sauvain, J.J., Donaldson, K., and Riediker, M. 2009. Oxidative potential of a panel of Carbonaceous and Metallic Nanoparticles, European Aerosol Conference 2009, Karlsruhe, Abstract T043A10.

Hong, G., Diao, S., Antaris, A.L., and Dai, H. 2015. Carbon nanomaterials for biological imaging and nanomedicinal Therapy. Chem. Rev. 115: 10816–10906. doi: 10.1021/acs.chemrev.5b00008.

Huang, H., Yuan, Q., Shah, J.S., and Misra, R.D. 2011. A new family of folate-decorated and carbon nanotube-mediated drug delivery system: synthesis and drug delivery response. Adv. Drug Deliv. Rev. 63: 1332–1339. doi: 10.1016/j.addr.2011.04.001.

Huang, Z., and Liu, J. 2018. Length-dependent diblock DNA with poly-cytosine (Poly-C) as high-affinity anchors on graphene oxide. Langmuir. 34: 1171–1177. doi: 10.1021/acs.langmuir.7b02812.

Ibrahim, E., Luo, J., Ahmed, T., Wu, W., Yan, C., and Li, B. 2020. Biosynthesis of silver nanoparticles using onion endophytic bacterium and its antifungal activity against rice pathogen *Magnaporthe oryzae*. J. Fungi. 6(4): 294. doi:10.3390/jof6040294.

Jacobs, C.B., Peairs, M.J., and Venton, B.J. 2010. Review: carbon nanotube based electrochemical sensors for biomolecules. Anal. Chim. Acta. 662: 105–127. doi: 10.1016/j.aca.2010.01.009.

Jagtap, P., Nath, H., Kumari, P.B., Dave, S., Mohanty, P., Das, J., and Dave, S. 2021. Mycogenic fabrication of nanoparticles and their applications in modern agricultural practices & food industries. pp. 475–488. *In*: Sharma, V.K. et al. (eds.). Academic Press. doi: https://doi.org/10.1016/B978-0-12-821734-4.00007-1.

Fungi-Mediated Synthesis of Carbon-Based Nanomaterials   211

Jena, P.V., Roxbury, D., Galassi, T.V., Akkari, L., Horoszko, C.P., Iaea, D.B. et al. 2017. A carbon nanotube optical reporter maps endolysosomal lipid flux. ACS Nano. 11: 10689–10703. doi: 10.1021/acsnano.7b04743.

Jeong, C.J., Roy, A.K., Kim, S.H., Lee, J.E., Jeong, J.H., In, I., and Park, S.Y. 2014. Fluorescent carbon nanoparticles derived from natural materials of mango fruit for bio-imaging probes. Nanoscale. 6(24): 15196–15202. doi:10.1039/c4nr04805a.

Jiang, L., Gao, L. and Sun, J. 2003. Production of aqueous colloidal dispersions of carbon nanotubes. J. Coll. Interf. Sci. 260(1): 89–94.

Jiang, J., Oberdörster, G., and Biswas, P. 2008. Characterization of size, surface charge, and agglomeration state of nanoparticle dispersions for toxicological studies. J. Nanoparticle Res. 11(1): 77–89. doi: 10.1007/s11051-008-9446-4.

Kang, J.H., Kima, H.S., and Shin, U.S. 2017. Thermo conductive carbon nanotube-framed membranes for skin heat signal-responsive transdermal drug delivery. Polym. Chem. 8: 3154–3163. doi: 10.1039/c7py00570a.

Khalifa, N.S., and Hasaneen, M.N. 2018. The effect of chitosan–PMAA–NPK nanofertilizer on *Pisum sativum* plants. 3 Biotech. 8(4). doi:10.1007/s13205-018-1221-3.

Kim, J., Park, S.Y., Kim, S., Lee, D.H., Kim, J.H., Kim, J.M., Kang, H., Han, J.S., Park, J.W., Lee, H. and Choi, S.H. 2016. Precise and selective sensing of DNA-DNA hybridization by graphene/Si-nanowires diode-type biosensors. Sci. Rep. 6: 31984. doi: 10.1038/srep31984.

Kim, S.H., Lee, J.E., Sharker, S.M., Jeong, J.H., In, I., and Park, S.Y. 2015. *In vitro* and *in vivo* tumor targeted photothermal cancer therapy using functionalized graphene nanoparticles. Biomacromolecules 16: 3519–3529. doi: 10.1021/acs.biomac.5b00944.

Kim, S.W., Kim, T., Kim, Y.S., Choi, H.S., Lim, H.J., Yang, S.J., and Park, C.R. 2012. Surface modifications for the effective dispersion of carbon nanotubes in solvents and polymers. Carbon. 50: 3–33. doi: 10.1016/j.carbon.2011.08.011.

Kim, S.W., Lee, Y.K., Lee, J.Y., Hong, J.H., and Khang, D. 2017. PEGylated anticancer-carbon nanotubes complex targeting mitochondria of lung cancer cells. Nanotechnology. doi: 10.1088/1361-6528/aa8c31.

Kowshik, M., Ashtaputre, S., Kharrazi, S., Vogel, W., Urban, J., Kulkarni, S.K., and Paknikar, K.M. 2003. Extracellular synthesis of silver nanoparticles by a silver-tolerant yeast strain MKY3. Nanotech. 14(1): 95–100.

Kumar, S., Ahlawat, W., Kumar, R., and Dilbaghi, N. 2015. Graphene, carbon nanotubes, zinc oxide and gold as elite nanomaterials for fabrication of biosensors for healthcare. Biosens. Bioelectron. 70: 498–503. doi: 10.1016/j.bios.2015.

Kumar, S., Amala, G., and Gowtham, S.M. 2017. Graphene based sensors in the detection of glucose in saliva—a promising emerging modality to diagnose diabetes mellitus. RSC Adv. 7: 36949–36976. doi: 10.1039/C6AY01023G.

Kundu, A., Nandi, S., Das, P., and Nandi, A.K. 2015. Fluorescent graphene oxide via polymer grafting: an efficient nanocarrier for both hydrophilic and hydrophobic drugs. ACS Appl. Mater. Interfaces. 7: 3512–3523. doi: 10.1021/am507110r.

Kung, J.C., Tseng, I.T., Chien, C.S., Lin, S.H., Wang, C.C., and Shih, C.J. 2020. Microwave assisted synthesis of negative-charge carbon dots with potential antibacterial activity against multi-drug resistant bacteria. RSC Adv. 10(67): 41202–41208. doi:10.1039/d0ra07106d.

Landry, M.P., Ando, H., Chen, A.Y., Cao, J., Kottadiel, V.I., Chio, L., Yang, D., Dong, J., Lu, T.K., and Strano, M.S. 2017. Single-molecule detection of protein efflux from microorganisms using fuorescent single-walled carbon nanotube sensor arrays. Nat. Nanotechnol. 12: 368–377. doi: 10.1038/nnano.2016.284

Lee, P.C., Lin, C.Y., Peng, C.L., and Shieh, M.J. 2016. Development of a controlled-release drug delivery system by encapsulating oxaliplatin into SPIO/MWNT nanoparticles for effective colon cancer therapy and magnetic resonance imaging. Biomater. Sci. 4: 1742–1753. doi: 10.1039/C6BM00444J.

Li, K., Liu, W., Ni, Y., Li, D., Lin, D., Su, Z., and Wei, G. 2017. Technical synthesis and biomedical applications of graphene quantum dots. J. Mater. Chem. B. 5: 4811–4826. doi: 10.1039/C7TB01073G.

Lin, G., Mi, P., Chu, C., Zhang, J., and Liu, G. 2016. Inorganic nanocarriers overcoming multidrug resistance for cancer theranostics. Adv. Sci. 3: 1600134. doi: 10.1002/advs.201600134.

212 Mycosynthesis of Nanomaterials: Perspectives and Challenges

Lin, Y., Lu, F., Tu, Y., and Ren, Z. 2004. Glucose biosensors based on carbon nanotube nanoelectrode ensembles. Nano Lett. 4: 191–195. doi: 10.1021/nl0347233.

Liu, J., Guo, S., Han, L., Wang, T., Hong, W., Liu, Y., and Wang, E. 2012. Synthesis of phospholipid monolayer membrane functionalized graphene for drug delivery. J. Mater. Chem. 22: 20634–20640.

Liu, J., Liu, K., Feng, L., Liu, Z., and Xu, L. 2017. Comparison of nanomedicine-based chemotherapy, photodynamic therapy and photothermal therapy using reduced graphene oxide for the model system. Biomater. Sci. 5: 331–340. doi: 10.1039/c6bm00526h.

Liu, Z., Chen, K., Davis, C., Sherlock, S., Cao, Q., Chen, X., and Dai, H. 2008. Drug delivery with carbon nanotubes for *in vivo* cancer treatment. Cancer Res. 68: 6652–6660. doi: 10.1158/0008-5472.CAN-08-1468.

Long, C., Chen, X., Jiang, L., and Zhuangjun, F. 2015. Porous layerstacking carbon derived from in-built template in biomass for high volumetric performance supercapacitors. Nano Ener. 12: 141–151.

Marella, S., Kumar, A.R.N., and Prasad, T.N.V.K.V. 2021. Nanotechnology-based innovative technologies for high agricultural productivity: Opportunities, challenges, and future perspectives. In, Recent Developments in Applied Microbiology and Biochemistry. http://dx.doi.org/10.1016/B978-0-12-821406-0.00019-9.

Mehra, N.K., and Jain, N.K. 2015. Optimization of a pretargeted strategy for the PET imaging of colorectal carcinoma via the modulation of radioligand pharmacokinetics. Mol. Pharm. 12: 630–643. doi: 10.1021/acs.molpharmaceut.5b00294.

Mejri, A., Vardanega, D., Tangour, B., Gharbi, T., and Picaud, F. 2015. Substrate temperature to control moduli and water uptake in thin films of vapor deposited N,N-Di(1-naphthyl)-N,N-diphenyl-(1,1'-biphenyl)-4,4'-diamine (n.d.). J. Phys. Chem. B. 119: 604–611. doi: 10.1021/acs.jpcb.5b05814.

Monthioux, M., and Kuznetsov, V.L. 2006. Who should be given the credit for the discovery of carbon nanotubes? Carbon. 44: 1621–1623. doi: 10.1016/j.carbon.2006.03.019.

Mourato, A., Gadanho, M., Lino, A.R., and Tenreiro, R. 2011. Biosynthesis of crystalline silver and gold nanoparticles by extremophilic yeasts. Bioinorg. Chem. Appl. 546074.

Odom, T.W., Huang, J.L., Kim, P., and Lieber, M.C. 1998. Atomic structure and electronic properties of single-walled carbon nanotubes. Nature. 391: 62–64. doi: 10.1038/34145.

Ping, J., Vishnubhotla, J.R., Vrudhula, A., and Johnson, A.T.C. 2016. Scalable production of high-sensitivity, label-free DNA biosensors based on back-gated graphene field Effect transistors. ACS Nano. 10: 8700–8704. doi: 10.1021/acsnano.6b04110.

Prasad, R. 2017. Fungal Nanotechnology: Applications in Agriculture, Industry, and Medicine. doi: 10.1007/978-3-319-68424-6.

Rai, M., Yadav, A., Bridge, P., and Gade, A. 2009. Myconanotechnology: a new and emerging science. pp. 258–267. *In*: Rai, M.K., and Bridge, P.D. (eds.). Applied Mycology. CAB International Publishers, New York. doi: 10.1079/9781845935344.0258.

Rai, M., Bonde, S., Golinska, P., Wemcel, J.T., Gade, A., Abd-Elsalam, K.A., Shende, S., Gaikwad, S., and Ingle, A.P. 2021. Fusarium as a novel fungus for the synthesis of nanoparticles: mechanism and applications. J. Fungi. (Basel, Switzerland). 7(2). doi: 10.3390/jof7020139.

Raza, K., Kumar, D., Kiran, C., Kumar, M., Guru, S.K., Kumar, P. et al. 2016. Enhanced antitumor efficacy and reduced toxicity of docetaxel loaded estradiol functionalized stealth polymeric nanoparticles. Mol. Pharm. 13: 2423–2432. doi: 10.1021/acs.molpharmaceut.5b00281.

Rigodanza, F., Burian, M., Arcudi, F., Đorđević, L., Amenitsch, H., and Prato, M. 2021. Snapshots into carbon dots formation through a combined spectroscopic approach. Nat. Comm. 12(1). doi:10.1038/s41467-021-22902-w.

Roy, P., Chen, P.C., Periasamy, A.P., Chen, Y.N., and Chang, H.T. 2015. Photoluminescent carbon nanodots: Synthesis, physicochemical properties and analytical applications. Mater. Today 18: 447–458.

Saad, E.D. Fauda, A., Saied, E. et al. 2021. *Rhizopus oryzae*-mediated green synthesis of magnesium oxide nanoparticles (MgO-NPs): A promising tool for antimicrobial, mosquitocidal action, and tanning effluent treatment. J. Fungi. 8: 372. doi: 10.3390/jof7050372.

Saha, S., Sarkar, J., Chattopadhyay, D., Patra, S., Chakraborty, A., and Acharya, K. 2010. Production of silver nanoparticles by a phytopathogenic fungus *Bipolaris nodulosa* and its antimicrobial activity. Dig. J. Nanomater. Bios. 5(4): 887–895.

Saxena, J., Sharma, M.M., Gupta, S., and Singh, A. 2014. Emerging role of fungi in nanoparticle synthesis and their applications. World J. Pharma. Pharmaceut. Sci. 3(9): 1586–1613.

Shah, R.M., Malherbe, F., Eldridge, D., Palombo, E.A., and Harding, I.H. 2014. Physicochemical characterization of solid lipid nanoparticles (SLNs) prepared by a novel microemulsion technique. J. Colloid Interface Sci. 428: 286–294.

Shi, X., Zheng, Y., Wang, C., Yue, L., Qiao, K., Wang, G., Wang, L., and Quan, H. 2015. Dual stimulus responsive drug release under the interaction of pH value and pulsatile electric field for a bacterial cellulose/sodium alginate/multi-walled carbon nanotube hybrid hydrogel. RSC Adv. 5: 41820–41829. doi: 10.1039/C5RA04897D.

Singh, S., Mehra, N.K., and Jain, N.K. 2016. Development and characterization of the paclitaxel loaded riboflavin and thiamine conjugated carbon nanotubes for cancer treatment. Pharm. Res. 33: 1769–1781. doi: 10.1007/s11095-016-1916-2.

Song, E., Han, W., Li, C., Cheng, D., Li, L., Liu, L., Zhu, G., Song, Y., and Tan, W. 2014. Hyaluronic acid-decorated graphene oxide nanohybrids as nanocarriers for targeted and pH-responsive anticancer drug delivery. ACS Appl. Mater. Interfaces. 6: 11882–11890. doi: 10.1021/am502423r.

Song, J., Yang, X., Jacobson, O., Lin, L., Huang, P., Niu, G., Ma, Q., and Chen, X. 2015. Sequential drug release and enhanced photothermal and photoacoustic effect of hybrid reduced graphene oxide-loaded ultrasmall gold nanorod vesicles for cancer therapy. ACS Nano. 22; 9(9): 9199–209. doi: 10.1021/acsnano.5b03804. Epub 2015 Aug 28. PMID: 26308265; PMCID: PMC5227595.

Song, J., Wang, F., Yang, X., Ning, B., Harp, M.G., Culp, S.H., Hu, S., Huang P., Nie, L., Chen, J., and Chen, X. 2016. Gold nanoparticle coated carbon nanotube ring with enhanced Raman scattering and photothermal conversion Property for theranostic applications. J. Am. Chem. Soc. 138: 7005–7015. doi: 10.1021/jacs.5b13475.

Su, S., Wang, J., Vargas, E., Wei, J., Zaguilain, R.M., Sennoune, S.R., Pantoya, M.L., Wang, S., Chaudhuri, J., and Qiu, J. 2016. Porphyrin immobilized nanographene oxide for enhanced and targeted photothermal therapy of brain cancer. ACS Biomater. Sci. Eng. 2: 1357–1366. doi: 10.1021/nl100996u.

Su, Y., Hu, Y., Wang, Y., Xu, X., Yuan, Y., Li, Y., Wang, Z., Chen, K., Zhang, F., Ding, X., Li, M., Zhou, J., Liu, Y., and Wang, W. 2017. A precision-guided MWNT mediated reawakening the sunk synergy in RAS for anti-angiogenesis lung cancer therapy. Biomaterials 139: 75–90. doi: 10.1016/j.biomaterials.2017.05.046.

Sunkar, S., and Nachiyar, V. 2013. Endophytes as potential nanofactories. Int. J. Chem. Environ. Biol. Sci. 1: 488–491.

Suvarnaphaet, P., and Pechprasarn, S. 2017. Graphene-based materials for biosensors: a review. Sensors 17: E2161. doi: 10.3390/s17102161.

Tang, L., Wang, Y., and Li, J. 2015. The graphene/nucleic acid nanobiointerface. Chem. Soc. Rev. 44: 6954–6980. doi: 10.1039/C4CS00519H.

Travlou, N.A., Giannakoudakis, D.A., Algarra, M., Labella, A.M., Rodríguez-Castellón, E., and Bandosz, T.J. 2018. S- and N-doped carbon quantum dots: Surface chemistry dependent antibacterial activity. Carbon. 135: 104–111. doi:10.1016/j.carbon.2018.04.018.

Trushkevych, O., Collings, N., Wilkinson, T.D., Crossland, W.A., Milne, W.I., Geng, J., Johnson, B.F.G., and Macaulay, S. 2007. Characterization of carbon nanotube - nematic liquid crystal composite materials. CNT 2007, Cambridge, UK.

Trushkevych, O., Collings, N., Hasan, T., Scardaci, V., Ferrari, A. C., Wilkinson, T.D., Crossland, W.A., Milne, W.I., Geng, J., Johnson, B.F.G., and Macaulay, S. 2008. Characterization of carbon nanotube–thermotropic nematic liquid crystal composites. J. Phy. D: App. Phy. 41(12): 125106. doi:10.1088/0022-3727/41/12/125106.

Ulissi, Z.W., Sen, F., Gong, X., Sen, S., Iverson, N., Boghossian, A.A., Godoy, L.C., Wogan, G.N., Mukhopadhyay, D., and Strano, M.S. 2014. Spatiotemporal intracellular nitric oxide signaling captured using internalized, near-infrared fluorescent carbon nanotube nanosensors. Nano Lett. 14: 4887–4894. doi: 10.1021/nl502338y.

Vainshtein, M., Belova, N., Kulakovskaya, T., Suzina, N., and Sorokin, V. 2014. Synthesis of magneto-sensitive iron-containing nanoparticles by yeasts. J. Ind. Microbiol. Biotechnol. 41(4): 657–663.

214 *Mycosynthesis of Nanomaterials: Perspectives and Challenges*

Wang, X., Wang, C., Cheng, L., Lee, S.T., and Liu, Z. 2012. Noble metal coated single-walled carbon nanotubes for applications in surface enhanced Raman scattering imaging and photothermal therapy. J. Am. Chem. Soc. 134: 7414–7422. doi: 10.1021/ja300140c.

Wang, Y., Zhang, L., Liang, R.P., Bai, J.M., and Qiu, J.D. 2013. Using graphene quantum dots as photoluminescent probes for protein kinase sensing. Anal. Chem. 85: 9148–9155. doi: 10.1021/ac401807b.

Wang, Y.F., and Hu, A. 2014. Carbon Quantum Dots: synthesis, properties and applications. J. Mater. Chem. C. 2: 6921–6939.

Wu, J. 2022. Understanding the Electric Double-Layer Structure, Capacitance, and Charging Dynamics. Chem. Rev. 122(12): 10821–10859. doi: https://doi.org/10.1021/acs.chemrev.2c00097.

Xuan, C.T., Thuy, N.T., Luyen, T.T., Huyen, Tran T.T., and Tuan, M.A. 2017. Carbon Nanotube Field-Effect Transistor for DNA Sensing. J. Electronic Mater. 46(6): 3507–3511. doi: 10.1007/s11664-016-5238-2.

Yang, J., Zhang, X., Ma, Y.-H., Gao, G., Chen, X., Jia, H.-R., Li, Y.H., Chen, Z., and Wu, F.G. 2016. Carbon dot-based platform for simultaneous bacterial distinguishment and antibacterial applications. ACS Appl. Mater. Interf. 8(47): 32170–32181. doi: 10.1021/acsami.6b10398.

Yu, J., Grossiord, N., Koning, C.E., and Loos, J. 2007. Controlling the dispersion of multi-wall carbon nanotubes in aqueous surfactant solution. Carbon. 45(3): 618–623.

Yu, X., Gao, D., Gao, L., Lai, J., Zhang, C., Zhao, Y., Zhong, L., Jia, B., Wang, F., Chen, X., and Liu, Z. 2017. Inhibiting metastasis and preventing tumor relapse by triggering host immunity with tumor-targeted photodynamic therapy using photosensitizer-loaded functional nanographenes. ACS Nano. 11: 10147–10158. doi: 10.1021/acsnano.7b04736.

Zhang, M., Wang, W., Zhou, N., Yuan, P., Su, Y., Shao, M., Chi, C., and Pan, F. 2017. Near-infrared light triggered photo-therapy, in combination with chemotherapy using magnetofluorescent carbon quantum dots for effective cancer treating. Carbon 118: 752–764. doi: 10.1016/j.carbon.2017.03.085.

Zhang, P., Zhao, X., Ji, Y., Ouyang, Z., Wen, X., Li, J. et al. 2015. Electrospinning graphene quantum dots into a nanofibrous membrane for dual-purpose fluorescent and electrochemical biosensors. J. Mater. Chem. B. 3: 2487–2496. doi: 10.1039/C4TB02092H.

Zhao, Q., Gan, Z., and Zhuang, Q. 2002. Electrochemical sensors based on carbon nanotubes. Electroanalysis. 14: 1609–1613. doi: 10.1002/elan.200290000.

Zhao, X., Liu, L., Li, X., Zeng, J., Jia, X., and Liu, P. 2014. Biocompatible graphene oxide nanoparticle-based drug delivery platform for tumor microenvironment-responsive triggered release of doxorubicin. Langmuir. 30: 10419–10429. doi: 10.1021/la502952f.

Zhao, X., Yang, L., Li, X., Jia, X., Liu, L., Zeng, J. et al. 2015. Functionalized graphene oxide nanoparticles for cancer cell specific delivery of antitumor drug. Bioconjug. Chem. 26: 128–136. doi: 10.1021/bc5005137.

Zheng, X.T., Ananthanarayanan, A., Luo, K.Q., and Chen, P. 2015. Glowing graphene quantum dots and carbon dots: properties, syntheses, and biological applications. Small. 11: 1620–1636. doi: 10.1002/smll.201402648.

Zhou, Q., Zheng, J., Qing, Z., Zheng, M., Yang, J., Yang, S. et al. 2016. Detection of circulating tumor DNA in human blood via DNA-mediated surface-enhanced Raman spectroscopy of single-walled carbon nanotubes. Anal. Chem. 88: 4759–4765. doi: 10.1021/acs.analchem.6b00108.

Zhu, L., Deng, C., Chen, P., Dong, X., Su, Y. H., Yuan, Y. et al. 2014. Glucose oxidase biosensors based on carbon nanotube non-woven fabrics. Carbon. 67: 795–796. doi: 10.1016/j.carbon.2013.10.046.

Zhu, W., Li, Y., Dai, L., Li, J., Li, X., Li, W., Duan, T., and Lei, J. 2018. Bioassembly of fungal hyphae/carbon nanotubes composite as a versatile adsorbent for water pollution control. Chem. Eng. J. 339: 214–222. doi: 10.1016/j.cej.2018.01.134.

Zuo, P., Lu, X., Sun, Z., Guo, Y., and He, H. 2016. A review on syntheses, properties, characterization and bioanalytical applications of fluorescent carbon dots. Microchim. Acta. 183: 519–542.

Chapter 11

Fungi-Based Synthesis of Nanoparticles and its Large-Scale Production Possibilities

Nadun H. Madanayake[1] and *Nadeesh M. Adassooriya*[2,*]

Introduction

Nanomaterials (NMs) have gradually transformed into an inevitable requirement for industries in a wide range of applications. NMs can be simply defined as particles having at least one of its dimensions within the size range of 1–100 nm (Madanayake et al. 2019). Different methods including chemical, physical and biological approaches have been studied for the synthesis of NMs. Furthermore, chemical and physical methods have been widely explored for scaling up applications. Concerns about nanoparticle (NPs) synthesis using physical and chemical-based techniques have emerged due to the risks associated with human and environmental health. Chemical-based approaches utilize larger quantities of toxic chemicals to synthesize and stabilize NMs. Also, wastes generated from synthesis can be highly toxic unless they are released into the environment after proper treatments. Furthermore, the energy required for the synthesis of NMs using chemical and physical-based approaches is high and it utilizes a bulk of fossil fuels to fulfill this demand. This can lead to generating tons of greenhouse gases which is one of the serious global issues (Dorcheh and Vahabi 2016). Therefore, it is imperative to have greener and environmentally friendly approaches to synthesize NMs in large-scale production.

---

[1] Department of Botany, Faculty of Science, University of Peradeniya, Peradeniya 20400, Sri Lanka.

[2] Department of Chemical & Process Engineering, Faculty of Engineering, University of Peradeniya, Peradeniya 20400, Sri Lanka.

* Corresponding author: nadeeshm@eng.pdn.ac.lk

Green synthesis of NMs using biogenic materials has been identified as a feasible approach. Generally, plants or plant parts, bioreduction compounds from bacteria, fungi, yeasts, algae, and actinomycetes are used in the biogenic synthesis of NMs. These approaches are simpler, economic, and environmentally friendly. Therefore, scientists have shown a preference to use greener methods to synthesize NMs (Boroumand et al. 2015; Pal et al. 2019). Fungi have a long history of being used as cell factories in a wide range of industrial applications. Fungi are usually grown on simple nutrient media and their biomass is easier to handle. Hence, they can be easily used in scaling-up applications by converting cheap, readily available renewable biomass into useful end products (Syed and Ahmad. 2012; Cairns et al. 2021). Currently, fungi are used to produce enzymes, vitamins, polysaccharides, polyhydric alcohols, pigments, lipids, glycolipids, and many other solvents required for different industrial applications. In addition, molecular biology and genetic engineering tools allow scientists to easily manipulate fungi to over-express homologous and heterologous proteins and other metabolites such as antibiotics, pigments, and fatty acids (Adrio and Demain 2003). This makes them cellular factories with higher efficiency for large-scale purposes.

Studies have reported that protein and enzymes synthesized by fungi can reduce metal ions into their elemental forms (Kumar et al. 2007; Syed and Ahmad 2012; Fernández et al. 2016; Seetharaman et al. 2018). The extracellular and intracellular enzymes produced by different fungal species at given conditions can successfully catalyze the NM synthesis. Therefore, mycosynthesis of NMs has become a promising approach in large-scale production. Hence, this chapter elaborates on the fungal-based synthesis of NPs and their potential to be used in large-scale production. Also, how myconanotechnology can be utilized in agricultural applications will be further discussed.

Fungi in green synthesis of nanoparticles

Fungi are an industrially important group of eukaryotic microorganisms which have been widely used for biotechnological and microbiological applications. Both primary and secondary metabolic pathways are used industrially for the synthesis of different products from these organisms. Proteins and different oxidoreductases (class of enzymes that can catalyze oxidation-reduction reactions) synthesized by fungi are used to synthesize NMs. It has been reported under different growth conditions fungi are capable of reducing metallic ions such as silver, copper, iron, platinum, zinc, and gold into their elemental forms (Riddin et al. 2006; Nithya and Ragunathan 2009; Thakker et al. 2013; Raliya and Tarafdar 2014; El-Batal et al. 2020). Table 1 summarizes some of the studies reported on the green synthesis of different NMs using fungal-based approaches.

Mechanisms involved in fungal-based nanomaterial synthesis

Ahmad et al. (2002) reported that extracellularly secreted sulfate reductase enzymes by *F. oxysporum* can reduce SO_4^{2-} groups in $CdSO_4$ to form CdS NPs. However, the green synthesis approach to fabricate CdS NPs was slow and it was highlighted

Myco-synthesis of Nanomaterials 217

Table 1. Mycosynthesis of different nanoparticles.

Organisms	Nanoparticles	Size/nm	References
Aspergillius species	Zinc	25	Raliya and Tarafdar (2014)
Coriolus versicolor	Cadmium sulphide	20	Sanghi and Verma (2009)
Fusarium oxysporum	Zirconia	3–11	Bansal et al. (2004)
F. oxysporum f. sp. lycopersici	Platinum	10–50	Riddin et al. (2006)
Pleurotus sajorcaju	Silver	5–50	Nithya and Ragunathan (2009)
Aspergillus flavus (KF934407)	Silver	50	Fatima et al. (2016)
Fusarium solani	Silver	5–35	Ingle et al. (2009)
F. oxysporum f. sp. cubense JT1	Gold	22	Thakker et al. (2013)
Botrytis cinerea and *Penicillium expansum*	Zinc oxide	70	He et al. (2011)
Saccharomyces cerevisiae	Cadmium sulphide	2.5–5.5	Prasad and Jha (2010)
A. flavus	Titanium dioxide	62–74	Rajakumar et al. (2012)
Alternaria alternata	Selenium	13–15	Sarkar et al. (2011)
F. oxysporum	Platinum	5–30	Syed and Ahmad (2012)
F. oxysporum	Silver	50	Gholami-Shabani et al. (2014)
Phomopsis liquidambaris.	Silver	18.7	Seetharaman et al. (2018)
Penicillium chrysogenum	Copper oxide		El-Batal et al. (2020).

that the activation barrier to converting SO_4^{2-} acts as the rate-limiting step for the slow rate of conversion to form semiconducting CdS NPs. Seetharaman et al. (2018) manufactured spherical or near to spherical polydispersed Ag NPs (18.7 nm) using aqueous extracts of endophytic fungi *P. liquidambaris*. It was stated that the reduction of Ag^+ ions to Ag NPs occurs via free amino groups, cysteine residues, and carboxylate groups in proteins. Also, stabilization of Ag NPs was achieved by the same compounds as capping agents during the synthesis. Syed and Ahmad (2012) fabricated Pt NPs (5–30 nm, spherical) using extracellular extracts of *F. oxysporum*. Fourier transformed infra-red spectroscopy (FTIR) confirmed that amide groups from proteins present in aqueous extracts were involved in the reduction and stabilization of $PtCl_6^-$ to form Pt NPs. More importantly, Pt NPs were stabilized by surface-bound proteins resulting in longer-term stability. This is additional merit that does not require to application of separate capping and stabilizing agents as in chemical-based approaches.

Nitrate reductases from *F. oxysporum*, *Cryptococcus laurentii*, and *Rhodotorula glutinis* were reported in the literature for the bioreduction of Ag^+ ions to form Ag NPs (Kumar et al. 2007; Fernández et al. 2016). Nitrate reductase is a multidomain enzyme-containing prosthetic group such as molybdopterin, Fe-heme, and flavin adenine dinucleotide moieties that facilitate the transfer of electrons from NADH (Nicotinamide adenine dinucleotide) to nitrate (Gholami-Shabani et al. 2014). This is a widely accepted mechanism to formulate Ag NPs. These enzymes can catalyze the transformation of NO_3^- to NO_2^- in their biological pathways. Gholami-Shabani et al. (2014) synthesized Ag NPs using nitrate reductases extracted from *F. oxysporum*. Accordingly, the $AgNO_3$ was allowed to react with a mixture containing phosphate

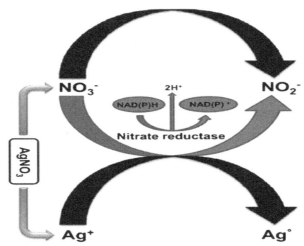

Figure 1. Nitrate reductase based Ag NPs synthesis (Gholami-Shabani et al. 2014).

buffer (0.2 M, pH 7.2), gelatin, 4-hydroxyquinoline, NADPH (Nicotinamide adenine dinucleotide phosphate) as enzyme cofactor alongside with fungal nitrate reductase for 5 h at 25 °C. Here, gelatin was used as the capping agent for synthesized Ag NPs, NADPH as a cofactor to activate the enzyme and 4-hydroxyquinoline as an electron carrier. It was found that stable and non-aggregated spherical-shaped Ag NPs with an average size of 50 nm were obtained. Figure 1 explains the general principle behind the bioreduction of Ag⁺ to form Ag NPs using nitrate reductase. In addition, nitrate reductase-based synthesis of Au NPs was also reported in the literature (Singh et al. 2019).

Factors affecting the fungal-based nanomaterial synthesis

pH

Effect of pH on the mycosynthesis of Au NPs using the fungus *Penicillium brevicompactum* at variable pH values of 2, 4, 6, 8 and 10 were evaluated by Mishra et al. (2011). It was found that ascending the pH in the medium significantly alters the shape or size of Au NPs. Lower pH values such as pH 2 do not promote the synthesis of Au NPs whereas pH 6 and 8 favor the synthesis of relatively well-dispersed Au NPs. A slight increment in the size of NPs was observed when the pH of the medium was above 8 while reducing the yield obtained. Tidke et al. (2014) synthesized Au NPs (spherical shaped with a mean particle size of 17 nm) using fungal cell filtrates of *Fusarium acuminatum*. It was evidenced that acidic pH favors the synthesis of Au NPs. Alkaline pH (6 to 10) favors the synthesis of Ag NPs using the endophytic fungus *Epicoccum nigrum* (Qian et al. 2013). Balakumaran et al. (2015) reported that the synthesis of monodispersed and stable Ag NPs was obtained at neutral pH (pH 7) using *Guignardia mangiferae*. Also, it was highlighted that aggregation of Ag NPs was observed at pH 3 and 4 proving that neutral and slightly alkaline pH are favorable for NM synthesis.

Temperature

A decrease in particle size was observed by Fayaz et al. (2009) when Ag NPs were synthesized using aqueous extracts of *Trichoderma viride* at different temperatures. It was observed that the particle size of Ag NPs decreased from 80–100 nm to 2–4 nm when the reaction temperature was increased from 10 to 40 °C. Generally, when the reaction rate gets increased at higher temperatures, reactants start to consume faster and deplete quickly. This leads to forming NPs with narrow size distribution at higher temperatures.

Stage of the fungal growth phase

The age of a fungal culture plays an important role in the bioreduction of metallic ions to form NPs (Zhang et al. 2011). Gericke and Pinches (2006) evaluated the effect of culture age on the synthesis of Au NPs by harvesting fungal biomass after 24, 48, and 72 h of growth. It was reported that the age of the *Verticillium* sp. at the time of exposure to Au^+ ions does not show any impact on the shape of the synthesized Au NPs. However, there was a reduction in the number of Au NPs per cell as the age of fungal cells was increased. This might be attributed to the reduction of bioreduction enzymes and proteins in *Verticillium* sp. at the latter stages of their growth.

Effect of growth media

Molnár et al. (2018) investigated the effect of fungal growth media on the synthesis of Au NPs using potato dextrose broth (PDB) and a modified Czapek-Dox medium. It was found that PDB itself is capable of synthesizing Au NPs showing a maximum absorption peak at 555 nm. Furthermore, electron microscopic analysis confirms that the Au NPs synthesized using PDB are spherical and hexagonal in shape with particle sizes ranging from 1 to 80 nm. However, the modified Czapek-Dox medium did not synthesize Au NPs which is mainly due to the absence of capping agents in the media. Sometimes antibiotics are used in growing media to avoid bacterial contaminations. Antibiotics are also capable of reducing certain metal ions by behaving as reducing agents. Moreover, they can act as capping agents to stabilize NPs synthesized. Therefore, growth media and its components have a significant impact on the control of size and shape thus affecting the uniformity of NPs synthesized through fungi-based green approaches. In addition, metal ion concentration and metal type too have a significant impact on the biogenesis of NMs using fungi.

Applications of myconanotechnology in agriculture

It has been reported that myconanotechnological applications have been widely utilized in pathogen detection and control, wound curing, food preservation, textiles, etc. Also, NPs synthesized using fungi have concerned the researchers to apply them in agricultural applications as well. Nanofungicides, nanopesticides and nanofertilizers are major areas where myconanotechnology has been targeted for utilization in agriculture (Hnamte et al. 2017).

Plant pathogens cause detrimental effects on agricultural crops by reducing their crop yield and productivity. Agrochemicals such as fungicides have been widely applied to control the damages caused by fungal phytopathogens. However, conventional fungicides can be lethal to beneficial organisms (Akther and Hemalatha 2019). Nanotechnology is one such alternative that can be used to mitigate the limitations of conventional approaches to control them. Furthermore, myconanotechnological-based approaches have been studied for their potential to synthesize antimicrobial agents. Win et al. (2020) evaluated the potential of utilizing Ag NPs fabricated using *Alternaria* sp. showing a greater antifungal activity against phytopathogens such as *Fusarium* spp., and *Alternaria* sp. A similar kind of results was obtained by Roy et al. (2013) where Ag NPs were synthesized using extracellular filtrates of *Aspergillus foetidus* MTCC8876 shows antifungal activities for some fungal strains proving its antifungal potential. Gamma radiation-assisted synthesis of CuO NPs using *Penicillium chrysogenum* showed antimicrobial activity against plant pathogens including *F. oxysporum, Alternaria solani, Ralstonia solanacearum,* and *Erwinia amylovor* (El-Batal et al. 2020). Therefore, fungi-mediated CuO NPs thus synthesized can be utilized as an antimicrobial agent in agriculture against phytopathogenic fungi and bacteria.

Nanobiofertilizers are emerging plant growth-promoting approaches in agriculture. Here application of beneficial microorganisms such as mycorrhizae can enhance plant growth and its resistance to biotic and abiotic stress. Rajak et al. (2017) studied the effect of Cu NPs applied in combination with the growth-promoting fungus *Piriformospora indica* on *Cajanus cajan* seedlings. It was found that host plants have enhanced the survival strength of *C. cajan* due to their synergistic effects on growth. Hence, *P. indica* plays a crucial role in Cu NPs absorption to *C. cajan* and improves the available Cu content in the plants. Therefore, the combined application of Cu NPs and *Piriformospora indica* on *Cajanus cajan* can be used as a potential plant growth promoter in agriculture.

Large-scale production of nanomaterials using fungi

The general approach to synthesizing bioactive compounds using microorganisms is fermentation. Fermentation techniques have greater importance economically and in biotechnological approaches. Fermentation technology has been advanced to a greater extent to apply in scaling up applications. Solid-state fermentation (SSF) and submerged fermentation (SmF) are used in industrial applications to synthesize bioactive compounds. Fungal fermentation utilizes the above fermentation techniques for the synthesis of different metabolites. Since most of the myconanotechnological-based approaches are used to synthesize NMs using enzymes, understanding the upstream and downstream processing of fungal fermentation is imperative. Upstream processing deals with the initial stage of fermentation in biotechnology. This includes the selection of fungal strains, appropriate substrates for the optimal growth of selected fungal strain; fermentation strategies, and parameters for optimized production desired product (pH, temperature, aeration, agitation rate). Downstream processing deals with the extraction, purification, and formulation methods of the target product obtained from fermentation (Jonathan et al. 2021).

Upstream processing for fungal fermentation

Obtaining the desired organism and its improvement

Generally, fungi are capable of adopting and tolerating heavy metal stresses via different mechanisms. These mechanisms include transport across cell membranes, biosorption to cell walls, entrapment in extracellular capsules, precipitation, and biotransformation via enzymatic approaches (Dhillon et al. 2012). It is a well-known fact that fungi are reported to eliminate heavy metals from contaminated areas through bioaccumulation and biosorption for higher tolerance. Therefore, fungi have been used in bioremediation applications as well (Joshi et al. 2011). Fungal species such as *Trichoderma* spp., *Fusarium* spp., *Aspergillus* spp., and *Rhizopus* spp. are reported to scavenge heavy metals from contaminated sites (Joshi et al. 2011; Kumar et al. 2011; Al-Hagar et al. 2020). Hence, the same principle can be applied in the synthesis of NMs using fungi for scaling up approaches. In addition, optimizing the selected strains for industrial applications is required to identify the most appropriate media to cultivate, optimum temperature, pH, and aeration rates, etc. However, for scaling up sometimes it is required to go for strain improvement. Genetic engineering tools are one of the premium strategies applied by biotechnologists for strain improvement in order to enhance the expression and secretion of metabolites of interest.

Genetic engineering approaches such as replacing original signal peptides with more efficient peptide targets, the fusion of heterologous proteins to naturally secreting ones, regulation of unfolded protein response (UPR) and endoplasmic-reticulum-associated protein degradation (ERAD) to promote protein secretion, optimization of the intracellular transport routes, constructing protease-deficient strains, regulate mycelium morphology and to regulate sterol regulatory element-binding proteins (SREBP) (Wang et al. 2020) are some of the common strategies for strain improvement.

Xylanases are renowned as one of the widely applied enzymes in many industrial applications. Several studies have reported that xylanases can be used to synthesize different NMs (Elegbede et al. 2019). Xylanases from *Penicillium citrinum* are promising candidates which show higher enzymatic activity at a wide pH range. Therefore, the manipulation of xylanase expression systems can enhance the production of xylanases for industrial applications. Ouephanit et al. (2019) reported that, xynA gene encoding xylanase A of *P. citrinum* was expressed in *Yarrowia lipolytica* expressing hosts under the control of the strong constitutive thyrotrophic embryonic factor (TEF) promoter. Higher xylanase A production was achieved from this system using preproLIP2 secretion signal than native *P. citrinum* proving the higher heterologous expression of xylanase A.

Signal peptide sequence in a given expression system plays a critical role in the secretion of proteins in prokaryotes and eukaryotes (Peter et al. 2019). Therefore, replacing an existing signal peptide sequence with a strong signal peptide sequence can enhance the secretion efficiency of metabolites (Wang et al. 2020). Xu et al. (2018) developed a genetically modified *A. niger* to enhance the extracellular secretion of α-galactosidase by replacing the native signal peptide AglB gene with glucoamylase signal peptide. It was observed that the new recombinant strain had

222 Mycosynthesis of Nanomaterials: Perspectives and Challenges

Table 2. Genetic manipulations enhance the protein production by fungi.

Target protein and its origin	Host	Strategy	References
α-Galactosidase	A. niger	Interchanging the original signal peptide sequence with a glucoamylase signal peptide	Xu et al. (2018)
β-Glucuronidase	A. niger	Regulating the UPR and ERAD by removing the ERAD factor doaA and overexpression of the oligosaccharyl transferase sttC	Jacobs et al. (2009)
Glucose oxidase	T. reesei	Overexpression of snc1 to optimize the intracellular transport process	Wu et al. (2017)
Cellulase	Myceliophthora sp.	Constructing a protease-deficient strain by deletion of res-1, cre-1, gh1-1, and alp-1 using CRISPR/Cas9 system in M. thermophila and M. heterothallica	Liu et al. (2017)
Laccase	A. niger	Protease-deficient strain were constructed by deleting aspartic protease genes pepAa, pepAb, or pepAd	Wang et al. (2008)

significantly increased the secretion of α-galactosidase in *A. niger.* The construction of protease deficient strains is another strategy that can be implemented to enhance the production of certain endogenous proteins. For instance, cellulase production in *Trichoderma reesei* was significantly decreased due to the secretion of proteases. Qian et al. (2019) deleted three target proteases (tre81070, tre120998, and tre123234) by one-step genetic transformation. It was observed that the strains developed by deleting the following protease expressions had decreased the protease activity by 78% while increasing the cellulase production during fermentation. Therefore, the following strategies can be implemented during upstream processing in order to improve the strains for efficient production of NM by fungi. Table 2. Summarizes some of the genetic manipulations to upregulate the protein expression by fungi.

Selection of bioreactors for large-scale production of fungal metabolites

Filamentous fungi are capable of synthesizing larger quantities of target metabolites such as proteins and enzymes that can be used to synthesize NMs. However, the production of different metabolites from fungi depends on the method of cultivation and its growing conditions. Fungi can grow as free mycelia or as pellets in liquid media. Hence, based on these features different bioreactors have been developed to stimulate the growth of fungi (Musoni et al. 2015). SSF and SmF are basic strategies used by biotechnologists in industrial applications to synthesize desired compounds. Both these fermentation strategies have been reported to use in the extraction of chemicals to assist in NM synthesis (Vetchinkina et al. 2014; Jini and Rohiniraj 2017; Tandon et al. 2017). For scaling up applications, bioreactors are utterly important for the fermentation processes. Conventional bioreactors for SSF includes tray, drum, and packed bed bioreactors whereas continuous stirred-bioreactors, continuous flow stirred-tank reactors, plug-flow reactor, and fluidized-bed reactors are used in SmF (Musoni et al. 2015).

SSF is suited for fermentation applications with fungi having lower water activity. Substrates such as bran, bagasse, lingo cellulosic materials and paper pulp can be utilized as substrates for fermentation. Here, the substrates are utilized slowly and steadily, hence substrates used here can be applied for long fermentation periods. Also, substrates used in SSF can vary in terms of composition, size, mechanical resistance, porosity, and water-holding capacity. However, it cannot be used in fermentation processes involving organisms that require high water activity (Subramaniyam and Vimala 2012; Musoni et al. 2015).

In SmF, fungi are growing in liquid substrates with necessary nutrient sources which it promotes the secretion of bioactive compounds to the media. Here, nutrients in the media will be utilized quickly by growing organisms. Hence, nutrients are required to be continuously supplied to the fermenters. It is easier to purify the products obtained from SmF (Musoni et al. 2015). The oxygen transfer capacity is a critical factor in SmF and it depends on the size and shape of the bioreactors used. In addition, the mode of agitation and aerations are required to be considered in scale-up studies. In addition to oxygen transfer capacity, designs of SSF require maintaining other important parameters such as temperature, pH, and water content of the solid medium. Besides these factors, fungal morphology's effect on the mechanical agitation and the requirement of sterile conditions also plays a critical role in deciding the bioreactor type (Musoni et al. 2015). Figure 2 shows the typical structure of fermenters used in SmF.

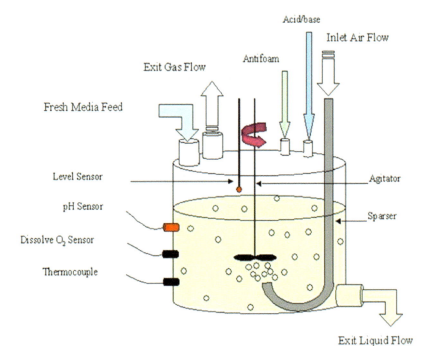

Figure 2. Structure of a typical fermentor (Stirred Tank Batch Bioreactor) (Al-Maqtari et al. 2019).

224 *Mycosynthesis of Nanomaterials: Perspectives and Challenges*

Most industrial enzymes produced by fungi are extracellular (secreted by cells into the external environment), they remain in the fermented broth after the biomass has been removed. Enzymes and proteins secreted from fungi are used as reducing and stabilizing agents in the synthesis of NPs (Ahmad et al. 2002; Boroumand et al. 2015; Guilger-Casagrande and Lima 2019). Both solid-state and SmF are used for enzyme production in large-scale production. However, most of the industrial enzymes are synthesized using SmF Several enzymes of industrial importance have been extracted from the fungi. Also, they can be genetically modified in order to upregulate the expression of heterologous molecules (Subramaniyam and Vimala 2012). They also secrete large quantities of extracellular proteins that contribute to the stability of the NPs (Guilger-Casagrande and Lima 2019).

Downstream processing for fungal fermentation

Downstream processing describes the stages that deal with extraction, purification, and packaging of metabolites. In myconanotechnology, this can be attributed as the isolation, extraction and purification of reducing agents involved in NM synthesis. This can be followed by synthesis of NPs and its purification under scaling-up approaches. Conventionally, downstream processes play a crucial role in biopharmaceutical (antibiotics, vitamin, vaccine) and industrial enzymes, as these products require higher purity and quality (Kumar and Murthy 2011; Jungbauer 2013; Jonathan et al. 2021).

The enzymes in the remaining broth are concentrated by evaporation, membrane filtration or crystallization depending on their intended application. If pure enzyme preparations are required for the synthesis, they are usually isolated by gel or ion-exchange chromatography. Sometimes crude powder enzymes are made into granules to make them more convenient to use. Enzymes are immobilized on surfaces of inert granules in order to ensure their prolonged activeness (Al-Maqtari et al. 2019). One approach to fabricating NMs using fungi is by obtaining the extracellular fractions of fungal biomass. El-Moslamy et al. (2017) evaluated the scale-up production of Ag NPs using *Trichoderma harzianum. T. harzianum* was cultivated in a stirred tank bioreactor (7.0 L BioFlo 310 fermenter) equipped with automatic control of temperature, pH, agitation rate, and aeration rate. The fungi were inoculated with 10% (v/v) inoculums from the fungal pre-culture and fungi were allowed to grow under the following conditions; aeration rate > 20% O_2 level in the culture medium, pH (5.5–6.0), polypropylene glycol antifoaming agent (5.0 ml/l). The fungal biomass was periodically removed and washed with distilled water, dried and the extracellular fractions were used to synthesize Ag NPs (0.01 M $AgNO_3$ and diluted supernatant (10 v/v pH 5) at 30 °C and 200 rpm for 24 hr).

El Domany et al. (2017) biosynthesized Ag NPs using cell filtrate from SmF of *F. oxysporum*. The fungal biomass was grown aerobically in MYPG broth (0.3 g. malt extract, 0.3 g. yeast extract, 0.5 g. peptone and 1 g. glucose) in an orbital shaker at 25 ± 2 °C at 120 rpm for 96 h. The cell-free filtrate (20 ml) was mixed 150 ml of 1 mM $AgNO_3$ 1mM and incubated for 48 h at 30 °C by agitating at 120 rpm. Spherical Ag NPs with particle sizes 10–25 nm were obtained and NPs thus synthesized have higher stability due to stabilization from biological agents released from the fungus.

El-Batal et al. (2015) used laccases produced by *Pleurotus ostreatus* to synthesize Au NPs. Laccases were produced using SSF with appropriate inducers. Enzymes were extracted using 1 mM citrate phosphate buffer (pH 5) for 2 h in a shaker at 200 rpm. Extracts from the fungi were obtained by tincture press and centrifuged to separate the crude extract. The partial purification was done in order to apply the enzyme extracts for the synthesis of Au NPs. Ammonium sulphate was introduced into the crude extract to achieve 80% saturation and kept at 4 °C for 48 h. Then the saturated mixture was centrifuged at 2415 g for 15 min at 4 °C and the supernatant was removed. The obtained pellet was dissolved in 1 mM citrate phosphate buffer pH 5. Here the precipitate was desalted by dialysis bag to eliminate low molecular weight substances and other ions that interfere with the enzyme activity. Then partially purified laccases were used to synthesize Au NPs by reducing tetrachloroauric acid by stirring the mixture for 90 min with a magnetic stirrer. Therefore, there is a greater potential for the synthesis of different NMs especially metallic NPs for large-scale productions. Proper maintenance of culture conditions and optimized approaches during upstream and downstream processing will provide NMs with higher monodispersity having well-controlled sizes and shapes for industrial applications.

Conclusion and future perspectives

The application of NMs in different industrial applications has a drastic increment due to their unique physicochemical properties. Generally, the required NMs for these applications can be fabricated using chemical physical, and biological methods. Chemical and physical approaches have been frequently used to synthesize NPs. However, these approaches have serious concerns with respect to their potential risks to health and environment due to the use of hazardous reagents and costly requirements for the synthesis. The biological synthesis of NMs has recently been highly explored to mitigate the limitations of chemical and physical approaches. Enzymes secreted by fungi are used to synthesize different NMs. Several factors affect the biogenic synthesis of NPs using fungi. It has been reported that pH, temperature, metal type, and concentration, growth media, and the stage of fungi in its growth phase affects particle size, shape and monodispersity. Upstream and downstream processing are important stages that are required to be addressed for scaling up approaches. Hence, the development of reliable experimental protocols for the synthesis of NMs with well-controlled sizes, shapes, durability and monodispersity is one of the primary challenges for large-scale production. Also, it should be noted that the nanobiotechnological approaches thus implemented should be harmless and environmentally friendly which can be fulfilled by biological sources. Therefore, myconanotechnology has a greater potential to implement environmentally friendly approaches to synthesize NMs to fulfill the global requirement in the future.

Abbreviations

ERAD	-	Endoplasmic-reticulum-associated protein degradation
FTIR	-	Fourier transformed infra-red spectroscopy
NADH	-	Nicotinamide adenine dinucleotide

226 *Mycosynthesis of Nanomaterials: Perspectives and Challenges*

NADPH - Nicotinamide adenine dinucleotide phosphate
NMs - Nanomaterials
NPs - Nanoparticles
PDB - Potato dextrose broth
SSF - Solid-state fermentation
SREBP - Sterol regulatory element-binding proteins
SmF - Submerged fermentation
TEF - Thyrotrophic embryonic factor
UPR - unfolded protein response

References

Adrio, J.L., and Demain, A.L. 2003. Fungal biotechnology. Int. Microbiol., 6(3): 191–199.

Ahmad, A., Mukherjee, P., Mandal, D., Senapati, S., Khan, M.I., Kumar, R., and Sastry, M. 2002. Enzyme mediated extracellular synthesis of CdS nanoparticles by the fungus, *Fusarium oxysporum*. J. Am. Chem. Soc. 124(41): 12108–12109.

Akther, T., and Hemalatha, S. 2019. Mycosilver nanoparticles: Synthesis, characterization and its efficacy against plant pathogenic fungi. Bionanoscience 9(2): 296–301.

Al-Hagar, O.E., Bayoumi, R.A., Abdel Aziz, O.A., and Mousa, A.M., 2020. Biosorption and adsorption of some heavy metals by *Fusarium* sp. F6c isolate as affected by gamma irradiation and agricultural wastes. Sci Asia 46(1): 1–37.

Al-Maqtari, Q.A., Waleed, A.A., and Mahdi, A.A. 2019. Microbial enzymes produced by fermentation and their applications in the food industry-A review. Int. J. Agri. Inn. Res. 8: 62–82.

Balakumaran, M.D., Ramachandran, R., and Kalaichelvan, P.T. 2015. Exploitation of endophytic fungus, *Guignardia mangiferae* for extracellular synthesis of silver nanoparticles and their *in vitro* biological activities. Microbiol. Res. 178: 9–17.

Bansal, V., Rautaray, D., Ahmad, A., and Sastry, M. 2004. Biosynthesis of zirconia nanoparticles using the fungus *Fusarium oxysporum*. J. Mater. Chem. 14(22): 3303–3305.

Boroumand, M.A., Namvar, F., Moniri, M., Azizi, S., and Mohamad, R. 2015. Nanoparticles biosynthesized by fungi and yeast: a review of their preparation, properties, and medical applications. Molecules 20(9): 16540–16565.

Cairns, T.C., Zheng, X., Zheng, P., Sun, J., and Meyer, V. 2021. Turning inside out: filamentous fungal secretion and its applications in biotechnology, agriculture, and the clinic. J. Fungi 7(7): 535.

Dhillon, G.S., Brar, S.K., Kaur, S., and Verma, M. 2012. Green approach for nanoparticle biosynthesis by fungi: current trends and applications. Crit. Rev. Biotechnol. 32(1): 49–73.

Dorcheh, S.K., and Vahabi, K. 2016. Biosynthesis of nanoparticles by fungi: large-scale production. Fungal Metabolites: 395–414.

El-Batal, A.I., ElKenawy, N.M., Yassin, A.S., and Amin, M.A. 2015. Laccase production by *Pleurotus ostreatus* and its application in synthesis of gold nanoparticles. Biotechnol. Rep. 5: 31–39.

El-Batal, A.I., El-Sayyad, G.S., Mosallam, F.M., and Fathy, R.M. 2020. *Penicillium chrysogenum*-mediated mycogenic synthesis of copper oxide nanoparticles using gamma rays for in vitro antimicrobial activity against some plant pathogens. J. Clust. Sci. 31(1): 79–90.

El Domany, E.B., Essam, T.M., Ahmed, A.E., and Farghli, A.A. 2017. Biosynthesis, characterization, antibacterial and synergistic effect of silver nanoparticles using *Fusarium oxysporum*. J. Pure Appl. Microbiol. 11(3): 1441–1446.

El-Moslamy, S.H., Elkady, M.F., Rezk, A.H., and Abdel-Fattah, Y.R. 2017. Applying Taguchi design and large-scale strategy for mycosynthesis of nano-silver from endophytic *Trichoderma harzianum* SYA. F4 and its application against phytopathogens. Sci. Rep. 7(1): 1–22.

Elegbede, J.A., Lateef, A., Azeez, M.A., Asafa, T.B., Yekeen, T.A., Oladipo, I.C., Hakeem, A.S., Beukes, L.S., and Gueguim-Kana, E.B. 2019. Silver-gold alloy nanoparticles biofabricated by fungal xylanases exhibited potent biomedical and catalytic activities. Biotechnol. Prog. 35(5): e2829.

Fatima, F., Verma, S.R., Pathak, N., and Bajpai, P. 2016. Extracellular mycosynthesis of silver nanoparticles and their microbicidal activity. J. Glob. Antimicrob. Resist. 7: 88–92.

Fayaz, A.M., Balaji, K., Kalaichelvan, P.T., and Venkatesan, R. 2009. Fungal based synthesis of silver nanoparticles—an effect of temperature on the size of particles. Colloids Surf. 74(1): 123–126.

Fernández, J.G., Fernández-Baldo, M.A., Berni, E., Camí, G., Durán, N., Raba, J., and Sanz, M.I. 2016. Production of silver nanoparticles using yeasts and evaluation of their antifungal activity against phytopathogenic fungi. Process Biochem. 51(9): 1306–1313.

Gericke, M., and Pinches, A. 2006. Biological synthesis of metal nanoparticles. Hydrometallurgy 83(1-4): 132–140.

Gholami-Shabani, M., Akbarzadeh, A., Norouzian, D., Amini, A., Gholami-Shabani, Z., Imani, A., Chiani, M., Riazi, G., Shams-Ghahfarokhi, M., and Razzaghi-Abyaneh, M. 2014. Antimicrobial activity and physical characterization of silver nanoparticles green synthesized using nitrate reductase from *Fusarium oxysporum*. Appl. Biochem. Biotechnol. 172(8): 4084–4098.

Guilger-Casagrande, M., and Lima, R.D. 2019. Synthesis of silver nanoparticles mediated by fungi: a review. Front. Bioeng. Biotechnol. 7: 287.

He, L., Liu, Y., Mustapha, A., and Lin, M. 2011. Antifungal activity of zinc oxide nanoparticles against *Botrytis cinerea* and *Penicillium expansum*. Microbiol. Res. 166(3): 207–215.

Hnamte, S., Siddhardha, B., and Sarma, V.V. 2017. Myconanotechnology in Agriculture. In Fungal Nanotechnology (pp. 77–88). Springer, Cham.

Ingle, A., Rai, M., Gade, A., and Bawaskar, M. 2009. *Fusarium solani*: a novel biological agent for the extracellular synthesis of silver nanoparticles. J. Nanoparticle Res. 11(8): 2079–2085.

Jacobs, D.I., Olsthoorn, M.M., Maillet, I., Akeroyd, M., Breestraat, S., Donkers, S., van der Hoeven, R.A., van den Hondel, C.A., Kooistra, R., Lapointe, T., and Menke, H. 2009. Effective lead selection for improved protein production in *Aspergillus niger* based on integrated genomics. Fungal Genet. Biol, 46(1): S141–S152.

Jini, J., and Rohiniraj, N.R. 2017. Production and partial purification of extracellular laccase from mushroom *Pleurotus* sp., and its application in green synthesis of silver nanoparticle. IJRASET 5(12): 2611–2619.

Jonathan, J., Tania, V., Tanjaya, J.C., and Katherine, K. 2021. Recent advancements of fungal xylanase upstream production and downstream processing. IJLS., ISSN: 2656-0682 (online) 3(1): 37–58.

Joshi, P.K., Swarup, A., Maheshwari, S., Kumar, R., and Singh, N. 2011. Bioremediation of heavy metals in liquid media through fungi isolated from contaminated sources. Indian J. Microbiol. 51(4): 482–487.

Jungbauer, A. 2013. Continuous downstream processing of biopharmaceuticals. Trends Biotechnol. 31(8): 479–492.

Kumar, D., and Murthy, G.S. 2011. Impact of pretreatment and downstream processing technologies on economics and energy in cellulosic ethanol production. Biotechnol. Biofuels 4(1): 1–19.

Kumar, R., Bhatia, D., Singh, R., Rani, S., and Bishnoi, N.R. 2011. Sorption of heavy metals from electroplating effluent using immobilized biomass *Trichoderma viride* in a continuous packed-bed column. Int. Biodeterior. Biodegradation 65(8): 1133–1139.

Kumar, S.A., Abyaneh, M.K., Gosavi, S.W., Kulkarni, S.K., Pasricha, R., Ahmad, A., and Khan, M.I. 2007. Nitrate reductase-mediated synthesis of silver nanoparticles from $AgNO_3$. Biotechnol. Lett. 29(3): 439–445.

Liu, Q., Gao, R., Li, J., Lin, L., Zhao, J., Sun, W., and Tian, C. 2017. Development of a genome-editing CRISPR/Cas9 system in thermophilic fungal *Myceliophthora* species and its application to hyper-cellulase production strain engineering. Biotechnol. Biofuels 10(1): 1–14.

Madanayake, N.H., Rienzie, R., and Adassooriya, N.M. 2019. Nanoparticles in nanotheranostics applications. In Nanotheranostics (pp. 19–40). Springer, Cham.

Mishra, A., Tripathy, S.K., Wahab, R., Jeong, S.H., Hwang, I., Yang, Y.B., Kim, Y.S., Shin, H.S., and Yun, S.I. 2011. Microbial synthesis of gold nanoparticles using the fungus *Penicillium brevicompactum* and their cytotoxic effects against mouse mayo blast cancer C 2 C 12 cells. Appl. Microbiol. Biotechnol. 92(3): 617–630.

Molnár, Z., Bódai, V., Szakacs, G., Erdélyi, B., Fogarassy, Z., Sáfrán, G., Varga, T., Kónya, Z., Tóth-Szeles, E., Szűcs, R., and Lagzi, I. 2018. Green synthesis of gold nanoparticles by thermophilic filamentous fungi. Sci. Rep. 8(1): 1–12.

Musoni, M., Destain, J., Thonart, P., Bahama, J.B., and Delvigne, F. 2015. Bioreactor design and implementation strategies for the cultivation of filamentous fungi and the production of fungal metabolites: from traditional methods to engineered systems. BASE.

Nithya, R., and Ragunathan, R. 2009. Synthesis of silver nanoparticle using *Pleurotus sajor* caju and its antimicrobial study. Dig. J. Nanomater. 4(4): 623–629.

Ouephanit, C., Boonvitthya, N., Theerachat, M., Bozonnet, S., and Chulalaksananukul, W. 2019. Efficient expression and secretion of endo-1, 4-β-xylanase from *Penicillium citrinum* in non-conventional yeast Yarrowia lipolytica directed by the native and the preproLIP2 signal peptides. Protein Expr. Purif. 160: 1–6.

Pal, G., Rai, P., and Pandey, A. 2019. Green synthesis of nanoparticles: A greener approach for a cleaner future. In Green synthesis, characterization and applications of nanoparticles (pp. 1–26). Elsevier.

Peter, S.C., Dhanjal, J.K., Malik, V., Radhakrishnan, N., Jayakanthan, M., Sundar, D., Sundar, D., and Jayakanthan, M. 2019. Encyclopedia of bioinformatics and computational biology. Ranganathan, S., Grib-skov, M., Nakai, K., Schönbach, C. (Eds.). pp. 661–676.

Prasad, K., and Jha, A.K. 2010. Biosynthesis of CdS nanoparticles: an improved green and rapid procedure. J. Colloid Interface Sci., 342(1): 68–72.

Roy, S., Mukherjee, T., Chakraborty, S., and Das, T.K. 2013. Biosynthesis, characterisation & antifungal activity of silver nanoparticles synthesized by the fungus *Aspergillus foetidus* MTCC8876. Dig. J. Nanomater. 8(1): 197–205.

Qian, Y., Yu, H., He, D., Yang, H., Wang, W., Wan, X., and Wang, L. 2013. Biosynthesis of silver nanoparticles by the endophytic fungus *Epicoccum nigrum* and their activity against pathogenic fungi. Bioprocess Biosyst. Eng. 36(11): 1613–1619.

Qian, Y., Zhong, L., Sun, Y., Sun, N., Zhang, L., Liu, W., Qu, Y., and Zhong, Y. 2019. Enhancement of cellulase production in *Trichoderma reesei* via disruption of multiple protease genes identified by comparative secretomics. Front. Microbiol. 10: 2784.

Rajak, J., Bawaskar, M., Rathod, D., Agarkar, G., Nagaonkar, D., Gade, A., and Rai, M. 2017. Interaction of copper nanoparticles and an endophytic growth promoter *Piriformospora indica* with *Cajanus cajan*. J. Sci. Food Agric. 97(13): 4562–4570.

Rajakumar, G., Rahuman, A.A., Roopan, S.M., Khanna, V.G., Elango, G., Kamaraj, C., Zahir, A.A., and Velayutham, K. 2012. Fungus-mediated biosynthesis and characterization of TiO_2 nanoparticles and their activity against pathogenic bacteria. *Spectrochimica* Acta Part A: Molecular and Biomolecular Spectroscopy 91: 23–29.

Raliya, R., and Tarafdar, J.C. 2014. Biosynthesis and characterization of zinc, magnesium and titanium nanoparticles: an eco-friendly approach. Int. Nano Lett. 4(1): 93.

Riddin, T.L., Gericke, M., and Whiteley, C.G. 2006. Analysis of the inter-and extracellular formation of platinum nanoparticles by *Fusarium oxysporum* f. sp. *Lycopersici* using response surface methodology. Nanotechnology 17(14): 3482.

Sanghi, R., and Verma, P. 2009. A facile green extracellular biosynthesis of CdS nanoparticles by immobilized fungus. J. Chem. Eng. 155(3): 886–891.

Sarkar, J., Dey, P., Saha, S., and Acharya, K. 2011. Mycosynthesis of selenium nanoparticles. Micro Nano Lett. 6(8): 599–602.

Seetharaman, P.K., Chandrasekaran, R., Gnanasekar, S., Chandrakasan, G., Gupta, M., Manikandan, D.B., and Sivaperumal, S. 2018. Antimicrobial and larvicidal activity of eco-friendly silver nanoparticles synthesized from endophytic fungi *Phomopsis liquidambaris*. Biocatal. Agric. Biotechnol. 16: 22–30.

Singh, S., Dev, A., Gupta, A., Nigam, V.K., and Poluri, K.M. 2019. Nitrate reductase mediated synthesis of surface passivated nanogold as broad-spectrum antibacterial agent. Gold Bull. 52(3): 197–216.

Subramaniyam, R., and Vimala, R. 2012. Solid state and submerged fermentation for the production of bioactive substances: a comparative study. Int. J. Sci. Nat. 3(3): 480–486.

Syed, A., and Ahmad, A. 2012. Extracellular biosynthesis of platinum nanoparticles using the fungus *Fusarium oxysporum*. Colloids Surf. B, 97: 27–31.

Tandon, U.S., Habte, T., Adgo, A., and Kebede, A. 2017. Solid State Fermentation: Extraction of Enzyme from Fungus and its use in the synthesis of Silver Nanoparticles.

Thakker, J.N., Dalwadi, P., and Dhandhukia, P.C. 2013. Biosynthesis of gold nanoparticles using *Fusarium oxysporum* f. sp. *cubense* JT1, a plant pathogenic fungus. Int. Sch. Res. Notices.

Tidke, P.R., Gupta, I., Gade, A.K., and Rai, M. 2014. Fungus-mediated synthesis of gold nanoparticles and standardization of parameters for its biosynthesis. IEEE Transactions on Nanobioscience 13(4): 397–402.

Vetchinkina, E.P., Loshchinina, E.A., Burov, A.M., Dykman, L.A., and Nikitina, V.E. 2014. Enzymatic formation of gold nanoparticles by submerged culture of the basidiomycete *Lentinus edodes*. J. Biotechnol. 182: 37–45.

Wang, Q., Zhong, C., and Xiao, H. 2020. Genetic engineering of filamentous fungi for efficient protein expression and secretion. *Front.* Bioeng. Biotechnol. 8: 293.

Wang, Y., Xue, W., Sims, A.H., Zhao, C., Wang, A., Tang, G., Qin, J., and Wang, H. 2008. Isolation of four pepsin-like protease genes from *Aspergillus niger* and analysis of the effect of disruptions on heterologous laccase expression. Fungal Genet. Biol. 45(1): 17–27.

Win, T.T., Khan, S., and Fu, P. 2020. Fungus-(*Alternaria* sp.) Mediated Silver Nanoparticles Synthesis, Characterization, and Screening of Antifungal Activity against Some Phytopathogens. J. Nanotechnol.

Wu, Y., Sun, X., Xue, X., Luo, H., Yao, B., Xie, X., and Su, X. 2017. Overexpressing key component genes of the secretion pathway for enhanced secretion of an *Aspergillus niger* glucose oxidase in *Trichoderma reesei*. Enzyme Microb. Technol. 106: 83–87.

Xu, Y., Wang, Y.H., Liu, T.Q., Zhang, H., Zhang, H., and Li, J. 2018. The GlaA signal peptide substantially increases the expression and secretion of α-galactosidase in *Aspergillus niger. Biotechnol.* Let. 40(6): 949–955.

Zhang, X., Yan, S., Tyagi, R.D., and Surampalli, R.Y. 2011. Synthesis of nanoparticles by microorganisms and their application in enhancing microbiological reaction rates. Chemosphere 82(4): 489–494.

Section II
Characterization Techniques of Mycosynthesised Nanoparticles and Mechanism of Synthesis

Chapter 12

Techniques for Characterization of Biologically Synthesized Nanoparticles by Fungi

Pramod Ingle,[1] *Kapil Kamble,*[2] *Patrycja Golinska,*[3] *Mahendra Rai*[1,3,*] and *Aniket Gade*[1,4]

Introduction

A number of methods are being employed today, Electron microscopy, XRD, probe microscopy like atomic force microscopy, dynamic light scattering, laser light diffraction, and particle sedimentation by gravitational and centrifugal techniques to fully understand the particle size, size distribution, state of aggregation, particle morphology and surface area, bulk density, surface chemistry and powder chemistry (purity and dopant level and uniformity) (Saltiel and Giesche 2000; Jastrzębska and Olszyna 2015).

The poor knowledge available about the actual composition and characteristics of nanoparticles has attracted researchers towards the development of analytical methodologies for extracting quality information from nanoparticles or nanostructured compounds (Albuquerque et al. 2016). Therefore, developing effective analytical methods for accurate and rapid characterization of nanoparticles is mandatory with a view to nanotechnological advancements. Most of the methods used to characterize nanoparticles at present are based on atomic force microscopy or tunneling electron

---

[1] Nanobiotechnology Lab, Department of Biotechnology, Sant Gadge Baba Amravati University, Amravati, Maharashtra, India – 444602.

[2] Department of Microbiology, Sant Gadge Baba Amravati University, Amravati, Maharashtra, India – 444602.

[3] Department of Microbiology, Nicolaus Copernicus University, 87-100 Torun, Poland.

[4] Department of Biological Science and Biotechnology, Institute of Chemical Technology,Matunga, Mumbai,Maharshtra, India-400019.

* Corresponding author: mahendra.rai@v.umk.pl

234 *Mycosynthesis of Nanomaterials: Perspectives and Challenges*

microscopy, and physical measurements from electron microscopy (Gommes et al. 2003). These techniques, however, provide only limited information about nanoparticles so there is a need for chemical methods that can provide the missing information. Accurate characterization of nanostructured materials is currently another hot topic (Valcárcel et al. 2008). Complementary analysis methods are routinely followed for the accurate measurements and characterization of nanoparticles. The influence of time and environment on nanoparticle properties entail the need for timely sample analysis, for a reasonable understanding of the physical and chemical meaning of each type of measurement (Baer et al. 2010; Abe et al. 2018).

Characterization methods for analysis and measurement of nanophenomena are indispensable in the development of nanotechnology. Especially, the development of precise analysis with atomic level for local nanostructures such as chemical composition and bonding state, defects, and impurities, is a key to explicating the mechanism of nanophenomena (Naito et al. 2018).

Fungi are generally metal-tolerant and simple to handle, these are interesting agents for biogenic nanoparticle synthesis. Furthermore, the extracellular proteins generated by fungi aid in the stability of nanoparticles. For large-scale nanoparticle synthesis, fungal cultures are preferable over bacterial cultures because the biomass can be easily removed using traditional procedures such as centrifugation. Fungi are also favored over plant-mediated nanoparticle synthesis because they can withstand agitation and pressure throughout the synthesis process. Large scale nanoparticles synthesis conditions like incubation time, temperature and media composition can be modified to obtain nanoparticles of desired size and properties, which is an additional advantage of fungi over other agents employed for nanoparticle synthesis (Guilger-Casagrande and Lima 2019).

Need for characterization of nanoparticles

Materials with novel and significantly improved physical, chemical, and biological properties are a result of nanoscale fabrication (Godale and Sharon 2019). Quality, affordability, and reproducibility are essential if the nanoparticle-based industry is to become a large-scale commercial reality. Critical to these objectives is the development of characterization methods that can efficiently determine particle size, distribution, and morphologies. Verification tools are necessary for process understanding and certification. Real-time data analysis and rapid response instruments that can be employed during synthesis, processing (densification), and device production will be needed for process and quality control (Saltiel and Giesche 2000; Gagea and Rajkovic 2008).

Data reproducibility can provide greater confidence in the measurement data. Different analysis methods can provide important information about the sample (Baer et al. 2010). A number of technical and non-technical roadblocks in characterization methods that deter advances in Nanotechnology include—Electron microscopy is very slow and cumbersome for statistically reliable sampling. Also, a high level of technical skill is required to operate the instrument and conduct the analysis, aggregation of particles can make identification of true size distributions problematic,

Characterization Techniques for Biogenic Nanoparticles   235

lack of standards and common points of reference often lead to confusion and even with high-resolution microscopy. It is very ambiguous to distinguish if a particle is thin and disk-like or a 3-D structure (Saltiel and Giesche 2000).

Therefore, improvement of present characterization technology and the development of new techniques are needed to keep up with processing advances. Moreover, improved characterization tools are also essential as the nanotechnology industry matures and development towards large-scale commercial production and growth of new applications will depend on increased research and development which in turn depends on better characterization techniques.

Considerations for analysis of nanoparticles

Some basic information associated with nanoparticles storage, and their preparation is useful during analyzing nanoparticles. One of the important properties of the nanoparticles is their stability concern at ambient conditions thereby significantly increasing the attention that analysts need to pay to the impacts that an analysis probe, environmental conditions, and time can have on the materials analyses (Phillips and Quake 2006; Sarathy et al. 2008; Baer et al. 2010). Secondly, the portion of other atoms or molecules associated with the surface may increase the probability of surface impurities, surface modification, and surface contamination (Grainger and Castner 2008) on nanoparticle properties and complicate accurate measurements. Surface coatings and impurities can significantly alter many aspects of nanoparticle behavior. The vitalities associated with probes used for particle analysis may equal or exceed those required to transform the particles in several ways (Baer et al. 2008). These general issues can have bearing on the care taken for sample handling, the time consumed and nature of the analyses, and results in comparison from a variety of methods (Baer et al. 2010).

The issues about contamination, impurities, and environmental effects necessarily raise concerns about the handling and mounting of specimens for analysis. For nanoparticles, a variety of guides like ASTM (Standard Guide for Specimen Preparation and Mounting in Surface Analysis) and ISO have been developed. Different nanoparticle characterization techniques may produce a variable particle size and distribution data, sometimes nanoparticle properties and measurement data can be influenced by agglomeration or aggregation of nanoparticles with the supporting substrate (Baer and Engelhard 2010; Baer et al. 2010). Such effects include the coupling of plasmon modes in metal nanoparticles within close proximity, the effect of particle spacing on electronic and magnetic properties of the composite, and the effect of corona layers on optical properties of nanocrystal superlattices (Glover and Meldrum 2005).

There are different challenges faced in the characterization of nanoparticles using many of the currently available techniques along with their limitations. The considerations for nanoparticle characterization may include particle stability, contamination, environmentally-impacted changes, contamination, and technique-specific constraints related to the characterization of nanoparticles (Baer et al. 2010).

Characterization techniques

The role of nanoparticles in research and product development is growing rapidly in the application areas related to medicine, diagnosis, environmental science, agriculture, and consumer products. Making it mandatory to understand their properties and behaviors as they are synthesized, applied, and evolve in a particular environment, process, or application. There is also a need to understand the health, safety, and environmental impacts of nanoparticles in both their synthesized form and as a part of the consumer product (Baer et al. 2010).

Nanoparticles can be fabricated biologically using various biological materials like plants, bacteria, fungi, algae, and yeast. Either the whole organism is used or their extracts are applied for the synthesis of nanoparticles. Different methods used for the characterization of these nanoparticles are given in Figure 1. Biogenic nanoparticles have shown their applications in numerous fields including antimicrobials, antioxidants, antiangiogenic, antitumors, etc. (Tarannum et al. 2019; Galatage et al. 2020). These biogenic nanoparticles can be further detected and characterized by various techniques.

Varied information can be obtained from the analysis of nanoparticles using different characterization techniques, information about the biomolecules' role in the synthesis and stabilization of nanoparticles can be obtained from FTIR, XRD will reveal information about the crystalline nature, Electron microscopy (SEM, and TEM) analysis will provide information about the size and shape of the nanoparticles, EDX will provide with the elemental composition of the nanoparticles, NTA estimates size, size distribution and concentration of nanoscale particles in colloids and Zeta

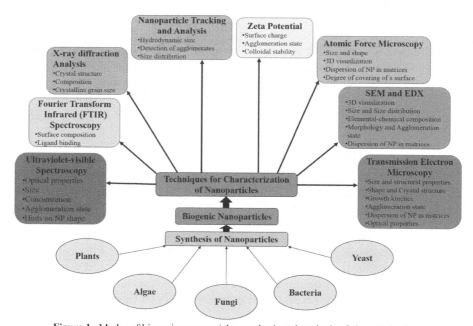

Figure 1. Modes of biogenic nanoparticles synthesis and methods of characterization.

Characterization Techniques for Biogenic Nanoparticles 237

sizer by measuring zeta potential will give insights into the stability of nanoparticles (Kurzydłowski et al. 2012; Gade et al. 2013).

Ultraviolet-visible (UV-Vis) spectroscopy

Ultraviolet-visible (UV-Vis) spectroscopy has been widely used for the detection of nanoparticles. The UV-Vis Spectroscopy shows the absorbance of nanoparticles at a particular wavelength and also determines the electronic structures through exciting electrons from the ground to excited states (absorption). Color transitions in nanomaterials arise due to molecular and structural changes in the substances being examined, leading to corresponding changes in the ability to absorb light in the visible region of the electromagnetic spectrum. The metallic nanoparticles are known to exhibit different characteristic colors (Banu et al. 2011; Suresh 2012; Elnashaie et al. 2015; López-Lorente and Valcárcel 2016). Colloidal AgNPs exhibit a deep-brown color due to the well-known surface plasmon absorption in the range of ca. 400–450 nm, whereas in the case of gold it absorbs in the range of ca. 500–550 nm. This absorption of electromagnetic radiation by metallic nanoparticles originates from the coherent oscillation of the valence band electrons induced by an interaction with the electromagnetic field (Deshmukh et al. 2012). These resonances are known as surface plasmons, which occur only in the case of nanoparticles and not in the case of bulk metallic particles (Papavassiliou 1979). Hence, the UV-Vis spectrum can be utilized to study the unique optical properties exhibited by nanoparticles and can serve as a preliminary method for the detection of nanoparticle synthesis. In UV- Vis spectroscopy the absorption will be observed in the form of the peak when plotted as the function of wavelength with respect to the intensity of absorbed radiation. UV-Vis spectroscopy proves to be one of the important and simplest ways to confirm the formation of nanoparticles. The shifting of absorption peak towards higher energy wavelengths indicates a decrease in the size of the nanoparticles (Gade et al. 2014; Verma et al. 2017). The effect of various physical parameters like light source and intensity, the time course of the reaction, and pH can be demonstrated with the help of a shift in the absorption spectrum of nanomaterials (Gade et al. 2013).

The setup of instrument plays a key role in spectrometric measurements. The transmittance of undesired wavelengths during analysis may cause serious errors due to imperfect light selectors or monochromators or light compartments. Unnecessary scattering of light due to suspended solids, bubbles in cuvettes or samples may result in measurement errors and irreproducible results. Samples with multicomponent systems having a close range of absorption wavelength may result in ambiguous quantitative analysis. Misalignment of measurement and detection components yields irreproducible and inaccurate results (Harris 2007; Tomaszewska et al. 2013; Quevedo et al. 2021).

Fourier transform infrared (FTIR) spectroscopy

The region of the electromagnetic spectrum which extends from the red end to the microwave region is called as Infrared region. The region includes the radiation of wavelength ranging from 0.7 to 500 µm. and the energy corresponding to these frequencies corresponds to the infrared region (4000–400 cm^{-1}) of the electromagnetic spectrum. By using the infra-red spectroscopy twisting, bending, rotating and vibration motions of atoms in a molecule are examined. When a vibrating molecule produces the dipole movement, it interacts with the electric field of IR radiation (Pettinari and Santini 1999; Suresh 2012).

The term Fourier transformation (FT) is a mathematical conversion that refers to a recent development in the manner in which the data are collected and converted from an interference pattern to an infrared absorption spectrum that is like a molecular "fingerprint". The FTIR measurement can be utilized to study the presence of protein molecules in the solution, as the FTIR spectra in the 1000–2000 cm^{-1} region provide information about the presence of different functional groups (Aroui et al. 2012; Pushkar and Mungray 2020; https://agta.org/advantages-and-disadvantages-of-raman-fourier-transform-infrared-spectroscopy-ftir-in the-gemological-field/).

The versatility of FTIR is based on the long wavelength of the radiation, which minimizes the scattering problem. In addition to that, a wide range of sampling methods has been developed using transmission, reflectance, or emission technique. The choice of the sampling method is mostly based upon the sample preparation and type of sample (Amritsar et al. 2006). FTIR spectroscopy is designed to measure vibration spectra, yield information pertaining to chemical bonds, and allow the measurement of substances in any state (gas, liquid, or solid), with only a small amount of sample required and minimal interference from coexisting substances. FTIR spectroscopy is based on molecular vibrations, accompanied by changes in the bipolar moment in the middle IR region (2.5–25 µm) and the wave number between 4,000 and 400 cm^{-1}. This mode of vibration includes bending/stretching, rotating, pinching, twisting, oscillating vibration, and so on, which collectively cause intense IR absorption. IR activity is not seen during the bending/stretching vibration of CO_2 or diatomic molecules, which is also not accompanied by any change in the bipolar moment. The FTIR spectrum is often used for the direct identification of certain specific functional groups constituting organic molecules (Figure 2A). Among inorganic solids, the hydroxide group, bound water, oxyanions (carbonates, nitrates, and sulfates), and so on are known to cause intense IR peaks (Rai et al. 2015).

In the transmission method, the light transmitted through a disk-shaped sample is measured with the latter usually prepared by pressure forming. If disk forming is difficult, an alternative method available is pellet forming with KBr and spraying of the sample onto the IR ray-transmitting material board. With the diffuse reflection method, i.e., diffuse reflectance IR Fourier transform spectroscopy (DRIFT), spectra are obtained by measuring the diffuse reflective light, which returns after diffusion within the sample. If the light is applied to the powder layer, it diffuses in various directions, while repeating reflection on the particle surface and is eventually released out of the sample from the uppermost layer. During this process, light is repeatedly transmitted through the superficial layer of the particles. For this reason, the diffuse reflective light contains information about the IR absorptive property of the material in the vicinity of the particle surface. In research related to fine particles, this method is often used for the identification of the surface and observation of adsorbed species (Genty-Vincent et al. 2015).

Characterization Techniques for Biogenic Nanoparticles 239

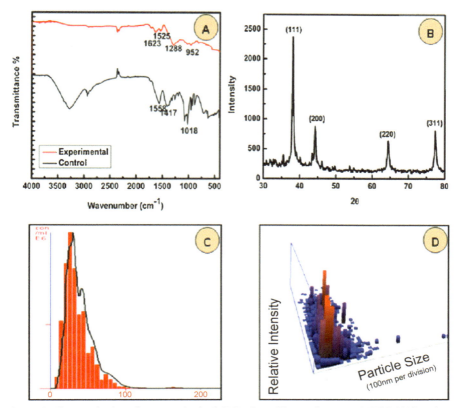

Figure 2. Characterization of mycosynthesized SNPs from *Phoma glomerata* (MTCC-2210), where (A) FTIR spectrum of SNPs, (B) XRD graph of SNPs, (C) NTA Particle size distribution and (D) NTA 3D plot of particle size distribution (Gade et al. 2014).

The sample analysis chambers are maintained at specific relative humidity higher than the external environment. The change in internal humidity results in the non-functionality of instrumental components. Sampling chambers of FTIR spectroscopes are usually smaller in size, which limits the amount of sample analyzed. Many times samples absorb all incident infrared radiations and only small number of samples in the form of rings are analyzed. This may provide inconsistent and unreliable results (Fels et al. 2015; Faghihzadeh et al. 2016; https://www.bruker.com/en/products-and-solutions/infrared-and-raman/ft-ir-routine-spectrometer/what-is-ft-ir-spectroscopy.html; www.agta.org).

X-ray diffraction analysis

X-ray diffraction (XRD) is one of the most widely used indirect methods for determining particle size. It also provides information regarding the structures or phases of a crystal. In the powder method, the crystal to be examined is in the form of fine powder placed in the beam of monochromatic X-ray. Each particle will behave like a tiny crystal or a group of small crystals oriented in a random direction with respect to the incident beam (Sommariva et al. 2014; Kawaguchi et al. 2017; Sharma et al. 2017). Some particles will be oriented such that their one particular plane will diffract the beam at one particular angle. Other particles will diffract in other directions and so on. Thus, every set of the plane is capable of diffraction. In the case of small crystals, broadening of the diffraction peak occurs due to the nonexistence of the atomic planes which cancel the scattered X-rays at angles deviating slightly from the exact Bragg's angle (Momose 2000; Janssens 2004; Bellucci et al. 2013).

The X-ray diffraction patterns are obtained by measurement of the angles at which an X-ray beam is diffracted by the crystalline phases in the specimen. Bragg's equation relates the distance between two hkl planes (d) and the angle of diffraction (2θ) as $n\lambda = 2d\sin\theta$, where, λ = wavelength of X-rays, n = an integer known as the order of reflection (h, k, and l represent Miller indices of the respective planes) (Suresh 2012; Bragg and Bragg 1913). X-ray diffraction broadening analysis has been widely used to determine the crystal size of nanoscale materials (Figure 2B). The average size of the nanoparticles can be estimated using the Debye–Scherrer equation: $D = k\lambda/\beta\cos\theta$, where D = thickness of the nanocrystal, k is a constant, λ = wavelength of X-rays, β = width at half maxima of (111) reflection at Bragg's angle 2θ (Kruh 1962; Suresh 2012; Venkateswarlu et al. 2014).

XRD is the basic method for characterizing crystal structures and the amorphous nature of nanomaterials. It gives an idea about the direction of crystal growth as well. While the crystal structures are specified by the arrangement of atoms separated by 1 angstrom (Liu and Wang 2017). The crystal structure of nanoparticles can be more appropriately characterized by electron beam diffraction (ED) method using a TEM rather than XRD. However, the XRD method has several advantages like XRD allows measurements in the air or necessary atmosphere, sample preparation is easier and average crystal structures can be quantitatively evaluated (Gong et al. 2008).

The XRD method is based on the measurements of X-ray intensities scattered by the statistically distributed electrons belonging to the atoms in the material. The arrangement of atoms or the population of electrons is determined by analysis of angular dependence of scattered X-rays in both methods. The XRD method is suitable to determine the crystal structures by analyzing the positions and intensities of diffraction peaks typically observed for the well-crystallized material in the range of diffraction angles from 20 to 80°. The method is also used for evaluating the crystallite size by analyzing the width and the shape of the peak profiles (Chen 2011; Kharat and Mendhulkar 2016).

Since the most stable structure of a pure material is crystal, where the atoms are periodically arranged, the pattern of the positions and intensities of XRD peaks can be uniquely assigned to the material. Therefore, XRD measurement is important to identify the main component of materials. The most commonly used database for the identification is by Joint Committee on powder diffraction, standard file (JCPDS). JCPDS database consists of sets of data about interplanar distances and intensities of the significant diffraction peaks. Evaluation of crystallite size by peak profile analysis is one of the important applications of the powder XRD method to study nanoparticles. The range of crystallite size that can be evaluated by the powder XRD method depends on the accuracy of the instrument and analytical methods (Uvarov and Popov 2013; Crovetto et al. 2015; Pilloni et al. 2018).

Nanoparticle tracking and analysis (NTA)

Important criteria for the use of synthesized nanoparticles for the formulation of any product are an idea about its toxicity and for the nanotoxicological study. It is necessary to know the state of the nanoparticles to be used particularly their size and size distribution in the appropriate test media. Nanoparticles Tracking and Analysis (NTA) technique allows nanoparticles to be sized in suspension on a particle-by-particle basis (Figure 2C and D). NTA allows individual nanoparticles in a suspension to be microscopically visualized (though not, of course, imaged) and their Brownian motion to be separately but simultaneously analyzed and from which the particle size distribution (and changes therein) can be obtained on a particle-by-particle basis allowing higher resolution and, therefore, a better understanding of aggregation than other ensemble methods like Dynamic light scattering (DLS) and Differential centrifugation sedimentation (DCS) (Montes-Burgos et al. 2009; Mahmoudi et al. 2011; Gade et al. 2013).

NTA technique can be extended to multi-parameter analysis, allowing for characterization of particle size and light scattering intensity on an individual basis. This multi-parameter measurement capability allows subpopulations of particles with varying characteristics to be resolved in a complex mixture. Changes in one or more of such properties can be followed both in real-time and *in situ* (Montes-Burgos et al. 2009).

The statistical accuracy of results obtained by the NTA nanoparticle tracking system is dependent on a number of factors; primarily particle concentration analyzed, the length of time over which the sample is analyzed, and the size of the particles present. While the sample chamber is approximately 250 µl in volume, the section of the laser beam visualized (the scattering volume) is only small (a section approx 80 µm diameter x 100 µm length x 25 µm deep). Accordingly, for a statistically significant number of particles to be present on the beam, sample concentrations should lie between 10^7 and 10^9 particles/ml. Lower magnifications improve measurement accuracy for larger particles but at the expense of being ability to visualize smaller particles (Malloy and Carr 2006; Carr et al. 2008).

NTA allows a wide range of particle types to be analyzed in various solvent types with the only user input required being that of sample temperature and solvent viscosity. Multi-angle measurements are superfluous. The technique of NTA is robust and low cost representing an attractive alternative or complement to higher cost and more complex methods of nanoparticle analysis such as photon correlation spectroscopy (PCS) or electron microscopy that are currently employed in a wide range of technical and scientific sectors. It uniquely allows the user a simple and direct qualitative view of the sample under analysis (perhaps to validate data obtained from other techniques such as PCS) and from which an independent quantitative estimation of sample size, size distribution, and concentration can be immediately obtained (Carr et al. 2008; Carr and Malloy 2008).

242 *Mycosynthesis of Nanomaterials: Perspectives and Challenges*

The nanoparticle tracking and analysis (NTA) technique is a light-scattering technique and allows nanoparticles to be sized in suspension on a particle-by-particle basis. NTA uses a laser light source to illuminate nanoscale particles. Enhanced by a near-perfect black background, the particles appear individually as point-scatterers moving under Brownian motion. The NTA image analysis software automatically tracks and sizes many particles simultaneously (Kharat and Mendhulkar 2016; Fernando et al. 2017). Results are displayed as frequency size distributions or may be output to a spreadsheet format. Supplementary scatter intensity data is combined with these size distributions to provide 3D plots which highlight disparate populations. The video clip of each sample studied is recorded and retained for further analysis (Malloy 2011).

For the measurement, samples are prepared in an appropriate liquid (Nuclease free water or Milli Q) at a concentration level of 10^7–10^9 parts/ml and placed in the sample chamber which has a volume of 0.3 ml. The laser then illuminates the samples in the chamber with the dispersed light being captured by a high sensitivity camera, via the microscope. Particles are individually tracked and visualized on the screen of the control computer. The smallest particles appear as fast-moving dots of light while larger particles diffuse more slowly. The direct observation of particle motion and scattering behavior provides a wealth of additional information beyond particle size. These real-time observations validate the reported particle size distributions and provide instant insight into polydispersity and the state of aggregation. Measurements take just minutes allowing time-based changes and aggregation kinetics to be quantified (Malloy 2011).

The capabilities of NTA measurements include-

* Direct and live view of particles in suspension.
* Particle-by-particle, high-resolution particle sizing.
* Especially applicable to polydisperse systems.
* Measurement of concentration and particle count.
* Scattering intensity distribution.
* Evidence of non-sphericity and of aspect ratio.
* Sizing down to 10 nm, material dependent.
* Number vs intensity vs size is provided for each particle size class.

NTA allows a wide range of particle types to be analyzed in a wide range of solvent types with the only user input required being that of sample temperature and solvent viscosity. NTA is robust and low cost representing an attractive alternative or complement to higher cost and more complex methods of nanoparticle analysis. NTA analysis is limited due to the strong decrease in scattered light intensity with particle diameter, scaling with diameter to the 6th power. Sample composition also plays a vital role in measurement. Smaller particles even below 100 nm are measured with accuracy when samples have high purity. The accuracy may be affected when samples are in the form of mixtures or with impurities (http://www.nanoparticleanalyzer. com/; Kim et al. 2019).

SEM and EDX

Scanning electron microscopy (SEM) is one of the most widely used techniques for the characterization of nanoparticles. The popularity of the SEM can be attributed to many factors: the versatility of its various modes of imaging, the excellent spatial resolution now achievable, the very modest requirement on sample preparation and condition, the relatively straightforward interpretation of the acquired images, the accessibility of associated spectroscopy and diffraction techniques (Lukaszkowicz 2011; Erdemoğlu and Baláž 2012). With the recent generation of SEM instruments, high-quality images can be obtained with an image magnification as low as about 5X and as high as > 1,000,000X (as shown in Figure 3A); this wide range of image magnifications bridges our visualization ability from naked eyes to nanometer dimensions (Yao and Wang 2005).

In an SEM instrument, a fine electron probe, formed by using a strong objective lens to de-magnify a small electron source, is scanned over a specimen in a two-dimensional raster. Signals generated from the specimen are detected, amplified, and used to modulate the brightness of a second electron beam that is scanned synchronously with the SEM electron probe across a cathode-ray-tube (CRT) display. SEM provides topographical information like the shape and size of the nanoparticles. A scanning electron microscope generates an electron beam scanning back and forth over a sample. The interaction between the beam and the sample produces different

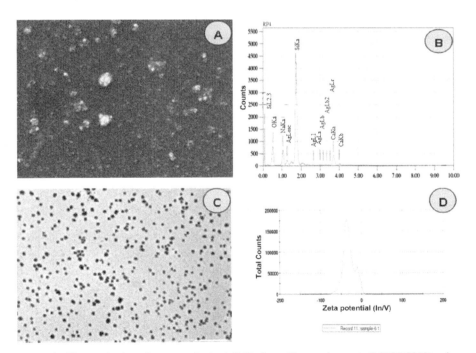

Figure 3. Characterization of mycosynthesized SNPs from *Phoma glomerata* (MTCC-2210), where (A) SEM micrograph (scale bar 100 nm), (B) EDX, (C) TEM micrograph (scale bar 100 nm) and (D) Zeta potential graph (Gade et al. 2014).

types of signals providing detailed information about the surface structure and morphology of the sample (Yao and Wang 2005; Suresh 2012).

SEM can provide both morphological information at the submicron scale and elemental information at the micron scale. Recent developments in terms of electron source (field emission) have led to the development of high-resolution SEM. Using a secondary or backscattering electron image one can look at particles as small as 10–20 nm. Since the scattering angle depends on the atomic number of the nucleus, the primary electrons arriving at a given detector position can be used to produce images containing topological and compositional information (Lawes 1987). Chemical information using Energy-dispersive X-ray spectroscopy (EDX), however, is obtained at the micron scale and not for individual particles (https://www.texaspowerfulsmart.com/oxidative-stress/sem-afm.html).

SEM uses a focused beam of high-energy accelerated electrons to generate a variety of signals at the surface of generally solid specimens. The signals generated from the interaction by electron and sample reveal information about the sample like external morphology (texture), chemical composition, crystalline structure, and orientation of materials making up the sample. The electron-sample interaction signal includes secondary electrons (SE), backscattered electrons (BSE), diffracted backscattered electrons (EBSD), photons, visible light, and heat (Saravanan and Rani 2012; Anwar et al. 2015; Hafez et al. 2015). SE and BSE are commonly used for imaging samples: SE are most valuable for showing morphology and topography on samples and BSE is most valuable for illustrating contrasts in composition in multiphase samples. EBSD is used to determine crystal structures and orientations of minerals and X-rays that are used for elemental analysis and continuum X-rays (Busca 2014). In SEM analysis the characteristic X-rays are produced for each element in a mineral that is "excited" by the electron beam. SEM analysis is considered to be "non-destructive"; i.e., x-rays generated by electron interactions do not lead to volume loss of the sample, so it is possible to analyze the same materials repeatedly. The main problem with the application of SEM to nanoparticle characterization analysis is that sometimes it is not possible to clearly differentiate the nanoparticles from the substrate. Problems become even more exacerbated when the nanoparticles under study have a tendency to adhere strongly to each other, forming agglomerates (Saravanan and Rani 2012; Pathak and Thassu 2016).

In EDX X-rays emitted from the atoms represent the characteristics of the element and their intensity distribution represents the thickness-projected atom densities in the specimen. EDX plays an important role in microanalysis, particularly for heavier elements and it is a key tool for identifying the chemical composition of a specimen (Figure 3B). A modern SEM and TEM are capable of producing a fine electron probe, allowing direct identification of the local composition of an individual nanocrystal (Wang 2001).

There are different technological variants of EDX that can even be used for the compositional analysis of nanoparticles (Hall et al. 2007). When low-energy electrons rather than X-rays are detected, the technique is referred to as auger electron spectroscopy (AES) (Unterhalt et al. 2002). A higher-resolution technique to EDX is the complementary technique of electron energy loss spectroscopy (EELS) (Park et al. 2005), which can be used to differentiate even different forms of the same element (e.g., diamond and graphite can usually be differentiated by EELS). Another variant includes X-ray photoelectron spectroscopy (XPS) which, uses an input beam of x-rays rather than electrons and detects the kinetic energies of emitted electrons (Craparo et al. 2006).

The limitations of the SEM and EDX methods are attributed to the moisture content of samples under examination. If the samples are not polished, complete quantitative chemical analyses become impossible to obtain, but only semi-quantitative analyses are possible (Eric 2017). SEMs are large, expensive, and must be installed in an area devoid of any possible electric, magnetic or vibration interference. The maintenance involves keeping steady current to electromagnetic coils and steady voltage, along with circulation of cool water (Choudhary and Priyanka 2017). EDX is a non-destructive method of analysis in which samples can be analyzed in situ with very less sample preparations. In commonly used detectors of EDX, nitroges generate weak, minor peaks which are hard to be detected. The bulk analysis is not possible, limiting the smaller quantities of samples to be analyzed (https://www.intertek.com/analysis/microscopy/edx/; Nasrazadani and Hassani 2016).

TEM

Transmission electron microscopy (TEM) analyzes the transmitted or forward-scattered electron beam. Here the electron beam is passed through a series of lenses to determine the image resolution and obtain the magnified image. The highest structural resolution possible (point resolution) is achieved upon the use of high-voltage instruments (acceleration voltages higher than 0.5 MeV). Enhanced radiation damage, which may have stronger effects on nanostructured materials, must however be considered in these cases. With corrections it is possible to achieve sub-angstrom resolution with microscopes operating at lower voltages (typically, 200 keV), allowing the oxygen atoms to be resolved in oxides materials (Fernandez-Garcia et al. 2004; 2006). On the other hand, as high resolution is achieved in TEM as the result of electron wave interference among diffracted peaks and transmitted beams in the absence of deflection, a limitation to structural resolution can arise from nanoparticles with a very low number of atoms (Figure 3C). Nevertheless, conventional TEM is the most common tool used to investigate the crystal structure of materials at the sub-nanometer scale. The topographic information obtained by TEM in the vicinity of atomic resolution can be utilized for structural characterization and identification of various phases of nanomaterials, *viz.*, hexagonal, cubic, or lamellar (Wang 2000). One shortcoming of TEM is that the electron scattering information in a TEM image originates from a three-dimensional sample, but is projected onto a two-dimensional detector. Therefore, structural information along the electron beam direction is superimposed at the image plane (Murty et al. 2013; Bhaumik et al. 2015).

In TEM Selected area diffraction (SAD) offers a distinctive advantage to determine the crystal structure of individual nanomaterials, such as nanocrystals and nanorods, and the crystal structures of different parts of the sample. In SAD, the condenser lens is defocused to produce parallel illumination at the specimen and a selected-area aperture is used to limit the diffracting volume. SAD patterns are often used to determine the Bravais lattices and lattice parameters of crystalline

246 *Mycosynthesis of Nanomaterials: Perspectives and Challenges*

materials by the same procedure used in XRD. TEM has been also explored for other applications in nanotechnology, which include the determination of melting points of nanocrystals (Goldstein et al. 1992) and the measurement of mechanical and electrical properties of individual nanowires and nanotubes (Wang 2000; Murty et al. 2013).

There exist a number of different TEM techniques that may be used to obtain structural images with atomic-level resolution; two of these techniques are detailed below: high-resolution TEM (HRTEM) and high angle annular dark-field (HAADF) scanning transmission electron microscopy (STEM) (Serin et al. 2013). The progress made in TEM has enabled the direct imaging of atomic structures in solids and surfaces. Nanometer-sized particles are commonly present in many different types of materials and the use of TEM allows for gathering information about particle size, shape, and any surface layers or absorbates (Henry 2005; Brydson and Brown 2008). There are three types of transmitted electrons observed by TEM; they are unscattered electrons, elastically scattered electrons, and in-elastically scattered electrons. The bright-field imaging technique is the most common method for the imaging method using TEM (formation of images only with the transmitted electron beam). The electrons used for imaging are selected only via the aperture diaphragm according to their scattering angles. Therefore, only electrons with a large scattering angle contribute to the contrast generation. The energy of the electrons and the difference in their energies remain unaccounted for, despite the fact that the bandwidth of the energy differences caused by the chromatic aberration of the objective lens has considerable influence on contrast and resolution (Reimer 1984).

Nanoparticle-containing samples can be susceptible to the highly energetic electron beam of the TEM instrument (Bentley et al. 2006). Beam susceptibility makes it very difficult sometimes to carry out electron diffraction studies on nanoparticles that are prone to beam damage. In this case, by using low electron beam currents, it is possible to obtain lattice fringe images and electron diffraction (Sharma et al. 2019). In spite of many advantages, TEM imaging still presents a series of challenges. For instance, image overlap is a typical problem during observation. When this occurs, the surrounding matrix usually tends to mask the supported nanoparticles. In some special cases, however, the existence of an epitaxial relationship between the nanoparticles and their support can be used to obtain size and shape information (Smith 2007). Moreover, nanoparticles can be susceptible to damage under the electron beam irradiation conditions normally used for high-resolution imaging. The disadvantage of both SEM and TEM in the characterization of nanoparticles is that one can never be sure that the observed image is truly representative of the bulk nanoparticle sample. Consequently, bulk-sensitive methods that provide information regarding the quality, size, and structural properties of a given sample must be employed (Pathak and Thassu 2016).

Zeta potential measurements

Zeta potential is a measure of the magnitude of the repulsion or attraction between particles. Its measurement brings detailed insight into the dispersion mechanism and is the key to electrostatic dispersion control. In colloidal systems, zeta potential is used to represent the electro-kinetic potential (Figure 3D). Zeta potential can give a better understanding and control over colloidal suspensions. Colloidal particles dispersed in a solution are electrically charged due to their ionic characteristics and dipolar attributes. Each particle dispersed in a solution is surrounded by oppositely charged ions called the fixed layer. Outside the fixed layer, there are varying compositions of ions of opposite polarities, forming a cloud-like area. This area is called the diffuse double layer, and the whole area is electrically neutral. When a voltage is applied to the solution in which particles are dispersed, particles are attracted to the electrode of the opposite polarity, accompanied by the fixed layer and part of the diffuse double layer, or internal side of the "sliding surface" (Franks 2002; Ghasemi-Kahrizsangi et al. 2015; Rai et al. 2015; Uragami 2017; Xu et al. 2018).

Atomic force microscopy (AFM)

Atomic force microscopy (AFM) is a powerful technique that has enabled major advances in the field of nanotechnology. The development of these "sharp" tip-based probes has made it possible for almost any laboratory around the world to examine surfaces of many types of materials with spatial resolutions approaching 1 nm for sample topography. The AFM tip size and shape, influences on the acquired images must be properly accounted for and corrected. A variety of images collected can be used to extract information about particle-size distributions and volumes (www.nanoparticles.pacificnanotech.com) (Baer et al. 2010). The interactions of living cells can be studied using the AFM technique. The effectiveness of various ions, molecules, and nanomaterials can be visualized with ease at the interface as shown in Figure 4 (Gade et al. 2016).

AFM is ideally suited for characterizing nanoparticles. It offers the capability of 3D visualization and both qualitative and quantitative information on many physical properties including size, morphology, surface texture, and roughness (Nagvekar et al. 2009). Statistical information, including size, surface area, and volume distributions, can be determined as well. A wide range of particle sizes can be characterized in the same scan, from 1 nm to 8 μm. In addition, the AFM can characterize nanoparticles in multiple mediums including ambient air, controlled environments, and even liquid dispersions. Obviously, there is not one single "best technique" for all situations. Determining the best technique for a particular situation requires knowledge of the particles being analyzed, the ultimate application of the particles, and the limitations of techniques being considered (Ahmed et al. 2015). AFM is able to scan nano-sized images of 150×150 nm and only one image at a time. Thermal drifts in the sample are possible due to the short scanning time, which may damage the sample. Besides this, AFM has limited magnification and vertical range.

Conclusion

The chapter describes the facts or information obtained by several techniques used for the characterization of biogenic nanomaterials. Revealing the complete features of nanoscale materials requires the use of several characterization techniques, the

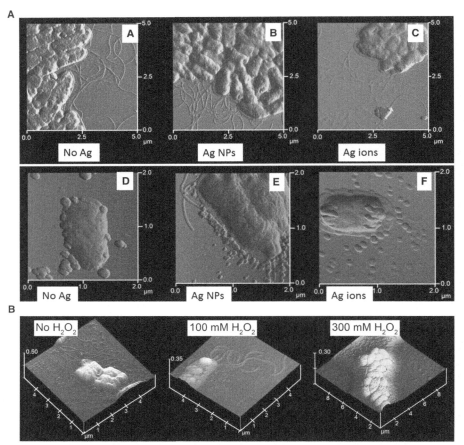

Figure 4. AFM imaging of wild type *Pseudomonas chlororaphis* O6 cells with and without treatments of Ag NPs, Ag ions and hydrogen peroxide. Figure 4a Representative images are shown for wild-type cells without treatments (A, B) and after treatment with acetone-purified Ag NPs (C, D) or Ag ions (E, F) at lethal levels (2 mg Ag/L) for 1 h. Images A, D, C, F show cells with and without flagella respectively for the control and Ag ion-treated cells (Source: Gade et al. 2016).

information provided by each characterization technique in bits and pieces can be useful for the acquisition of a complete picture about nanoscale materials providing the user with better guidance about the applicability of the synthesized nanoscale materials in a precise manner. Understanding of limitations, constraints, and advantages of each characterization technique provide guidance to researchers in the selection of suitable characterization techniques. In spite of several characterization techniques available today, still there are challenges faced by the researchers to improve the accuracy and resolution of several characterization techniques generating the need to develop new faster, simple, and more efficient characterization techniques.

Acknowledgements

MR thankfully acknowledges the financial support rendered by the Polish National Agency for Academic Exchange (NAWA) (Project No. PPN/ULM/2019/1/00117/U/00001) to visit the Department of Microbiology, Nicolaus Copernicus University, Toruń, Poland.

References

Abe, H., Adschiri, T., Aida, T., Akatsu, T., Akedo, J., Ando, M., Ando, Y. et al. 2018. Characterization methods for nanostructure of materials. Nanoparticle Technology Handbook, 255–300. doi:10.1016/b978-0-444-64110-6.00005-6.

Ahmed, M.A., Seddik, U., Okasha, N., and Imam, N.G. 2015. One-dimensional nanoferroic rods; synthesis and characterization. J. Mol. Str. 1099: 330–339. doi:10.1016/j.molstruc.2015.05.06.

Albuquerque, C.D.L., Nogueira, R.B., and Poppi, R.J. 2016. Determination of 17β-estradiol and noradrenaline in dog serum using surface-enhanced Raman spectroscopy and random Forest. Microchem. J. 128: 95–101. doi:10.1016/j.microc.2016.04.012.

Amritsar, J., Stiharu, I., and Packirisamy, M. 2006. Bioenzymatic detection of troponin C using micro-opto-electro-mechanical systems. J. Biomed. Opt. 11(2): 021010. doi:10.1117/1.2186326.

Anwar, S., Mubarik, M., Waheed, A., and Firdous, S. 2015. Polarization imaging and characterization of chitosan for applications in tissue engineering. Optik – Int. J. Light Electron Optics. 126(7-8): 871–876. doi:10.1016/j.ijleo.2015.02.027.

Aroui, H., Orphal, J., and Kwabia, F. 2012. Fourier transform infrared spectroscopy for the measurement of spectral line profiles. Fourier Trans. – Mater. Anal. doi:10.5772/36120.

Baer, D.R., Amonette, J.E., Engelhard, M.H., Gaspar, D.J., Karakoti, A.S., Kuchibhatla, S., Nachimuthu, P., Nurmi, J.T., Qiang, Y., Sarathy, V., Seal, S., Sharma, A., Tratnyek, P.G., and Wang, C.M. 2008. Characterization challenges for nanomaterials. Surf. Interface Anal. 40: 529–537.

Baer, D.R., and Engelhard, M.H. 2010. XPS analysis of nanostructured materials and biological surfaces. J. Elec. Spectro. Rel. Phen. 178–179.

Baer, D.R., Gaspar, D.J., Nachimuthu, P., Techane, S.D., and Castner, D.G. 2010. Application of surface chemical analysis tools for characterization of nanoparticles. Anal. Bioanal. Chem. 396(3): 983–1002. doi:10.1007/s00216-009-3360-1.

Banu, A., Rathod, V., and Ranganath, E. 2011. Silver nanoparticle production by *Rhizopus stolonifer* and its antibacterial activity against extended spectrum β-lactamase producing (ESBL) strains of Enterobacteriaceae. Mater. Res. Bulletin. 46(9): 1417–1423. doi:10.1016/j.materresbull.2011.0.

Bellucci, V., Camattari, R., and Guidi, V. 2013. Quasi-mosaicity as a powerful tool to investigate coherent effects. Optics for EUV, X-Ray, and Gamma-Ray Astronomy VI. doi:10.1117/12.2024137.

Bentley, J., Gilliss, S.R., Carter, C.B, Al-Sharab, J.F., Cosandey, F., Anderson, I.M., and Kotula, P.J. 2006. Nanoscale EELS analysis of oxides: composition mapping, valence determination and beam damage. J. Phy. Conference Series. 26: 69–72.

Bhaumik, J., Thakur, N.S., Aili, P.K., Ghanghoriya, A., Mittal, A.K., and Banerjee, U.C. 2015. Bioinspired nanotheranostic agents: synthesis, surface functionalization, and antioxidant potential. ACS Biomat. Sci. Eng. 1(6): 382–392. doi:10.1021/ab500171a.

Bragg, W.H., and Bragg, W.L. 1913. The reflection of X-rays by crystals. 88(605): Royal Society, London. doi: https://doi.org/10.1098/rspa.1913.0040.

Brydson, R., and Brown, A. 2008. An investigation of the surface structure of nanoparticulate systems using analytical electron microscopes corrected for spherical aberration. *In*: Turning Points in Solid-State, Material and Surface Science. London: RSC Publishing, 778–791.

Busca, G. 2014. Characterization of real catalytic materials. Heterogen. Catal. Mater. 23–35. doi:10.1016/b978-0-444-59524-9.00003-1.

Carr, B., Knowles, J., and Warren, J. 2008. A reagentless real-time method for the multiparameter analysis of nanoparticles as a potential "trigger" device. Optic. Based Biol. Chem. Det. Defence. IV. doi:10.1117/12.799627.

250 *Mycosynthesis of Nanomaterials: Perspectives and Challenges*

Carr, B., and Malloy, A. 2008. Nanoparticle Measurement Through Visualisation. Microscopy Today. 16(2): 22–25. doi:10.1017/s1551929500055887.

Chen, H. 2011. Electrical and material characterization of tantalum pentoxide (Ta2O5) charge trapping layer memory. Appl. Surf. Sci. 257(17): 7481–7485. doi:10.1016/j.apsusc.2011.03.055.

Choudhary, O.P., and Priyanka. 2017. Scanning electron microscope: advantages and disadvantages in imaging components. Int. J. Curr. Microbiol. App. Sci. 6(5): 1877–1882. https://doi.org/10.20546/ijcmas.2017.605.207.

Craparo, E.F., Cavallaro, G., Bondi, M.L., Mandracchia, D., and Giammona, G. 2006. PEGylated nanoparticles based on a polyaspartamide. preparation, physico-chemical characterization and intracellular uptake. Biomacromolecules. 7: 3083–3092.

Crovetto, A., Cazzaniga, A., Ettlinger, R.B., Schou, J., and Hansen, O. 2015. Optical properties and surface characterization of pulsed laser-deposited Cu 2 ZnSnS 4 by spectroscopic ellipsometry. Thin Solid Films, 582: 203–207. doi:10.1016/j.tsf.2014.11.075.

Deshmukh, S.D., Gade, A.K., and Rai, M. 2012. *Pseudomonas aeruginosa* mediated synthesis of silver nanoparticles having significant antimycotic potential against plant pathogenic fungi. J. Bionanosci. 6(2): 90–94. doi:10.1166/jbns.2012.1078.

Elnashaie, S., Danafar, F., and Hashemipour Rafsanjani, H. 2015. From nanotechnology to nanoengineering. Nanotech. Chem. Eng. 79–178. doi:10.1007/978-981-287-496-2_2.

Tomaszewska Emilia, Katarzyna Soliwoda, Kinga Kadziola, Beata Tkacz-Szczesna, Grzegorz Celichowski, Michal Cichomski, Witold Szmaja, and Jaroslaw Grobelny. 2013. Detection Limits of DLS and UV-Vis Spectroscopy in Characterization of Polydisperse Nanoparticles Colloids. Journal of Nanomaterials, vol. 2013, Article ID 313081, 10 pages, 2013. https://doi.org/10.1155/2013/313081.

Erdemoğlu, M., and Baláž, P. 2012. An overview of surface analysis techniques for characterization of mechanically activated minerals. Mineral Process. Extrac. Metal. Rev. 33(1): 65–88. doi:10.1080/08827508.2010.542582.

Eric, S. 2017. The application and limitations of the sem-eds method in food and textile technologies. Adv. Techno. 6(2): 05–10.

Faghihzadeh, F., Anaya, N.M., Schifman, L.A., and Oyanedel-Craver, V. 2016. Fourier transform infrared spectroscopy to assess molecular-level changes in microorganisms exposed to nanoparticles. Nanotechnol. Environ. Eng. 1, 1. https://doi.org/10.1007/s41204-016-0001-8.

Fels, L.E., Zamama, M., and Hafidi, M. 2015. Advantages and limitations of Using FTIR spectroscopy for assessing the maturity of sewage sludge and olive oil waste co-composts. In: Chamy, R., Rosenkranz, F., and Soler, L. (eds.). Biodegradation and Bioremediation of Polluted Systems - New Advances and Technologies. IntechOpen. https://doi.org/10.5772/60943.

Fernandez-Garcia, M., Martinez-Arias, A., Hanson, J.C., and Rodriguez, J.A. 2004. Nanostructured oxides in chemistry: characterization and properties. ChemInform. 35(47). doi:10.1002/chin.200447225.

Fernandez-Garcia, M., Rodriguez, J.A., Martinez-Arias, A., and Hanson, J.C. 2006. Techniques for the study of the structural properties. United States: N. doi:10.2172/893010.

Fernando, M.R., Jiang, C., Krzyzanowski, G.D., and Ryan, W.L. 2017. New evidence that a large proportion of human blood plasma cell-free DNA is localized in exosomes. PLOS ONE. 12(8): e0183915. doi:10.1371/journal.pone.0183915.

Franks, G.V. 2002. Zeta potentials and yield stresses of silica suspensions in concentrated monovalent electrolytes: isoelectric point shift and additional attraction. J. Coll. Interface Sci. 249(1): 44–51. doi:10.1006/jcis.2002.8250.

Gade, A., Gaikwad, S., Duran, N., and Rai, M. 2013. Screening of different species of Phoma for the synthesis of silver nanoparticles. Biotech. App. Biochem. 60(5): 482–493. doi:10.1002/bab.1141.

Gade, A., Gaikwad, S., Duran, N., and Rai, M. 2014. Green synthesis of silver nanoparticles by *Phoma glomerata*. Micron. 59: 52–59. doi:10.1016/j.micron.2013.12.005.

Gade, A., Adams, J., Britt, D.W., Shen, F.A., McLean, J.E., Jacobson, A., Kim, Y.C. and Anderson, A.J. 2016. Ag nanoparticles generated using bio-reduction and -coating cause microbial killing without cell lysis. BioMetals. 29(2): 211–223. doi: 10.1007/s10534-015-9906-0.

Gagea, L., and Rajkovic, I. 2008. Designing devices for avionics applications and the DO-254 guideline. International Semiconductor Conference. doi:10.1109/smicnd.2008.4703431.

Characterization Techniques for Biogenic Nanoparticles   251

Galatage, S.T., Hebalkar, A.S., Dhobale, S.V., Mali, O.R., Kumbhar, P.S., Nikade, S.V., and Killedar, S.G. 2020. Silver Nanoparticles: Properties, Synthesis, Characterization, Applications and Future Trends. DOI: 10.5772/intechopen.99173.

Genty-Vincent, A., Eveno, M., Nowik, W., Bastian, G., Ravaud, E., Cabillic, I., Uziel, J., Lubin-Germain, N., and Menu, M. 2015. Blanching of paint and varnish layers in easel paintings: contribution to the understanding of the alteration. Appl. Phy. A. 121(3): 779–788. doi: 10.1007/s00339-015-9366-y.

Ghasemi-Kahrizsangi, A., Neshati, J., Shariatpanahi, H., and Akbarinezhad, E. 2015. Effect of SDS modification of carbon black nanoparticles on corrosion protection behavior of epoxy nanocomposite coatings. Poly. Bull. 72(9): 2297–2310. doi:10.1007/s00289-015-1406-4.

Glover, M., and Meldrum, A. 2005. Effect of "buffer layers" on the optical properties of silicon nanocrystal superlattices. Opt. Mater. 27: 977–982.

Godale, C., and Sharon, M. 2019. Contemporary history of nanotechnology. Hist. Nanotech. 213–269. doi:10.1002/9781119460534.ch8.

Goldstein, A.N., Echer, C.M., and Alivisatos, A.P. 1992. Melting in semiconductor nanocrystals. Science. 256(5062): 1425and1427.

Gommes, C., Blacher, S., Masenelli-Varlot, K., Bossuot, C., McRae, E., Fonseca, A., Nagy, J.B., and Pirard, J.P. 2003. Image analysis characterization of multi-walled carbon Nanotubes. Carbon. 41: 2561–2572.

Gong, J., Liu, Y., Wang, L., Yang, J., and Zong, Z. 2008. Preparation and characterization of hexagonal close-packed Ni nanoparticles. Front. Chem. China. 3(2): 157–160. doi:10.1007/s11458-008-0030-3.

Grainger, D.W., and Castner, D.G. 2008. Nanobiomaterials and nanoanalysis: opportunities for improving the science to benefit biomedical technologies. Adv. Mater. 20: 867–877.

Guilger-Casagrande, M., and Lima, R.D. 2019. Synthesis of silver nanoparticles mediated by fungi: a review. Front. Bioeng. Biotech. 7: 287. https://doi.org/10.3389/fbioe.2019.00287.

Hafez, M., Yahia, I.S., and Taha, S. 2015. Study of the diffused reflectance and microstructure for the phase transformation of KNO_3. Acta Physica Polonica A. 127(3): 734–740. doi:10.12693/aphyspola.127.734.

Hall, J.B., Dobrovolskaia, M.A., Patri, A.K., and McNeil, S.E. 2007. Characterization of nanoparticles for therapeutics: Conclusion. Nanomed. 2(6): 789–803.

Harris DC. Quantitative Chemical Analysis. 7th ed, 3rd printing. W. H. Freeman; 2007.

Henry, C.R. 2005. Morphology of supported nanoparticles. Prog. Surf. Sci. 80(3-4): 92–116.

http://www.nanoparticleanalyzer.com/. Nanoparticle Tracking Analysis Limit of Detection Depends on Sample Composition.

https://agta.org/advantages-and-disadvantages-of-raman-fourier-transform-infrared-spectroscopy-ftir-in-the-gemological-field/.

https://www.bruker.com/en/products-and-solutions/infrared-and-raman/ft-ir-routine-spectrometer/what-is-ft-ir-spectroscopy.html.

https://www.texaspowerfulsmart.com/oxidative-stress/sem-afm.html.

Janssens, K. 2004. X-ray based methods of analysis. Comprehen. Anal. Chem. 129–226. doi:10.1016/s0166-526x(04)80008-4.

Jastrzębska, A.M., and Olszyna, A.R. 2015. The ecotoxicity of graphene family materials: current status, knowledge gaps and future needs. J. Nanopart. Res. 17(1). doi:10.1007/s11051-014-2817-0.

Kawaguchi, T., Fukuda, K., and Matsubara, E. 2017. Site- and phase-selective x-ray absorption spectroscopy based on phase-retrieval calculation. J. Phy.: Cond. Matt. 29(11): 113002. doi:10.1088/1361-648x/aa53bb.

Kharat, S.N., and Mendhulkar, V.D. 2016. Synthesis, characterization and studies on antioxidant activity of silver nanoparticles using *Elephantopus scaber* leaf extract. Mater. Sci. Eng. C. 62: 719–724. doi:10.1016/j.msec.2016.02.024.

Kim, A., Ng, W.B., Bernt, W., and Choo, N.J. 2019. Validation of size estimation of nanoparticle tracking analysis on polydisperse macromolecule assembly. Sci. Rep. 9, 2639. https://doi.org/10.1038/s41598-019-38915-x.

Kruh, R.F. 1962. Diffraction Studies of the Structure of Liquids. Chem. Rev. 62(4): 319–346. doi: https://doi.org/10.1021/cr60218a003.

Kurzydłowski, K.J., Lewandowska, M., and Wozniak, M.J. 2012. Experimental approach to the structure and properties of nanoparticles. In book: Towards Efficient Designing of Safe Nanomaterials:

252 *Mycosynthesis of Nanomaterials: Perspectives and Challenges*

Innovative Merge of Computational Approaches and Experimental Techniques. Edition: 1, Royal Society of Chemistry. Eds: Jerzy Leszczynski, Tomasz Puzyn.

Lawes, G. 1987. Scanning electron microscopy and x-ray microanalysis. United States: N. p., 1987. Web.

Liu, Z.K., and Wang, Y. 2017. Computational Thermodynamics of Materials. MRS Bulletin. 02: 162–163.

López-Lorente, Á.I., and Valcárcel, M. 2016. The third way in analytical nanoscience and nanotechnology: Involvement of nanotools and nanoanalytes in the same analytical process. TrAC Tren. Anal. Chem. 75: 1–9. doi:10.1016/j.trac.2015.06.011.

Lukaszkowicz, K. 2011. Review of nanocomposite thin films and coatings deposited by PVD and CVD Technology. Nanomaterials. doi:10.5772/25799.

Mahmoudi, M., Lynch, I., Ejtehadi, M.R., Monopoli, M.P., Bombelli, F.B., and Laurent, S. 2011. Protein-Nanoparticle Interactions: Opportunities and Challenges. Chem. Rev. 111(9): 5610–5637. doi:10.1021/cr100440g.

Malloy, A., and Carr, B. 2006. NanoParticle Tracking Analysis - The Halo™ System. Partic. Partic. Syst. Charact. 23(2): 197–204. doi:10.1002/ppsc.200601031.

Malloy, A. 2011. Count, size and visualized nanoparticles. Mater. Tod. 14(4): 170–173. doi:10.1016/s1369-7021(11)70089-x.

Momose, A. 2000. Design of X-ray interferometer for phase-contrast X-ray microtomography. AIP Conference Proceedings. doi:10.1063/1.1291125.

Montes-Burgos, I., Walczyk, D., Hole, P., Smith, J., Lynch, I., and Dawson, K. 2009. Characterization of nanoparticle size and state prior to nanotoxicological studies. J. Nanop. Res. 12(1): 47–53.

Murty B.S., Shankar P., Raj B., Rath B.B., and Murday J. 2013. Tools to characterize nanomaterials. in: textbook of nanoscience and nanotechnology. Springer, Berlin, Heidelberg. https://doi.org/10.1007/978-3-642-28030-6_5.

Nagvekar, A., Trickler, W., and Dash, A. 2009. Current analytical methods used in the *in vitro* evaluation of nano-drug delivery systems. Curr. Pharma. Anal. 5(4): 358–366. doi:10.2174/157341209789649122.

Naito, M., Yokoyama, T., Hosokawa, K, and Nogi, K. 2018. Nanoparticle Technology Handbook. Elsevier, UK. doi: https://doi.org/10.1016/C2017-0-01011-X.

Nasrazadani, S., and Hassani, S. 2016. Modern analytical techniques in failure analysis of aerospace, chemical, and oil and gas industries. Handbook of Materials Failure Analysis with Case Studies from the Oil and Gas Industry, 39–54. doi:10.1016/b978-0-08-100117-2.00010-8.

Papavassiliou, G.C. 1979. Optical properties of small inorganic and organic metal particles. Prog. Sol. St. Chem. 12: 185.

Park, S.H., Oh, S.G., Mun, J.Y., and Han, S.S. 2005. Effects of silver nanoparticles on the fluidity of bilayer in phospholipid liposome. Coll. Surf. B: Biointerf. 44: 117–122.

Pathak, Y., and Thassu, D. 2016. Drug Delivery Nanoparticles Formulation and Characterization. doi:10.3109/9781420078053

Pettinari, C., and Santini, C. 1999. IR and raman spectroscopy of inorganic, coordination and organometallic compounds. Encyclop. Spectrosco. Spectromet. 1021–1034. doi:10.1006/rwsp.2000.0374.

Phillips, R., and Quake, S.R. 2006. The biological frontier of physics. Phy. Tod. 59: 38–43.

Pilloni, M., Kumar, V.B., Ennas, G., Porat, Z., Scano, A., Cabras, V., and Gedanken, A. 2018. Formation of metallic silver and copper in non-aqueous media by ultrasonic radiation. Ultrason. Sonochem. 47: 108–113. doi:10.1016/j.ultsonch.2018.04.01.

Pushkar, P., and Mungray, A.K. 2020. Exploring the use of 3 dimensional low-cost sugar-urea carbon foam electrode in the benthic microbial fuel cell. Renew. En. 147: 2032–2042. doi:10.1016/j.renene.2019.09.142.

Quevedo, A.C., Guggenheim, E., Briffa, S.M., Adams, J., Lofts, S., Kwak, M., Lee, T.G., Johnston, C., Wagner, S., Holbrook, T.R., Hachenberger, Y.U., Tentschert, J., Davidson, N., and Valsami-Jones, E. 2021. UV-Vis spectroscopic characterization of nanomaterials in aqueous media. J. Vis. Exp. (176): e61764, doi:10.3791/61764.

Rai, M., Gade, A., Duran, N., and Ingle, A.P. 2015. Synthesis of silver nanoparticles by *Phoma gardeniae* and in vitro evaluation of their efficacy against human disease-causing bacteria and fungi. IET Nanobiotech. 9(2): 71–75. doi:10.1049/iet-nbt.2014.0013.

Reimer, L. 1984. Methods of detection of radiation damage in electron microscopy. Ultramicroscopy. 14(3): 291–303. doi:10.1016/0304-3991(84)90097-4.

Characterization Techniques for Biogenic Nanoparticles 253

Saltiel, C., and Giesche, H. 2000. Needs and opportunities for nanoparticle characterization. J. Nanopart. Res. 2: 325–6.

Sarathy, V., Tratnyek, P.G., Nurmi, J.T., Baer, D.R., Amonette, J.E., Chun, C.L., Penn, R.L., and Reardon, E.J. 2008. Aging of iron nanoparticles in aqueous solution: effects on structure and reactivity. J. Phy. Chem. A. 112: 2286–2293.

Saravanan, R., and Rani, M.P. 2012. Metal and Alloy Bonding—An Experimental Analysis. doi: 10.1007/978-1-4471-2204-3.

Serin, V., Zhang, J., Magén, C., Serra, R., Hÿtch, M.J., Lemort, L., Latroche, M., Ibarra, M.R., Knosp, B., and Bernard, P. 2013. Identification of the atomic scale structure of the La0.65Nd0.15Mg0.20Ni3.5 alloy synthesized by spark plasma sintering. Intermetallics. 32: 103–108. doi:10.1016/j.intermet.2012.09.00.

Sharma, P.K., Mishra, A., Kame, R., Malviya, V., and Malviya, P.K. 2017. X-ray diffraction and X-ray K-absorption studies of copper doped complex. J. Phy.: Conference Series. 836: 012042. doi:10.1088/1742-6596/836/1/012042.

Sharma, S., Jaiswal, S., Duffy, B., and Jaiswal, A. 2019. Nanostructured materials for food applications: spectroscopy, microscopy and physical properties. Bioengineering. 6(1): 26. doi:10.3390/bioengineering6010026.

Smith, D.J. 2007. Characterization of nanomaterials using transmission electron microscopy. pp. 1–27. *In:* Hutchison, J., Kirkland, A. (eds.). Nanocharacterisation. Cambridge, England: The Royal Society of Chemistry.

Sommariva, M., Gateshki, M., Gertenbach, J.A., Bolze, J., König, U., Vasile, B.Ş., and Surdu, V.A. 2014. Characterizing nanoparticles with a laboratory diffractometer: from small-angle to total X-ray scattering. Powder Diffraction. 29(S1): S47–S53. doi:10.1017/s0885715614001043.

Suresh, A.K. 2012. Introduction to nanocrystallites, properties, synthesis, characterizations, and potential applications. *In:* Metallic Nanocrystallites and their Interaction with Microbial Systems. SpringerBriefs in Molecular Science. Springer, Dordrecht. https://doi.org/10.1007/978-94-007-4231-4_1.

Tarannum, N., Divya, D., and Gautam, Y.K. 2019. Facile green synthesis and applications of silver nanoparticles: a state-of-the-art review. RSC Adv. 9(60): 34926–34948. doi:10.1039/c9ra04164h.

Unterhalt, H., Rupprechter, G., and Freund, H.J. 2002. Vibrational sum frequency spectroscopy on Pd (111) and supported Pd nanoparticles: CO adsorption from ultrahigh vacuum to atmospheric pressure. J. Phy. Chem. B. 106: 356–367.

Uragami, T. 2017. Science and Technology of Separation Membranes. doi:10.1002/9781118932551.

Uvarov, V., and Popov, I. 2013. Metrological characterization of X-ray diffraction methods at different acquisition geometries for determination of crystallite size in nano-scale materials. Mater. Character. 85: 111–123. doi:10.1016/j.matchar.2013.09.002.

Valcárcel, M., Simonet, B.M., and Cárdenas, S. 2008. Analytical nanoscience and nanotechnology today and tomorrow. Anal. Bioanal. Chem. 391(5): 1881–1887. doi:10.1007/s00216-008-2130-9.

Venkateswarlu, K., Sandhyarani, M., Nellaippan, T.A., and Rameshbabu, N. 2014. Estimation of crystallite size, lattice strain and dislocation density of nanocrystalline carbonate substituted hydroxyapatite by x-ray peak variance analysis. Proce. Mater. Sci. 5: 212–221. doi:10.1016/j.mspro.2014.07.260.

Verma, N., Bhatia, S., and Bedi, R.K. 2017. Role of pH on electrical, optical and photocatalytic properties of ZnO based nanoparticles. J. Mater. Sci.: Mater. Electron. 28(13): 9788–9797. doi:10.1007/s10854-017-6732-x.

Wang, Z.L. 2000. Characterizing the structure and properties of individual wire-like nanoentities. Adv. Mater. 12: 1295–1298.

Wang, Z.L. 2001. Characterization of nanophase materials. particle & particle systems characterization. 18(3): 142. doi:10.1002/1521-4117(200110)18:3<142::aid-ppsc142>3.0.co;2-n.

Xu, W., Li, Z., and Yin, Y. 2018. Colloidal assembly approaches to micro/nanostructures of complex morphologies. Small. 1801083. doi:10.1002/smll.201801083.

Yao, N., Wang, Z.L. (eds.). 2005. Handbook of Microscopy for Nanotechnology. Springer, Boston, MA. doi:10.1007/1-4020-8006-9.

CHAPTER 13

Mechanism of Synthesis of Metal Nanoparticles by Fungi

Nelson Durán,[1,2,*] *Marcelo B. De Jesus,*[3] *Ljubica Tasic,*[4] *Wagner J. Fávaro*[1] *and Gerson Nakazato*[5,*]

Introduction

Several fungi have been investigated to synthesize metallic nanoparticles, mainly silver, gold, and some metal oxides, and have attracted enormous interest because they exhibit advantages over using bacteria for the biogenic synthesis of nanoparticles. Biogenic syntheses have been reported since the early 1920s, and their commercial feasibility has also been published (Mukherjee et al. 2001a; Durán et al. 2005; 2007; Marcato et al. 2012a). At the beginning of this field, the research focus of the scientific community was on silver nanoparticles (AgNPs) due to their excellent antibacterial activities (Rai et al. 2009). Interestingly, *F. oxysporum* has been one of the most widely used in attempts to produce metallic nanoparticles, particularly silver (Ahmad et al. 2003a; Durán et al. 2005). Additionally, *F. oxysporum* has also been used to generate cadmium sulfide (CdS), zinc sulfide (ZnS), lead sulfide (PbS), and molybdenum sulfide (MoS) nanoparticles (Ahmad et al. 2002).

---

[1] Laboratory of Urogenital Carcinogenesis and Immunotherapy, Department of Structural and Functional Biology, Universidade Estadual de Campinas (UNICAMP), Campinas, SP, Brazil.
[2] Nanomedicine Research Unit (Nanomed), Center for Natural and Human Sciences (CCNH), Universidade Federal do ABC (UFABC), Santo André, SP, Brazil.
[3] Department of Biochemistry and Tissue Biology, Institute of Biology, University of Campinas, UNICAMP, Campinas, SP, Brazil.
[4] Institute of Chemistry, Biological Chemistry Laboratory, Universidade Estadual de Campinas, Campinas, SP, Brazil.
[5] Laboratory of Basic and Applied Bacteriology, Department of Microbiology, Biology Sciences Center, Universidade Estadual de Londrina (UEL), Londrina, PR, Brazil.
* Corresponding authors: nduran@unicamp.br; gnakazato@uel.br

Mechanism of Synthesis of Metal Nanoparticles by Fungi 255

Several other interesting fungi have also been used, such as *Aspergillus fumigatus* to produce AgNPs (Bhainsa and D'Souza 2006), *Trichoderma reesei* to obtain extracellular AgNPs (Vahabi et al. 2011), *A. fumigatus* and *F. oxysporum* (Ahmad et al. 2003a; Bhainsa and D'Souza 2006), and *Trichoderma harzianum* (Guilger-Casagrande et al. 2019). In the case of gold nanoparticles (AuNPs), there are fewer examples of their synthesis by fungi than those from silver. Mukherjee et al. published the synthesis of AuNPs by the fungus *Verticillium* sp., where intracellular AuNPs were found to be localized on the surface of mycelia via the biological reduction of $AuCl_4^-$ (Mukherjee et al. 2001b). Little is known about platinum nanoparticles. In this case, *Neurospora crassa* efficiently produced these nanoparticles (Castro-Longoria et al. 2012). *F. oxysporum* also produced platinum nanoparticles at low efficiency (Riddin et al. 2006). It is important to know that not all metallic nanoparticles are formed of elemental metals in their zero-valent form (Duran and Seabra 2012). Zinc oxide from *Aspergillus niger* (Shamim et al. 2019) is an antibacterial and magnetite (Fe_3O_4) with magnetic properties synthesized from *F. oxysporum* and the endophytic fungus *Verticillium* sp. (Bharde et al. 2006).

Many of the most studied fungi, such as *F. oxysporum,* are phytopathogenic; thus, they are potentially toxic (Spadaro and Gullino 2005). For example, *Trichoderma asperellum* and *Trichoderma reesei* produced AgNPs when exposed to silver salts (Vahabi et al. 2011; Mukherjee et al. 2008) and were found to be nonpathogenic, making them ideal for commercial use (Pantidos and Horsfall 2014). Unfortunately, the biogenic syntheses of selenium, ZnO, MgO, and cobalt oxide are mainly made from plant extracts. Many reviews have already been published on the most relevant metallic nanoparticle syntheses by fungi and other sources (Rubilar et al. 2013; Rai et al. 2014; Ingle et al. 2014; Schröfel et al. 2014; Yadav et al. 2015; Kitching et al. 2015; Siddiqi and Husen 2016; Poblete et al. 2013; Khan and Khan 2017; Sandhu et al. 2017; Khandel and Shahi 2018; Guilger-Casagrande and Lima 2019; Gahlawat and Choudhury 2019; Noorafsha et al. 2020; Qin et al. 2020; Bhardwaj et al. 2020; Das et al. 2021; Rai et al. 2021). To summarize the general biogenic synthesis in fungi, which exhibit biological activities, a table with the most relevant of several genera studied (Table 1).

This chapter examines the available methods to synthesize metal nanoparticles organized by fungal genera (*Fusarium, Aspergillus, Trichoderma, Penicillium, Alternaria, Rhizopus, Phoma, Pleorotus, Cladosporium, Colletotrichum, Schizophyllum*). In addition, recent advances in understanding biosynthetic mechanisms are discussed, and key knowledge gaps are identified. Finally, the concluding remarks provide recommendations for a better understanding of the underlying mechanisms of the biosynthesis of metallic nanoparticles.

Mechanism of biogenic metallic nanoparticles

In the next section, the most studied fungal strains are discussed in terms of biological activities and mechanistic aspects of synthesis.

256 *Mycosynthesis of Nanomaterials: Perspectives and Challenges*

Table 1. Selected biogenic syntheses from different genera of fungi with biological activities.

Nanoparticles	Fungus	Application	References
Silver	*Fusarium oxysporum*	Antimicrobial	Durán et al. 2007
Silver	*Aspergillus niger*	Antimicrobial	Gade et al. 2008
Silver	*Alternaria alternata*	Antifungal	Gajbhiye et al. 2009
Silver	*Phytophthora infestans*	Antimicrobial	Thirumurugan et al. 2009
Silver	*Raffaelea* sp.	Antimicrobial	Kim et al. 2009
Silver	*Aspergillus clavatus*	Antimicrobial	Saravanan and Nanda 2010
Silver	*Fusarium solani*	Antimicrobial	El-Rafie et al. 2010
Silver	*Tricholoma crassum*	Antimicrobial	Ray et al. 2011
Silver	*Lecanicillium lecanii*	Antimicrobial	Namasivayam and Avimanyu 2011
Silver	*Pleurotus ostreatus*	Antimicrobial	Devika et al. 2012
Silver	*Trichophyton rubrum, T. mentagrophytes* and *Microsporum canis*	Antifungal	Moazeni et al. 2012
Silver	*Fusarium oxysporum*	Antimicrobial	Rossi-Bergmann et al. 2012
Silver	*Fusarium oxysporum*	Antifungal	Marcato et al. 2012b
Silver	*Fusarium oxysporum*	Antifungal	Marcato et al. 2012a
Silver	*Penicillium diversum*	Antimicrobial	Ganachari et al. 2012
Silver	*Aspergillus tubingiensis* and *Bionectria ochroleuca*	Antimicrobial and antifungal	Rodrigues et al. 2013
Silver	*Aspergillus foetidus*	Antifungal	Roy et al. 2013
Silver	*Aspergillus flavus*	Antimicrobial	Naqvi et al. 2013
Silver	*Ganoderma neo-japonicum Imazeki*	Anticancer	Gurunathan et al. 2013
Silver	*Fusarium oxysporum*	Antifungal	Ishida et al. 2014
Silver	*Macrophomina phaseolina*	Antimicrobial	Chowdhury et al. 2014
Silver	*Inonotus obliquus*	Antimicrobial, antioxidant, anticancer	Nagajyothi et al. 2014
Silver	*Penicillium chrysogenum* and *Aspergillus oryzae*	Antifungal	Pereira et al. 2014
Silver	*Penicillium* sp.	Antimicrobial	Singh et al. 2014
Silver	*Schizophyllum radiatum*	Antimicrobial	Metuku et al. 2014
Silver	*Phoma gardeniae*	Antimicrobial and antifungal	Rai et al. 2015a
Silver	*Phoma capsulatum, Phoma putaminum,* and *Phoma citri*	Antimicrobial	Rai et al. 2015b
Silver	*Aspergillus flavus*	Antioxidant, antimicrobial	Sulaiman et al. 2015
Silver	*Guignardia mangiferae*	Anticancer	Balakumaran et al. 2015

Table 1 contd. ...

Mechanism of Synthesis of Metal Nanoparticles by Fungi   257

...Table 1 contd.

Nanoparticles	Fungus	Application	References
Silver	*Calocybe indica*	Anticancer	Gurunathan et al. 2015
Silver	*Pleurotus djamor* var. *roseus*	Anticancer	Raman et al. 2015
Silver	*Pleurotus cornucopiae* var. *citrinopileatus*	Antifungal	Owaid et al. 2015
Silver	*Fusarium oxysporum*	Antimicrobial and anticancer	Husseiny et al. 2015
Silver	*Colletotrichum* sp.	Antimicrobial	Azmath et al. 2016
Silver	*Sclerotinia sclerotiorum*	Antimicrobial	Saxena et al. 2016
Silver	*Arthroderma fulvum*	Antifungal	Xue et al. 2016
Silver	*Pestalotiopsis microspora*	Antioxidant and anticancer	Netala et al. 2016
Silver	*Fusarium oxysporum*	Antimicrobial	Rajput et al. 2016
Silver	*Metarhizium anisopliae*	Mosquitocidal	Amerasan et al. 2016
Silver	*Cunninghamella echinulata*	Anticancer	Anbazhagan et al. 2017
Silver	*Fusarium oxysporum*	Antifungal	Ballottin et al. 2017
Silver	*Fusarium verticillioides*	Antimicrobial	Mekkawy et al. 2017
Silver	*Penicillium aculeatum*	Antimicrobial	Ma et al. 2017
Silver	*Aspergillus oryzae*	Antimicrobial	Silva et al. 2017
Silver	*Rhizopus arrhizus, Trichoderma gamsii* and *Aspergillus niger*	Antimicrobial	Ottoni et al. 2017
Silver	*Aspergillus terreus*	Antimicrobial	Rani et al. 2017
Silver	*Fusarium oxysporum*	Antimicrobial	Hamedi et al. 2017
Silver	*Trichoderma harzianum*	Antifungal	Guilger et al. 2017
Silver	*Alternaria* sp.	Antimicrobial	Singh et al. 2017
Silver	*Fusarium oxysporum*	Antiparasitic	Fanti et al. 2018
Silver	*Fusarium oxysporum*	Antimicrobial	Stanisic et al. 2018
Silver	*Penicillium italicum*	Antimicrobial and antifungal	Nayak et al. 2018
Silver	*Trichoderma longibrachiatum*	Antifungal	Elamawi et al. 2018
Silver	*Cladosporium cladosporioides*	Antimicrobial	Hulikere and Joshi 2019
Silver	*Fusarium oxysporum*	Antimicrobial	Srivastava et al. 2019
Silver	*Fusarium oxysporum*	Antitumoral	Ferreira et al. 2019
Silver	*Fusarium oxysporum*	Antimicrobial on ruminants	Santos et al. 2019
Silver	*Trichoderma harzianum*	Antifungal	Guilger-Casagrande et al. 2019

Table 1 contd. ...

258 Mycosynthesis of Nanomaterials: Perspectives and Challenges

...Table 1 contd.

Nanoparticles	Fungus	Application	References
Silver	*Aspergillus fumigatus*	Antimicrobial, anticancer	Othman et al. 2019
Silver	*Talaromyces purpurogenus*	Antimicrobial, anticancer	Bhatnagar et al. 2019
Silver	*Trichoderma viride*	Anticancer, immunomodulatory activity	Adebayo-Tayo et al. 2019
Silver	*Penicillium citreonigrum*	Antimicrobial	Hamad 2019
Silver	*Beauveria bassiana*	Antimicrobial	Tyagi et al. 2019
Silver	*Fusarium oxysporum*	Anticancer	Ferreira et al. 2020
Silver	*Phomopsis helianthi*	Antimicrobial	Gond et al. 2020
Silver	*Fusarium scirpi*	Antimicrobial	Rodriguez-Serrano et al. 2020
Silver	*Aspergillus sydowii*	Antifungal. antiproliferative	Wang et al. 2021
Silver	*Aspergillus flavus*	Antimicrobial	Priyanka 2020
Silver	*Nigrospora oryzae*	Antifungal	Dawoud et al. 2021
Silver	*Aspergillus tubingensis*	Antifungal	Rodrigues et al. 2021
Silver	*Rhizopus stolonifera*	Anticancer	Banu et al. 2021
Silver	*Aspergillus terreus*	Antimocrobiil, anticancer	Lotfy et al. 2021
Platinum	*Fusarium oxysporum*	Antimicrobial	Gupta and Chundawat 2019
Copper	*Fusarium oxysporum*	Antimicrobial	Majumder 2012
Copper	*Aspergillus niger*	Antimicrobial	Naqvi et al. 2017
Copper	*Stereum hirsutum*	Potential antimicrobial	Cuevas et al. 2015
Copper	*Aspergillus niger*	Antioxidant, antimicrobial, anticancer	Noor et al. 2020
Copper	*Aspergillus flavus*	Potential antimicrobial	Saitawadekar and Kakde 2020
CuO	*Trichoderma asperellum*	Anticancer	Saravanakumar et al. 2019
Copper oxide (CuO)	*Penicillium chrysogenum*	Antimicrobial	El-Batal 2020
CuO	*Aspergillus terreus*	Antimicrobial, antifungal, and anticancer	Mani et al. 2021
ZnO	*Aspergillus fumigatus*	Improvement of plant biomass	Raliya and Tarafdar 2013
ZnO	*Aspergillus terreus*	Anticancer	Baskar et al. 2015
ZnO	*Aspergillus niger*	Antimicrobial	Kalpana et al. 2018

Table 1 contd. ...

...Table 1 contd.

Nanoparticles	Fungus	Application	References
ZnO	*Fusarium keratoplasticum* and *Aspergillus niger*	Antimicrobial	Mohamed et al. 2019
ZnO	*Aspergillus niger*	Potential antimicrobial	Shamim et al. 2019
ZnO	*Trichoderma* harzianum	Antimicrobial, anticancer	Saravanakumar et al. 2020
ZnO	*Alternaria tenuissima*	Antimicrobial, anticancer	Abdelhakim et al. 2020
Selenium	*Fusarium* sp. and *Trichoderma reesi*	Antifungal	Gharieb et al. 1995
Selenium	*Aspergillus oryzae*	Antimicrobial	Mosallam et al. 2018
Selenium	*Mariannaea* sp. HJ	Medicine and electronic	Zhang et al. 2019
Selenium	*Fusarium semitectum*	Antioxidant, antimicrobial, anticancer	Abbas and Baker 2020
Co_3O_4	*Aspergillus nidulans*	Potential antimicrobial, Energy storage.	Vijayanandan and Balakrishnan 2018
Co_3O_4	*Aspergillus brasiliensis*	Antimicrobial	Omran et al. 2020
MgO	*Aspergillus tereus*	Antimicrobial, eco removal of toxic materials	Saied et al. 2021
MgO	*Aspergillus niger*	Antimicrobial	Ibrahem et al. 2017
Al_2O_3	*Colletotrichum* sp.	Antimicrobial and antifungal	Suryavanshi et al. 2017
Fe_2O_3	*Alternaria alternata*	Antimicrobial	Mohamed et al. 2015
Gold	*Rhizopus oryzae*	Antifungal	Das et al. 2009; 2012
Gold	*Aspergillus niger*	Antiparasitic	Soni and Prakash 2012
Gold	*Fusarium oxysporum*	Antifungal	Thakker et al. 2013
God	*Nigrospora oryzae*	Anthelmintic	Kar et al. 2014
Gold	*Cladosporium oxysporum*	Potential antimicrobial, Catalysis	Bhargava et al. 2016
Gold	*Trichoderma viride* and *Hypocrea lixii*	Antimicrobial	Mishra et al. 2014
Gold	*Cladosporium cladosporioides*	Antioxidant, antimicrobial	Hulikere et al. 2017
Gold	*Trichoderma hamatum*	Antimicrobial	Abdel-Kareem and Zohri 2018
Gold	*Pleurotus ostreatus*	Antimicrobial and anticancer	El Domany et al. 2018
Gold	*Fusarium solani*	Anticancer	Clarance et al. 2020
Gold	*Aspergillus flavus*	Anticancer	Abu-Tahon et al. 2020
Gold	*Cantharellus* sp.	Antimicrobial	Jha et al. 2021

Fusarium

A major review of the biogenic synthesis of AgNPs by *Fusarium* (Rai et al. 2021) described that members of this genus could produce metal nanoparticles both inside and outside their cells. Metabolites are generally believed to play an essential role in forming metal nanoparticles. Enzymes, proteins, flavonoids, polysaccharides, alkaloids, phenolics, and organic acids are some of the metabolites produced by fungi such as *Fusarium* to survive in extreme environmental conditions and are mostly responsible for both the reduction of metal ions to metallic nanoparticles and for the stabilization of nanoparticles (Durán et al. 2011; Mahmoud et al. 2013; Yadav et al. 2015; Srivastava et al. 2019).

It was hypothesized that *F. oxysporum* secretes the enzyme NADH-dependent nitrate reductase, which should be in control for the reduction of silver ions to AgNPs (Ahmad et al. 2003a). A similar mechanism was suggested by Ingle et al. (2008) for *F. acuminatum*. Durán et al. (2005) and Kumar et al. (2007) proposed the same mechanisms for the biosynthesis of AgNPs produced by *F. oxysporum*. In this case, the electron shutter anthraquinone and NADPH nitrate reductase played vital roles in the biosynthesis of AgNPs (Figure 1). In the case of *F. oxysporum* f. sp. *cubense,* JT1 showed that this type of electron shutter also exhibited the capacity to reduce the ions of gold to AuNP (Thakker et al. 2013). This intracellular mycosynthesis followed a two-step mechanism. The process begins with the electrostatic interaction between the metallic ions and lysine residues on the surface of the fungal cell (Yadav et al. 2015), followed by the mycosynthesis of nanoparticles, characterized by the aggregation and formation of nanoparticles catalyzed by the enzymatic reduction of metal ions (Rajput et al. 2016).

Nevertheless, the synthesis of *Fusarium* nanoparticles has been published for platinum (Gupta and Chundawat 2019), copper (Majumder 2012), selenium (Gharieb et al. 1995; Abbas and Baker 2020), and gold (Thakker et al. 2013; Clarance et al. 2020), very little is known about the mechanistic aspects of their synthesis. The synthesis of gold nanoparticles by *F. oxysporum* depends on the reducing agent, which reduces Au^{3+} to the Au^0 state (Mukherjee et al. 2001a; 2002), but no details have been presented (Thakker et al. 2013); *F. oxysporum* extracellularly reduces hexachloroplatinic acid into platinum nanoparticles (Riddin et al. 2006; Govender et al. 2009; Syed and Ahmad 2012). The reduction process occurs through the electron shuttle quinine and the reductase enzyme (Gupta and Chundawat 2019). Proteins play an essential role in the biological synthesis of nanoparticles by stabilizing the nanoparticles and providing a uniform size distribution; these facts appear to be similar to the formations of AgNPs (Durán et al. 2005). Hypothetical mechanism for the reduction of selenium ions from sodium selenite by *F. semitectum* was the presence of multibranched polysaccharides of glucose (glycogen), and this was suggested by Zhang et al. (2004), who confirmed the role of different polysaccharides, such as chitosan, konjac glucomannan, acasia gum and carboxymethyl cellulose, in the reduction of selenium (Abbas and Baker 2020).

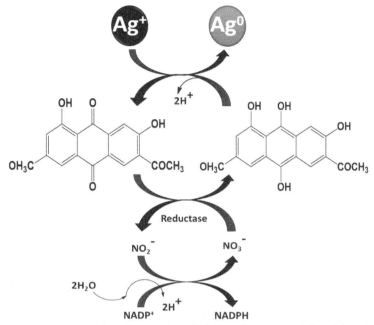

Figure 1. Hypothetical mechanism of AgNP biosynthesis from *F. oxysporum* (adapted from Durán et al. (2005) and Rai et al. (2021), an open-access article).

Aspergillus

In the case of *Aspergillus*, very little research has been done on biogenic synthesis. However, Gade et al. (2008) hypothesized that *A. niger* secretes an enzyme capable of reducing silver ions, thus forming AgNPs (such as nitrate reductase and anthraquinones), referring to previous work on *F. oxysporum* (Durán et al. 2005). Posteriorly, Ottoni et al. (2017) showed the nitrate reductase activity of culture supernatants for *A. niger* strains, similar to other researchers (Durán et al. 2005; Hamedi et al. 2013; Saifuddin et al. 2009).

A critical aspect reported by Rodrigues et al. (2013) on biogenic synthesis with *A. tubingiensis* is that they found the same protein bands on the fungal filtrate (FF) and during the formation of the AgNP (75, 122, 191, and 328 kDa). They speculated that these proteins stabilized AgNPs by binding to their surfaces.

Roy et al. (2013) used the extracellular filtrate of the fungal strain of *A. foetidus* to reduce Ag^+ to Ag^0 as reported by nitrate reductase, an extracellular enzyme leading to the formation of AgNPs (Kumar et al. 2007), similar to *F. oxysporum* (Ahmad et al. 2003) and *Bacillus licheniformis* (Kalimuthu et al. 2008).

Naqvi et al. (2013), using *A. flavus*, suggested that the mechanism underlying AgNP synthesis can also be predicted; the process heavily relies on the reducing

262 Mycosynthesis of Nanomaterials: Perspectives and Challenges

agents from the cellular environment. This report indicated that the specific position of reductase enzymes is typically cited (Ahmad et al. 2003a; Vigneshwaran et al. 2007). Next, enzymes from the cell surface reduce silver ions to AgNPs, naphthoquinones and anthraquinones, which seems to facilitate this process by acting as electron shutters (Mukherjee et al. 2001; Zhang et al. 2011; Jain et al. 2011).

It was found that in the presence of *A. terreus* (Rani et al. 2017) was possible to identify tannins, flavonoids, phenols, diterpenes, alkaloids, glycosides, and carbohydrates. Functional groups played an essential role in reducing and capping silver ions in AgNPs (Niraimathi et al. 2013; Prakash et al. 2013). For example, alcohols and phenols can be identified by characteristic peaks, suggesting that they can act when reducing silver ions to AgNPs (Ghosh et al. 2012). Studies also suggest that silver ion reduction and capping are due to NADH-dependent reductases in fungal endophytic extracts (Kathiresan et al. 2009; Govender et al. 2010). Unfortunately, for the biogenic synthesis of copper nanoparticles by *A. niger* and *A. terreus*, *A. flavus*, no information on its mechanism has been discussed (Naqvi et al. 2017; Saitawadekar and Kakde 2020; Mani et al. 2021).

Noor et al. (2020) stated that the mechanism involved in the extracellular synthesis of CuNPs by *A. niger* is the involvement of nitrate reductase. This was confirmed by studies using commercially available nitrate reductase discs that reduce Ag^+ ions to Ag^0 and the corresponding development of AgNPs. Similar to the nitrate reductase of *F. oxysporum* that has been used to produce AgNPs during *in vitro* analysis in the absence of oxygen and in the presence of a cofactor (NADPH), which is a stabilizer protein (phytochelatin), and an electron carrier (4-hydroxyquinoline) (Senapati et al. 2005; Kumar et al. 2007), the synthesis of CuNPs was investigated. These studies demonstrate that NADPH acts as a reducing agent for extracellular CuNP synthesis.

Zinc oxide generated by *A. terreus* by Baskar et al. (2015) did not present any mechanistic aspect of its synthesis. The same happened with the biogenic synthesis of Co3O4 from *A. nidulans* (Vijayanandan and Balakrishnan 2018).

Many biomolecules (enzymes/proteins, amino acids, polysaccharides, and also vitamins) have been identified in the fungal mycelial cell-free filtrate (MCFF) during the biotransformation mediated by *A. brasiliensis* of $CoSO_4$ to Co_3O_4 (Sarsar et al. 2016). These biomolecules cap and stabilize the nanoparticles through associations between the positively charged Co_3O_4 nanoparticles and the negatively charged groups of these biomolecules (Omran et al. 2020).

For *A. terreus*, it was found that during the biogenic synthesis of MgO nanoparticles, the activity of reducing agents in the biomass filtrate depended on the incubation temperature (Saied et al. 2021). The optimal temperature at which the metabolites of *A. terreus* acted as reducing agents was found to be 35°C, which was explained by the higher stability of enzymes and proteins at this temperature. The same optimal temperature was found for the synthesis of MgO nanoparticles using biomass filtrate from *Rhizopus oryzae* (Hassan et al. 2021).

Although *A. niger* can use both extracellular and intracellular pathways to produce ZnO nanoparticles, extracellular synthesis is more efficient (Shamim et al. 2019). Extracellular synthesis relies on secreted enzymes, such as the nitrate reductase enzyme, which acts by reducing metallic ions into nanoparticles. For example, during the synthesis of zinc oxide nanoparticles, the enzyme NADH reductase transfers electrons from NADH to Zn^{+2} and reduces them to Zn^0 nanoparticles. On the other hand, intracellular synthesis relies on enzymes that transport metals within the cell wall. Once in the periplasmic space and cytoplasm, the ions begin to be reduced to form nanoparticles. This pathway has been used, for example, by *Verticillium* sp. as a three-step process in which the ions were first trapped, then bioreduced, and finally capped (Yusof et al. 2019).

Raliya and Tarafdar (2013) showed that ZnO nanoparticles could be synthesized using *A. fumigatus* with the assistance of secreted proteins and enzymes, and they suggested that these molecules are important for the formation and capping of the nanoparticles. Kalpana et al. (2018) demonstrated that cell-free filtrate from *Aspergillus niger* performs the biosynthesis of ZnO nanoparticles. To obtain metal and metal oxide nanoparticles by fungi follow a mechanistic pathway similar to that described for green synthesis by bacteria; these mechanisms are outlined in Figure 2.

Figure 2 also shows the differences between intracellular and extracellular biosynthesis.

The content of the filtrate of *A. oryzae* metabolites plays an important role in all processes of reduction, capping, stabilization, and yielding of nanoparticles (Mosallam et al. 2018). The presence of metabolites, such as proteins, can chelate metal ions with their carbonyl, amine groups, or free electrons, which are also adsorbed onto the surface of selenium nanoparticles. This probably means that in addition to the bioreduction stage, they initiated the formation of selenium nanoparticles (Makarov et al. 2014). Then, the synthesis mechanism of selenium nanoparticles included three principal phases: the first activation phase, reduction of metal ions and nucleation; the second one is the growth phase, during which the small adjacent nanoparticles coalesced into particles of a larger size, leading to an increase in the thermodynamic

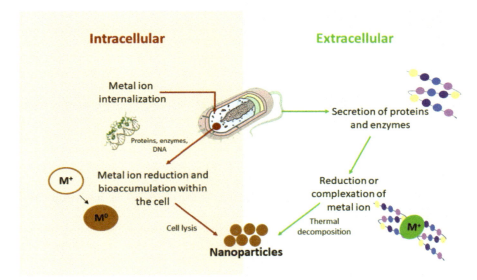

Figure 2. Green synthesis of nanoparticles using bacterial cultures following the same mechanism by fungi (extracted from Bandeira et al. 2020 by permission Elsevier B.V).

stability of nanoparticles; and the final step of the process, the determination of the final profile of the generated nanoparticles (Zhao 2010).

In the case of AuNPs synthesized from *A. niger* (Soni and Prakash 2012), the process occurred in the presence of reductases, which may be present in the cell-free extracts of this fungi, in a close way as in AgNP biosynthesis (Durán et al. 2005).

In *Aspergillus flavus* (Abu-Tahon et al. 2020), it has been proposed that AuNP production involves proteins or carbohydrates that may play a role in biosynthesis and cap biosynthesized nanoparticles (Pimprikar et al. 2009; Li et al. 2016).

Trichoderma

Similar to Ottoni et al. (2017), the detected nitrate reductase activity of culture supernatants for strains of *A. niger* was also seen in *T. gamsii* as the crucial enzyme for the synthesis of AgNPs.

T. harzianum, a filamentous mycoparasite used as a biological control agent, exhibits fungal reductase activity (Guilger et al. 2017), and this enzyme's concomitant role in reducing activity is to stabilize the AgNP in biogenic synthesis, as reported by Singh and Raja (2011) and Ahluwalia et al. (2014).

In AgNP biosynthesis by *T. longibrachiatum* (Elamawi et al. 2018), its mycelium was submitted to the metal salt solution, that induced the production of fungal enzymes and increased concentrations of some metabolites for survival, thus reducing the silver ion to AgNP. Furthermore, AgNP biosynthesis also relies on the catalytic activity of extracellular enzymes and fungal metabolites (Das and Marsili 2010; Vahabi et al. 2011).

During the biogenic synthesis of AgNPs by *T. viride*, metallic ions can act as stressors and induce the secretion of extracellular metabolites and enzymes (Adebayo-Tayo et al. 2019). These materials act as intermediates for fungal survival by reducing the concentration of silver ions. Then, exposure of the mycelium to the metal salt solution created osmotic stress that triggered the released enzymes and metabolites to survive. Therefore, the biosynthesis of nanoparticles could be described as a two-step process: Step I: metal ions are trapped near the fungal cells; Step II: fungal enzymes are secreted, and silver ions are reduced into the nanoparticles. Several naphthoquinones and anthraquinones, reported in some fungi, have unique redox properties capable of acting as electron shuttles during metal reductions (Durán et al. 2005). Most synthesis parameters (size, rate of synthesis, and efficiency) can be controlled by many physicochemical factors, such as pH, temperature, substrate concentration, and exposure time to the substrate (Gericke and Pinches 2006).

In the case of copper oxide of *T. asperellum* (Saravanakumar et al. 2019) or ZnO nanoparticles of *T. harzianum* (Saravanakumar et al. 2020), no information was given on the mechanistic aspects of their synthesis.

T. reesii was the only species tested capable of significant growth in the presence of 10 mmol of L^{-1} selenite (Gharieb et al. 1995). Growth inhibition was not markedly greater on sulfur-free Czapek-Dox agar than on the sulfur-containing medium for this fungus to produce selenium nanoparticles.

For *Trichoderma viride* (Mishra et al. 2014), which produces AuNPs, no information on the biosynthetic mechanism was provided.

Trichoderma species are a group of ubiquitous fungi and are capable of secreting enzymes and secondary metabolites, raising the expectation of using fungi as catalysts to synthesize AuNPs (Abdel-Kareem and Zohri 2018). For instance, the synthesis of biogenic AgNPs has been described for *T. harzianum* (Guilger et al. 2017), *T. atroviride* (Saravanakumar and Wang 2018), and *T. hamatum*; the latter being suitable for the synthesis of AgNPs and AuNPs (Saravanakumar et al. 2016).

Penicillium

The hypothetical pathway of the synthesis of AgNPs from *P. diversum* (Ganachari et al. 2012) assumed that the reductase enzymes present in the fungal extract were responsible for the reduction process. Then, a protein layer formed on the nanoparticles (characterized by FTIR, TEM, and AFM). However, the exact mechanism and chemical composition of the bioreduction process were not discussed.

The observation by Singh et al. (2014) on the biogenic synthesis of AgNPs from *Penicillium* sp. revealed that proteins in addition to acting as reducing agents, also com play a role as stabilizing agents. This stabilization could be mediated by proteins and enzymes present on the filtrate from fungal mycelium covering the AgNPs, which interact via covalent (free amino groups or cysteine residues) or non-covalent electrostatic interactions (negatively charged carboxylate groups) (Gole et al. 2001).

Pereira et al. (2014) on *P. chrysogenum* producing AgNPs suggested that different microorganisms have different production mechanisms (Li et al. 2011). However, in biological approaches, it has been suggested that the size of nanoparticles is determined by biomolecules excreted in large amounts by microorganisms, leading to the reduction of metal ions (Kumar et al. 2007; Thakkar et al. 2010).

AgNP from *P. aculeatum* (Ma et al. 2017) indicated a combination of AgNPs and protein components secreted from fungal biomass during the incubation period. These proteins can play a key role in reducing the number of silver ions. They may also bind to AgNPs to prevent agglomeration (Bhainsa ad D'Souza 2006). This combination of proteins can bind to nanoparticles via free amine groups, cysteine residues, or the electrostatic attraction of carboxylate residues with negative charges in the cell-free filtrate from fungal mycelia (Parikh et al. 2008).

Unfortunately, the biosynthesis mechanisms of AgNPs from *P. italicum have* not been described (Nayak et al. 2018).

Hamad (2019) based their biogenic synthesis of AgNPs by *P. citreonigrum* with the similarity to the synthesis by *F. oxysporum,* for which the enzyme nicotinamide adenine dinucleotide (NADH) was described as a mediator of the bioreduction of Ag+ to Ag (Ahmad et al. 2003a, b; Durán et al. 2005).

P. chrysogenum filtrate has been described as containing proteins, amino acids, pigments, and reduced enzymes that could cap and stabilize nanoparticles and play a role in reducing metal ions (Mosallam et al. 2018). In particular, the excess Cu^{2+} ions can be trapped in unique protein loops during the synthesis of CuO nanoparticles. Next, the –H and –OH groups of (–COOH) found in the filtrate proteins provide a network for synthesizing the metallic clusters of CuO nanoparticles (El-Batal et al. 2017).

Alternaria

Although the exact mechanism of the synthesis of AgNPs by *A. alternata* is not fully elucidated (Gajbhiye et al. 2009), it has been hypothesized that the fungus secretes NADH-dependent nitrate reductase in charge for silver ion reduction. Later, the presence of NADH-dependent nitrate reductase in the extracellular fungal cell filtrate was confirmed, and its role in the synthesis of nanoparticles was investigated (Ingle et al. 2008; Kumar et al. 2007). Unfortunately, Singh et al. (2017) did not discuss any aspect related to the mechanistic aspect of the biogenic synthesis of AgNPs from *Alternaria* sp.

Similar to *A. tenuissima* (Abdelhakim et al. 2020), they suggest that ZnO nanoparticle formation occurs in two steps: The first one is the reduction and then the capping. The second step starts with reducing zinc ions to the corresponding ZnO nanoparticles by the active metabolites of filtrate nitrate reductase (Kundu et al. 2014; Chauhan et al. 2015; Kalpana et al. 2018; Gao et al. 2019). Next, the NADH-dependent reductase transfer electrons from NADH to Zn(II) ions, this reduction leads to the consequent formation of ZnO nanoparticles (Hulkoti and Taranath 2014). Finally, ZnO nanoparticles are capped and stabilized by biomolecules from fungal cultures (El-Batal et al. 2020).

Mohamed et al. (2015) suggested that Fe_2O_3 nanoparticle synthesis from *A. alternata* follows a mechanism similar to AuNPs from *Rhizopus oryzae* (Das et al. 2012).

Rhizopus

Ottoni et al. (2017) demonstrated in an initial batch of strains of *R. arrhizus* cultured from sugar cane plantation soil their biogenic ability to produce AgNPs by reduction by nitrate reductase in a similar way to *T. gamsii* and *A. niger*. In the report of Banu et al. (2021) in the production of AgNPs by *R. stolonifera*, no information on biogenic synthesis was discussed. *R. oryzae* intracellularly generated AuNPs to act on gold detoxification mechanisms by reducing Au^{+3} to Au^0 nanoparticles (Das et al. 2009; 2012) (Figure 3). These authors suggested that at least two reduction routes occurred: (a) binding of Au(III) to the cell wall by electrostatic interaction tracked by reduction to AuNP by a reductase available on the cell wall, and (b) transport of Au(III) into the cytoplasm and then an enzymatic reduction to produce AuNP. The binding of Au(III) to the cell surface has been demonstrated by electrostatic interactions and the generation of Au^0 at the cell wall through Au(I) species. Gold ions (negatively charged) ($AuCl_4^-$) bind to mycelia of *R. oryzae* (positively charged) by electrostatic interaction with phosphoproteins and are then reduced to intermediate Au(I) species resulting from the high redox potential of Au(III) ($AuCl_4^- + 3e^- = Au^o + 4Cl^-$) (Usher et al. 2009).

This led to electron capture from a suitable electron donor, which reduced the number of Au(I) ions. The Au(I) species fromatiob was confirmed by the shift of C_{1s} and N_{1s} core-level peaks to higher energy in the XPS analysis, suggesting the possibility of gold(I) detoxification by a different mechanism. The active mechanism was probably responsible for reducing Au(I) species to metallic Au^0 particles, as was

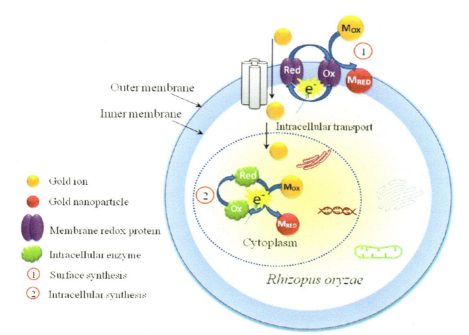

Figure 3. Proposed gold biomineralization mechanism in *Rhizopus oryzae* (extracted from Das et al. 2012 by permission of the American Chemical Society).

demonstrated in a study with a protein induction experiment. Binding of Au(III) ions and posterior reduction to Au(I) species occurred through a chemical interaction process, and reduction of Au(I) species to Au⁰ occurred via an enzymatic process reduction step, as shown by the XPS technique.

The presence of Au(I) species in the cytoplasm was probably an indication that the cytoplasmic reduction of Au(III) to Au⁰ occurred via intermediate Au(I) species. Together with the electrostatic interaction, it was suggested that the covalent binding of Au(III) with cytoplasmic proteins occurs since the protein exhibited a negative zeta potential value. Furthermore, protein expression was concentration-dependent at various concentrations of Au(III). This finding is supported by two upregulated proteins of molecular masses ~42 and ~45 kDa that promoted the biosynthesis of AuNPs. In contrast, the ~80 kDa protein bound to the surface of AuNPs led to a stable gold bioconjugate system. It is noteworthy that silver and AuNPs have been thoroughly studied, and the underlying mechanisms have been unequivocally described for all biogenic syntheses of metallic nanoparticles by fungi.

Phoma

The mycogenic synthesis of AgNPs by *Phoma gardenia* (Rai et al. 2015a), *Phoma capsulatum*, *Phoma putaminum*, and *Phoma citri* (Rai et al. 2015b) was found to be a simple and eco-friendly method. Several techniques demonstrated the presence of protein as a capping agent, resulting in high biocompatibility. This capping may occur between amino acid residues in the protein and synthesized AgNPs, resulting in well-known typical biogenic nanoparticles. The present finding corroborates Gole et al. (2001), who published that proteins could bind to nanoparticles through the electrostatic interaction of carboxylate groups (negatively charged) in the protein released from the fungus and then stabilize the AgNP. Unfortunately, no mechanistic aspects were discussed in these syntheses.

Pleurotus

Devika et al. (2012) hypothesized that the fungus secretes the NADH-dependent enzyme nitrate reductase, which is responsible for reducing silver ions during AgNP synthesis (Ahmad et al. 2002; Durán et al. 2005).

Raman et al. (2015) observed by spectroscopic analysis of protein in *Pleurotus djamor* var. *roseus* biosynthesis of AgNPs, the maximum absorbance of the amide bond of proteins with an alpha-helix structure was found in pure protein samples near 1639 cm^{-1}; after the reaction, this amide peak appeared as a single peak. This change in the spectra indicated that the secondary structure of the proteins was modified by binding to the AgNP. The spectral and electron microscopy data strongly support the use of *P. djamor* extract as a reducing agent in the biosynthesis of AgNPs.

The FT-IR spectrum of AgNPs was used to identify potential bioreductants of *Pleurotus cornucopiae* var. *citrinopileatus* aqueous extract involved in reducing the silver cation. The peaks at 1094 cm^{-1} and 1261 cm^{-1} peak in Owaid et al. (2015) disappeared after nanoparticle synthesis, indicating the involvement of the C–N group of the amine and the C–O group of the alcohol in the bioreduction process, respectively.

El Domany et al. (2018) suggested that fungi are more adequate than bacteria for the large-scale production of nanoparticles. In particular, for the biosynthesis of AuNPs, because the large fungi secretome contains proteins, enzymes, and small molecules that can play a role during the formation and stabilization of AuNPs (Siddiqi and Husen 2016; Mishra et al. 2014).

Cladosporium

The soil fungus *Cladosporium oxysporum* AJP03 produced AuNPs (approximately 72 nm in size). A series of experiments were managed to study the effect of distinct process parameters on particle size and yield. These nanoparticles exhibited excellent capacity as catalysts. Furthermore, coating material was found on the surface of the gold nanoparticle, suggesting that this capping of the protein was responsible for this remarkable catalytic efficiency. Unfortunately, no mechanistic aspects were discussed (Bhargava et al. 2016).

The FTIR spectra of the AuNP reduced by the *C. cladosporioides* extract showed a strong peak at 3307 cm^{-1}, characteristic of the O-H group of the phenolic and alcoholic molecules (Hulikere et al. 2017). The nanoparticles were stabilized by the hydrophilic flavonoids present in the extract, characterized by two important absorption bands at 2364 cm^{-1} from the O-H stretching of the carboxylic acids and 1650 cm^{-1} from the association of the amide-I bond of proteins and the secondary stretching vibrations in proteins. This study demonstrated the role of protein in coating and stabilizing AuNPs. The biogenic process was catalyzed by phenolic compounds associated with reductases present in the extract of *C. cladosporioides* (Hulikere et al. 2017).

Cladosporium cladosporioides biogenically synthesizes AgNPs (size –100 nm). The FTIR study highlighted the role of proteins, enzymes, and polyphenols from the extract of *C. cladosporioides* in the coating and biogenic reduction of AgNPs. An attempt has been made to understand the mechanism of the biogenic synthesis of AgNPs and the role of NADPH-dependent reductase in the formation of this nanoparticle (Hulikere and Joshi 2019).

Colletotrichum

Colletotrichum sp. ALF2-6 produced AgNPs (20–50 nm in size). FTIR analysis correlated reasonably well with previous scientific reports and another biogenic synthesis. Previously, this analysis demonstrated the role of proteins and phenolic compounds that reduce the silver salt, bind to the nanoparticles, and stabilize them, preventing aggregation. Unfortunately, no mechanistic aspects were discussed (Azmath et al. 2016).

Aluminum oxide (size 39 ± 35 nm) from *Colletotrichum* sp. (DBT 349) exhibited excellent antimicrobial activities (Suryavanshi et al. 2017). However, no characterization of the coated proteins or mechanistic discussion was done in this case.

Schizophyllum

Metuku et al. (2014) demonstrated that the FTIR spectra of biosynthesized AgNPs from *Schizophyllum radiatum* (10–49 nm in size) are in addition to peaks characteristic of stretching vibrations of aliphatic C–H bonds and CH_2-scissoring stretching vibrations in the planar region. Furthermore, various peaks of the C-N stretching vibration were observed at 1258 cm^{-1}, 1143 cm^{-1}, 1102 cm^{-1}, 1027 cm^{-1}, and 908 cm^{-1}. The appearance of bands at 1356 cm^{-1} and 1250 cm^{-1} in the spectra indicated that the capping agent of biogenic nanoparticles had an aromatic amine group with the specific subscription of amide linkages inside amino acid residues in the proteins. These data support the view that the presence of a protein cap on the surface of biogenically synthesized nanoparticles supports the hypothesis that metabolically generated proteins acted as stabilizing agents during nanoparticle generation and avoided them from agglomerating. However, no mechanistic aspects were discussed.

Conclusions

As disclosed, metallic, metal oxide, metal sulfide, and metal sulfate nanoparticles can be prepared using biosynthesis promoted by fungi. There have been significant advances in understanding the biosynthetic or biogenic synthesis of metallic and metal oxide nanoparticles. However, many studies are still carried out preferentially

270 *Mycosynthesis of Nanomaterials: Perspectives and Challenges*

exploring plant extracts for fungal synthesis. Regarding fungal biosynthesis of these nanoparticles, the mechanistic aspects have been neglected, and only occasionally few biosynthetic pathways have been studied in detail. Briefly, the syntheses of metal-containing nanoparticles are based on two- or three-step processes, with the reduction step followed by the nucleation step and stabilization step. The primary role in the first step is associated with reductases in almost all literature, where the role of the electron transfer is given to NADH and NAD(P)H cofactors, which must regenerate in the next step in another electron transfer step, this time promoted by different molecules that can oxidize, and where more details are needed, because of the variety of small molecules—secondary fungal metabolites that are specific for each fungal strain. Finally, the last step or steps (nucleation and stabilization) involve proteins that cap the formed NPs and stabilize them, influencing their shapes and sizes. In many cases, the molecular mechanisms of synthesis have been declared similar to those described for synthesis promoted by *F. oxysporum* and *Rhizopus oryzae*. However, it is known from the literature that the mechanistic aspects depend not only on the genera but also on the species, especially regarding secondary metabolites and their biosynthetic roles. It is clear from this overview that more careful mechanistic studies of the different fungal involvement and roles of biomolecules in metallic and metal oxide/sulfide/sulfate, among other nanoparticle syntheses, are needed in the literature compared to general studies with plant extracts and bacteria, which are beyond the scope of this chapter.

Acknowledgment

The authors would like to thank INOMAT/CNPq/UNICAMP, FAPESP, and CNPq for financial support.

References

Abbas, H.S., and Baker, D.H.A. 2020. Biological evaluation of selenium nanoparticles biosynthesized by fusarium semitectum as antimicrobial and anticancer agents. Egypt J. Chem. 63(4): 1119–1133.

Abdelhakim, H.K., El-Sayed, E.R., and Rashidi, F.B. 2020. Biosynthesis of zinc oxide nanoparticles with antimicrobial, anticancer, antioxidant, and photocatalytic activities by the endophytic *Alternaria tenuissima*. J. Appl. Microbiol. 128(6): 1634–1646.

Abdel-Kareem, M.M., and Zohri, A.A. 2018. Extracellular mycosynthesis of gold nanoparticles using Trichoderma hamatum: optimization, characterization and antimicrobial activity. Lett. Appl. Microbiol. 67(5): 465–475.

Abu-Tahon, M.A., Ghareib, M., and Abdallah, W.E. 2020. Environmentally benign rapid biosynthesis of extracellular gold nanoparticles using *Aspergillus flavus* and their cytotoxic and catalytic activities. Process Biochem. 95(0): 1–11.

Adebayo-Tayo, B.C., Ogunleye, G.E., and Ogbole, O. 2019. Biomedical application of greenly synthesized silver nanoparticles using the filtrate of Trichoderma viride: Anticancer and immunomodulatory potentials. Polim. Med. 49(2): 57–63.

Ahluwalia, V., Kumar, J., Sisodia, R., Shakil, N.A., and Walia, S. 2014. Green synthesis of silver nanoparticles by Trichoderma harzianum and their bio-efficacy evaluation against staphylococcus aureus and *Klebsiella pneumoniae*. Ind. Crops Prod. 55(0): 202–206.

Ahmad, A., Mukherjee, P., Mandal, D., Senapati, S., Khan, M.I., Kumar, R., and Sastry, M. 2002. Enzyme mediated extracellular synthesis of CdS nanoparticles by the fungus, *Fusarium oxysporum*. J. Am. Chem. Soc. 124(41): 12108–12109.

Ahmad, A., Mukherjee, P., Senapati, S., Mandal, D., Khan, M.I., Kumar, R., and Sastry, M. 2003a. Extracellular biosynthesis of silver nanoparticles using the fungus *Fusarium oxysporum*. Colloids and Surfaces B: Biointerf. 28(4): 313–318.

Ahmad, A., Senapati, S., Khan, M.I., Kumar, R., Ramani, R., Srinivas, V., and Sastry, M. 2003b. Intracellular synthesis of gold nanoparticles by a novel alkalotolerant actinomycete, *Rhodococcus* species. Nanotechnology. 14(3): 824–828.

Amerasan, D., Nataraj, T., Murugan, K., Panneerselvam, C., Madhiyazhagan, P., Nicoletti, M., and Benelli, G. 2016. Myco-synthesis of silver nanoparticles using *Metarhizium anisopliae* against the rural malaria vector *Anopheles culicifacies* Giles (Diptera: Culicidae) J. Pest Sci. 89(1): 249–256.

Anbazhagan, S., Azeez, S., Morukattu, G., Rajan, R., Venkatesan, K., and Thangavelu, K.P. 2017. Synthesis, characterization and biological applications of mycosynthesized silver nanoparticles. Biotech. 7(0): 333.

Azmath, P., Baker, S., Rakshith, D., and Satish, S. 2016. Mycosynthesis of silver nanoparticles bearing antibacterial activity. Saudi Pharm. J. 24(2): 140–146.

Ballottin, D., Fulaz, S., Cabrini, F., Tsukamoto, J., Durán, N., Alves, O.L., and Tasic, L. 2017. Antimicrobial textiles: Biogenic silver nanoparticles against *Candida* and *Xanthomonas*. Mat Sci Eng C. 75(0): 582–589.

Balakumaran, M.D., Ramachandran, R., and Kalaichelvan, P.T. 2015. Exploitation of endophytic fungus, *Guignardia mangiferae* for extracellular synthesis of silver nanoparticles and their *in vitro* biological activities. Microbiol. Res. 178(0): 9–17.

Bandeira, M., Giovanela, M., Roesch-Ely, M., Devine, D.M., and Crespo, J.S. 2020. Green synthesis of zinc oxide nanoparticles: A review of the synthesis methodology and mechanism of formation. Sustain Chem Phar. 15(0): 100223.

Banu, A., Gousuddin, M., and Yahya, E.B. 2021. Green synthesized monodispersed silver nanoparticles' Characterization and their efficacy against cancer cells. Biomed. Res. Ther. 8(8): 4476–4482.

Baskar, G., Chandhuru, J., Fahad, K.S., Praveen, A., Chamundeeswari, and M. Muthukumar, T. 2015. Anticancer activity of fungal L-asparaginase conjugated with zinc oxide nanoparticles. J. Mat. Sci: Mater Med. 26(1): 43.

Bhainsa, K.C., and D'Souza, S.F. 2006. Extracellular biosynthesis of silver nanoparticles using the fungus Aspergillus fumigatus. Colloids Surf B Biointerf. 47(2): 160–164.

Bharde, A., Rautaray, D., Bansal, V., Ahmad, A., Sarkar, I., Yasuf, M., Sanyal, M., and Sastry, M. 2006. Extracellular biosynthesis of magnetite using fungi. Small 2(1): 135–141.

Bhardwaj, K., Sharma, A., Tejwan, N., Bhardwaj, S., Bhardwaj, P., Nepovimova, E., Shami, A., Kalia, A., Kumar, A., Abd-Elsalam, K.A., and Kuca, K. 2020. *Pleurotus* macrofungi-assisted nanoparticle synthesis and its potential applications: A review. J. Fungi. 6(4): 351.

Bhargava, A., Navin, j., Azeem, K.M., Vikram, P., Venkataramana, D.R., and Jitendra, P. 2016. Utilizing metal tolerance potential of soil fungus for efficient synthesis of gold nanoparticles with superior catalytic activity for degradation of rhodamine B. J. Env. Manag. 183(1): 22–32.

Bhatnagar, S., Kobori, T., Ganesh, D., Ogawa, K., and Aoyagi, H. 2019. Biosynthesis of silver nanoparticles mediated by extracellular pigment from *Talaromyces purpurogenus* and their biomedical applications. Nanomater. 9(7): 1042.

Castro-Longoria, E., Moreno-Velázquez, S.D., Vilchis-Nestor, A.R., Arenas-Berumen, E., and Avalos-Borja, M. 2012. Production of platinum nanoparticles and nanoaggregates using *Neurospora crassa*. J. Microbiol. Biotechnol. 22(7): 1000–1004.

Chauhan, R., Reddy, A., and Abraham, J. 2015. Biosynthesis of silver and zinc oxide nanoparticles using *Pichia fermentans* JA2 and their antimicrobial property. Appl. Nanosci. 5(1): 63–71.

Chowdhury, S., Basu, A., and Kundu, S. 2014. Green synthesis of protein capped silver nanoparticles from phytopathogenic fungus *Macrophomina phaseolina* (Tassi) Goid with antimicrobial properties against multidrugresistant bacteria. Nano Res Lett. 9: 365.

Clarance, P., Luvankar, B., Sales, J., Khusro, A., Agastian, P., Tack, J.-C., Al Khulaifi, M.M., AL-Shwaiman, H.A., Elgorban, A.M., Syed, A., and Kim, H.-J. 2020. Green synthesis and characterization of gold nanoparticles using endophytic fungi *Fusarium solani* and its *in-vitro* anticancer and biomedical applications. Saudi. J. Biol. Sci. 27(0): 706–712.

272 Mycosynthesis of Nanomaterials: Perspectives and Challenges

Cuevas, R., Durán, N., Diez, M.C., Tortella, G., and Rubilar, O. 2015. Extracellular Biosynthesis of copper and copper oxide nanoparticles by *Stereum hirsutum*, a native white-rot fungus from Chilean forests. J. Nanomater. 16(1): 57–63.

Das, B.S., Das, A., Mishra, A., and Arakha, M. 2021. Microbial cells as biological factory for nanoparticle synthesis. Front Mater Sci. 15(2): 177–191.

Das, S.K., Das, A.R., and Guha, A.K. 2009. Gold nanoparticles: microbial synthesis and application in water hygiene management. Langmuir 25(14): 8192–8199.

Das, S.K., and Marsili, E. 2010. A green chemical approach for the synthesis of gold nanoparticles: characterization and mechanistic aspect. Rev. Environ. Sci. 9(0): 199–204.

Das, S.K., Liang, J., Schmidt, M., Laffir, F., and Marsili, E. 2012. Biomineralization mechanism of gold by *Zygomycete* Fungi *Rhizopous oryzae*. ACS Nano 6(7): 6165–6173.

Dawoud, T.M., Yassin, M.A., El-Samawaty, A.R.M., and Elgorban, A.M. 2021. Silver nanoparticles synthesized by *Nigrospora oryzae* showed antifungal activity. Saudi J. Biol. Sci. 28(3): 1847–1852.

Devika, R., Elumalai, S., Manikandan, E., and Eswaramoorthy, D. 2012. Biosynthesis of silver nanoparticles using the fungus *Pleurotus ostreatus* and their antibacterial activity. Open Access Sci Rep. 1(12): 557.

Durán, N., Marcato, P.D., Alves, O.L., De Souza, G.I.H., and Esposito, E. 2005. Mechanistic aspects of biosynthesis of silver nanoparticles by several *Fusarium oxysporum* strains. J. Nanobiotechnol. 3(0): 8.

Durán, N., Marcato, P.D., De Souza, D.I.H., Alves, O.L., and Esposito, E. 2007. Antibacterial effect of silver nanoparticles produced by fungal process on textile fabrics and their effluent treatment. J Biomed. Nanotechnol. 3(2): 203–208.

Durán, N., Marcato, P.D., Durán, M., Yadav, A., Gade, A., Rai, M. 2011. Mechanistic aspects in the biogenic synthesis of extracellular metal nanoparticles by peptides, bacteria, fungi and plants. Appl. Microbiol. Biotechnol. 90(5): 1609–1624.

Durán, N., and Seabra, A.B. 2012. Metallic oxide nanoparticles: state of the art in biogenic syntheses and their mechanisms. Appl. Microbiol. Biotechnol. 95(2): 275–288.

Elamawi, R.M., Al-Harbi, R.E., and Hendi, A.A. 2018. Biosynthesis and characterization of silver nanoparticles using *Trichoderma longibrachiatum* and their effect on phytopathogenic fungi. Egypt J. Biol. Pest Control. 28(0): 28.

El-Batal, A.I., El-Sayyad, G.S., Mosallam, F.M., and Fathy, R.M. 2020. *Penicillium chrysogenum*-mediated mycogenic synthesis of copper oxide nanoparticles using gamma rays for *in vitro* antimicrobial activity against some plant pathogens. J. Clust. Sci. 31(0): 79–90.

El-Batal, A.I., El-Sayyad, G.S., El-Ghamery, A., and Gobara, M. 2017. Response surface methodology optimization of melanin production by *Streptomyces cyaneus* and Synthesis of Copper Oxide Nanoparticles Using Gamma Radiation. J. Clust. Sci. 28(3): 1083–1112.

El Domany, E.B., Essam, T.M., Ahmed, A.E., and Farghali, A.A. 2018. Biosynthesis physico-chemical optimization of gold nanoparticles as anticancer and synergetic antimicrobial activity using *Pleurotus ostreatus* fungus. J. Appl. Pharm. Sci. 8(5): 119–128.

El-Rafie, M.H., Mohamed, A.A., Shaheen, T.H.I., and Hebeish, A. 2010. Antimicrobial effect of silver nanoparticles produced by fungal process on cotton fabrics. Carbo Polym. 80(3): 779–782.

Fanti, J.R., Tomiotto-Pellissier, F., Miranda-Sapla, M.M., Cataneo, A.H.D., Andrade, C.G.T.J., Panis, C., Rodrigues, J.H.S., Wowk, P.F., Kuczera, D., Costa, I.N., Nakamura, C.V., Nakazato, G., Durán, N., Pavanelli, W.R., and Conchon-Costa, I. 2018. Biogenic silver nanoparticles inducing *Leishmania amazonensis* promastigote and amastigote death *in vitro*. Acta Tropica. 178(0): 46–54.

Ferreira, L.A.B., Fóssa, F.G., Durán, N., de Jesus, M.B., and Fávaro, W.J. 2019. Cytotoxicity and antitumor activity of biogenic silver nanoparticles against non-muscle invasive bladder. J. Phys. Conf. Ser. 1323(0): 012020.

Ferreira, L.A.B., Garcia-Fossa, F., Radaic, A., Durán, N., Fávaro, W.J., and de Jesus, M.B. 2020. Biogenic silver nanoparticles: *in vitro* and *in vivo* antitumor activity in bladder cancer. Eur. J. Pharm. Biopharm. 151(0): 162–170.

Gade, A.K., Bonde, P., Ingle, A.P., Marcato, P.D., Durán, N., and Rai, M.K. 2008. Exploitation of *Aspergillus niger* for synthesis of silver nanoparticles. J. Biobased. Mater Bioener. 2: 243–247.

Gahlawat, G., and Choudhury, A.R. 2019. A review on the biosynthesis of metal and metal salt nanoparticles by microbes. RSC Adv. 9(23): 12944–12967.

Gao, Y., Anand, M.A.V., Ramachandran, V., Karthikkumar, V., Shalini, V., Vijayalakshmi, S., and Ernest, D. 2019. Biofabrication of zinc oxide nanoparticles from *Aspergillus niger*, their antioxidant, antimicrobial and anticancer activity. J. Clust. Sci. 30(4): 937–946.

Gajbhiye, M.B., Kesharwani, J.G., Ingle, A.P., Gade, A.K., and Rai, M.K. 2009. Fungus mediated synthesis of silver nanoparticles and its activity against pathogenic fungi in combination of fluconazole. Nanomed. 5(4): 382–386.

Ganachari, S.V., Bhat, R., Deshpande, R., and Venkataraman, A. 2012. Extracellular biosynthesis of silver nanoparticles using fungi *Penicillium diversum* and their antimicrobial activity studies. BioNanoScience 2(4): 316–321.

Gericke, M., Pinches, A. 2006. Biological synthesis of metal nanoparticles. Hydrometallurgy 83(1-4): 132–140.

Gharieb, M.M., Wilkinson, S.C., and Gadd, G.M. 1995. Reduction of selenium oxyanions by unicellular, polymorphic and filamentous fungi: cellular location of reduced selenium and implications for tolerance. J. Ind. Microbiol. 14(3,4): 300–311.

Gond, S.K., Mishra, A., Verma, S.K., Sharma, V.K., and Kharwar, R.N. 2020. Synthesis and characterization of antimicrobial silver nanoparticles by an endophytic fungus isolated from *Nyctanthes arbor-tristis*. Proc. Natl. Acad Sci. India Sect. B. Biol. Sci. 90(0): 641–645.

Ghosh, S., Patil, S., Ahire, M., Kitture, R., Kale, S., Pardesi, K. Cameotra, S.S., Bellare, J., Dhavale, D.D., Jabgunde, A., and Chopade, B.A. 2012. Synthesis of silver nanoparticles using *Dioscorea bulbifera* tuber extract and evaluation of its synergistic potential in combination with antimicrobial agents. Int. J. Nanomedicine. 7(0): 483–496.

Gole, A., Dash, C., Ramakrishnan, V., Sainkar, S.R., Mandalr, A.B., Rao, M., and Sastry, M. 2001. Pepsin-gold colloid conjugates: preparation, characterization, and enzymatic activity. Langmuir 17(5): 1674–1679.

Govender, Y., Riddin, T., Gericke, M., and Whiteley, C.G. 2009. Bioreduction of platinum salts into nanoparticles: a mechanistic perspective Biotechnol. Lett. 31(1): 95–100.

Govender, Y., Riddin, T.L., Gericke, M., and Whiteley, C.G. 2010. On the enzymatic formation of platinum nanoparticles. J. Nanopart. Res. 12: 261–271.

Guilger, M., Pasquoto-Stigliani, T., Bilesky-Jose, N., Grillo, R., Abhilash, P., Fraceto, L.F., and De Lima, R. 2017. Biogenic silver nanoparticles based on *Trichoderma harzianum*: synthesis, characterization, toxicity evaluation and biological activity. Sci. Rep. 7(0): 44421.

Guilger-Casagrande, M., and de Lima, R. 2019. Synthesis of silver nanoparticles mediated by fungi: A review. Front. Bioeng. Biotechnol. 7(0): 287.

Guilger-Casagrande, M., Germano-Costa, T., Pasquoto-Stigliani, T., Fraceto, L.F., and de Lima, R. 2019. Biosynthesis of silver nanoparticles employing *Trichoderma harzianum* with enzymatic stimulation for the control of *Sclerotinia sclerotiorum*. Sci. Rep. 9(0): 14351.

Gupta, K., and Chundawat, T.S. 2019. Bio-inspired synthesis of platinum nanoparticles from fungus *Fusarium oxysporum*: Its characteristics, potential antimicrobial, antioxidant and photocatalytic activities. Mater Res. Express. 6(10): 1–10.

Gurunathan, S., Raman, J., Malek, S.N.A., John, P.A., and Vikineswary, S. 2013. Green synthesis of silver nanoparticles using Ganoderma neo-japonicum Imazeki: A potential cytotoxic agent against breast cancer cells. Int. J. Nanomed. 8(0): 4399–4413.

Gurunathan, S., Park, J.H., Han, J.W., and Kim, J.H. 2015. Comparative assessment of the apoptotic potential of silver nanoparticles synthesized by *Bacillus tequilensis* and *Calocybe indica* in MDA-MB-231 human breast cancer cells: Targeting p53 for anticancer therapy. Int J Nanomed. 10(0): 4203.

Hamad, M.T. 2019. Biosynthesis of silver nanoparticles by fungi and their antibacterial activity. Int. J. Environ. Sci. Technol. 16(3): 1015–1024.

Hassan, S.E., Fouda, A., Saied, E., Farag, M.M.S., Eid, A.M., Barghoth, M.G., Awad, M.A., Hamza, M.F., and Awad, M.F. 2021. *Rhizopus oryzae* mediated green synthesis of magnesium oxide nanoparticles (MgO-NPs): A promising tool for antimicrobial, mosquitocidal action, and tanning effluent treatment. J. Fungi 7(5): 372.

Hamedi, S., Shojaosadati, S.A., Shokrollahzadeh, S., and Hashemi-Najafabadi, S. 2013. Extracellular biosynthesis of silver nanoparticles using a novel and nonpathogenic fungus, *Neurospora intermedia*: Controlled synthesis and antibacterial activity. World J. Microbiol. Biotechnol. 30(2): 693–704.

274   *Mycosynthesis of Nanomaterials: Perspectives and Challenges*

Hamedi, S., Ghaseminezhad, M., Shokrollahzadeh, S., and Shojaosadati, S.A. 2017. Controlled biosynthesis of silver nanoparticles using nitrate reductase enzyme induction of filamentous fungus and their antibacterial evaluation. Artif Cells Nanomed. Biotechnol., 45(8): 1588–1596.

Hulikere, M.M. , Joshi, C.G., Danagoudar, A., Poyya, J., Kudva, A.K., and Dhananjaya, B.L. 2017. Biogenic synthesis of gold nanoparticles by marine endophytic fungus-*Cladosporium cladosporioides* isolated from seaweed and evaluation of their antioxidant and antimicrobial properties. Process Biochem. 63(0): 137–144.

Hulikere M.M., and Joshi. C.G. 2019. Characterization, antioxidant and antimicrobial activity of silver nanoparticles synthesized using marine endophytic fungus- *Cladosporium cladosporioides*. Process Biochem. 82(0): 199–204.

Hulkoti, N.I., and Taranath, T.C. 2014. Biosynthesis of nanoparticles using microbes—A review. Colloids Surf. B Biointerf. 121(0): 474–483.

Husseiny, S.M., Salah, T.A., and Anter, H.A. 2015. Biosynthesis of size controlled silver nanoparticles by *Fusarium oxysporum*, their antibacterial and antitumor activities. Beni- Suef Univ. J. Basic Appl. Sci. 4(3): 225–231.

Ibrahem, E.J., Thalij, K.M., Amin, S., and Badawy, A.S. 2017. Antibacterial potential of magnesium oxide nanoparticles synthesized by *Aspergillus niger*. Biotechnol. J. Inter. 18(1): 1–7.

Ingle, A., Gade, A., Pierrat, S., Sonnichsen, C., and Rai, M. 2008. Mycosynthesis of silver nanoparticles using the fungus *Fusarium acuminatum* and its activity against some human pathogenic bacteria. Curr Nanosci. 4(2): 141–144

Ingle, I.P., Durán, N., and Rai, M. 2014. Bioactivity, Mechanism of action and cytotoxicity of copper-based nanoparticles: A review. Appl. Microbiol. Biotechnol. 98(3): 1001–1009.

Ishida, K., Cipriano, T.F., Rocha, G.M., Weissmüller, G., Gomes, F., Miranda, K., and Rozental, S. 2014. Silver nanoparticle production by the fungus *Fusarium oxysporum*: Nanoparticle characterization and analysis of antifungal activity against pathogenic yeasts. Mem. Inst. Oswaldo Cruz. 109(2): 220–228.

Jain, N., Bhargava, A., Majumdar, S., Tarafdar, J.C., and Panwar, J. 2011. Extracellular biosynthesis and characterization of silver nanoparticles using *Aspergillus flavus* NJP08: a mechanism perspective. Nanoscale. 3(2): 635–641.

Jha, P., Saraf, A., and Sohal, J.K. 2021. Antimicrobial activity of biologically synthesized gold nanoparticles from wild Mushroom *Cantharellus* species. Journal of Scientific Research 65(3).

Kalimuthu, K., Babu, R.S., Venkataraman, D., Bilal, M., and Gurunathan, S. 2008. Biosynthesis of silver nanocrystals by *Bacillus licheniformis*. Colloid Surf. B: Biointerf. 65(1): 150–153.

Kalpana, V.N., Kataru, B.Z.S,, Sravani, N., Vigneshwari, T., Panneerselvam, A., and Rajeswari, V.D. 2018. Biosynthesis of zinc oxide nanoparticles using culture filtrates of *Aspergillus niger*: Antimicrobial textiles and dye degradation studies, OpenNano 3(0): 48–55.

Kar, P.K., Murmu, S., Saha, X.S., Tandon, V., and Acharya, K. 2014. Anthelmintic efficacy of gold nanoparticles derived from a phytopathogenic fungus, *Nigrospora oryzae*. PLoS One. 9(1): e84693.

Kathiresan, K., Manivannan, S., Nabeel, M.A., and Dhivya, B. 2009. Studies on silver nanoparticles synthesized by a marine fungus, *Penicillium fellutanum* isolated from coastal mangrove sediment. Colloids Surf B Biointerf. 71(0): 133–137.

Khan, N.T., and Khan, M.J. 2017. Biogenic nanoparticles: An introduction to what they are and how they are produced. Int. J. Biotech Bioeng. 3(3): 66–70.

Khandel, P., and Shahi, S.K. 2018. Mycogenic nanoparticles and their bio-prospective applications: current status and future challenges. J. Nanostr. Chem. 8: 369–391.

Kim, S.W., Kim, K.S., Lamsal, K., Kim, Y.J., Kim, S.B., Jung, M., Sim, S.J., Kim, H.S., Chang, S.J., Kim, J.K., and Lee, Y.S. 2009. An *in vitro* study of the antifungal effect of silver nanoparticles on oak wilt pathogen *Raffaelea* sp. J. Microbiol. Biotechnol. 19(8): 760–764.

Kitching, M., Ramani, M., and Marsili, E. 2015. Fungal biosynthesis of gold nanoparticles: mechanism and scale up. Microb. Biotechnol. 8(6): 904–917.

Kumar, S.A., Abyaneh, M.K., Gosavi, S.W., Kulkarni, S.K., Pasricha, R., Ahmad, A., and Khan, M.I. 2007. Nitrate reductase-mediated synthesis of silver nanoparticles from $AgNO_3$. Biotechnol. Lett. 29(0): 439–445.

Kundu, D., Hazra, C., Chatterjee, A. Chaudhari, A., and Mishra, S. 2014. Extracellular biosynthesis of zinc oxide nanoparticles using *Rhodococcus pyridinivorans* NT2: Multifunctional textile finishing, biosafety evaluation and in vitro drug delivery in colon carcinoma. J. Photochem. Photobiol. B. Biol. 140(0): 194–204.

Li, J., Li, Q., Ma, X., Tian, B., Li, T., Yu, J., Dai, S., Weng, Y., and Hua, Y. 2016. Biosynthesis of gold nanoparticles by the extreme bacterium *Deinococcus radiodurans* and an evaluation of their antibacterial properties. Int. J. Nanomed. 11(0): 5931–5934.

Li, X., Xu, H., Chen, Z.-S., and Chen, G. 2011. Biosynthesis of nanoparticles by microorganisms and their applications. J. Nanomater. (8): 270974.

Lotfy, W.A., Alkersh, B.M., Sabry, S.A., and Ghozlan, H.A. 2021. Biosynthesis of silver nanoparticles by *Aspergillus terreus*: Characterization, optimization, and biological activities. Front Bioeng. Biotechnol. 9: 633468.

Ma, L., Su, W., Liu, J.-X., Zeng, X.X., Huang, Z., Li, W., Liu, Z.-C., and Tang, J.-X. 2017. Optimization for extracellular biosynthesis of silver nanoparticles by *Penicillium aculeatum* Su1 and their antimicrobial activity and cytotoxic effect compared with silver ions. Mater Sci. Eng C. 77(0): 963–971.

Mahmoud, M.A., Al-Sohaibani, S.A., Al-Othman, M.R., El-Aziz, A.M.A., and Eifan, S.A. 2013. Synthesis of extracellular silver nanoparticles using *Fusarium semitectum* (KSU-4) isolated from Saudi Arabia. Dig. J. Nanomat Biostruct. 8: 589–596.

Majumder, D.R. 2012. Bioremediation: Copper nanoparticles from electronic-waste. Inter J. Eng. Sci. Technol. 4(10): 4380–4389.

Mani, V.M., Kalaivani, S., Sabarathinam, S., Vasuki, M., Soundari, A.J.P.G., Das, M.P.A., Elfasakhany, A., and Pugazhendhi, A. 2021. Copper oxide nanoparticles synthesized from an endophytic fungus *Aspergillus terreus*: Bioactivity and anticancer evaluations. Environ Res. 201(0): 111502.

Makarov, V.V., Sinitsyna, O.V., Makarova, S.S., Yaminsky, I.V., Taliansky, M.E., and Kalinina, N.O. 2014. Green nanotechnologies: synthesis of metal nanoparticles using plants. Acta Naturae. 6(1): 35–44.

Marcato, P.D., Nakasato, G., Brocchi, M., Melo, P.S., Huber, S.C., Ferreira, I.R., Alves, O.L., and Durán, N. 2012a. Biogenic silver nanoparticles: Antibacterial and cytotoxicity applied to textile fabrics. J. Nano Res. 20(0): 69–76.

Marcato, P.D., Durán, M., Huber, S., Rai, M., Melo, P.S., Alves, O.L., and Durán, N. 2012b. Biogenic silver nanoparticles and its antifungal activity as a new topical transungual drug delivery. J. Nano Res. 20(0): 99–107.

Mekkawy, A.I., El-Mokhtar, M.A., Nafady, N.A., Yousef, N., Hamad, M.A., El-Shanawany, S.M., Ibrahim, E.H., and Elsabahy, M. 2017. *In vitro* and *in vivo* evaluation of biologically synthesized silver nanoparticles for topical applications: effect of surface coating and loading into hydrogels. Int. J. Nanomed. 12(0): 759–777.

Metuku, R.P., Pabba, S., Burra, S., Hima Bindu, N.S.V.S.S.S.L., Gudikandula, K., and Singara Charya, M.A. 2014. Biosynthesis of silver nanoparticles from *Schizophyllum radiatum* HE 863742.1: Their characterization and antimicrobial activity. Biotech. 4(0): 227–234.

Mishra, Z., Kumari, M., Pandey, S., Chaudhry, V., Gupta, K., and Nautiyal, C. 2014. Biocatalytic and antimicrobial activities of gold nanoparticles synthesized by *Trichoderma* sp. Bioresour. Technol. 166(0): 235–242.

Moazeni, M., Rashidi, N., Shahverdi, A.R., Noorbakhsh, F., and Rezaie, S. 2012. Extracellular production of silver nanoparticles by using three common species of dermatophytes: *Trichophyton rubrum*, *Trichophyton mentagrophytes* and *Microsporum canis*. Iran Biomed. J. 16(1): 52–58.

Mohamed, A.A., Fouda, A., Abdel-Rahman, M.A., Hassan, S.E.D., El-Gamal, M.S., Salem, S.S., and Shaheen, T.I. 2019. Fungal strain impacts the shape, bioactivity and multifunctional properties of green synthesized zinc oxide nanoparticles. Biocatal. Agrical. Biotechnol. 19(1): 101103.

Mohamed, Y.M., Azzam, A.M., Amin, B.H., and Safwat, N.A. 2015. Mycosynthesis of iron nanoparticles by *Alternaria alternata* and its antibacterial activity. Afr. J. Biotechnol. 14(14): 1234–1241.

Mosallam, F.M., El-Sayyad, G.S., Fathy, R.M., and El-Batal, A.I. 2018. Biomolecules-mediated synthesis of selenium nanoparticles using *Aspergillus oryzae* fermented Lupin extract and gamma radiation for hindering the growth of some multidrug-resistant bacteria and pathogenic fungi Microb. Pathog. 122(0): 108–116.

Mukherjee, P., Ahmad, A., Mandal, D., Senapati, S., Sainkar, S., Khan, M.I., Parishcha, R., Ajaykumar, P.V., Alam, M., Kumar, R., and Sastry, M. 2001a. Fungus-mediated syntheses of silver nanoparticles

276 *Mycosynthesis of Nanomaterials: Perspectives and Challenges*

and their immobilization in the mycelial matrix—a novel biological approach to nanoparticle synthesis. Nano Letters. 1(10): 515–519.

Mukherjee, P., Ahmad, A., Mandal, D., Senapati, S., Sainkar, S.R., Khan, M.I., Ramani, R., Parisha, R., Ajayakumar, V., Alam, M., Sastry, M., and Kumar, R. 2001b. Bioreduction of AuCl(4)(-) ions by the fungus, *Verticillium* sp. and surface trapping of the gold nanoparticles formed. Angew Chem. Int. Ed. Engl. 40(19): 3585–3588.

Mukherjee, P., Senapati, S., Mandal, D., Ahmad, A., Khan, M.I., Kumar, R., and Sastry, M. 2002. Extracellular synthesis of gold nanoparticles by the fungus *Fusarium oxysporum*. Chembiochem. 3(5): 461–463.

Mukherjee, P., Roy, M., Mandal, B.P., Dey, G.K., Mukherjee, P.K., Ghatak, J., Tyagi, A.K., and Kale, S.P. 2008. Green synthesis of highly stabilized nanocrystalline silver particles by a nonpathogenic and agriculturally important fungus *T. asperellum*. Nanotechnology 19(7): 075103.

Nagajyothi, P.C., Sreekanth, T.V.M., Lee, J., and Lee, K.D. 2014. Mycosynthesis: antibacterial, antioxidant and antiproliferative activities of silver nanoparticles synthesized from *Inonotus obliquus* (Chaga mushroom) extract. J. Photochem. Photobiol. B 130(0): 299–304.

Naqvi, S.Z.H., Kiran, U., Ali, M.I., Jamal, A., Hameed, A., Ahmed, S. et al. 2013. Combined efficacy of biologically synthesized silver nanoparticles and different antibiotics against multidrug-resistant bacteria. Int. J. Nanomed. 8(0): 3187–3195.

Naqvi, S.T.Q., Shah, Z., Fatima, N., Qadir, M.I., Ali, A., and Muhammad, S.A. 2017. Characterization and biological studies of copper nanoparticles synthesized by *Aspergillus niger*. J.` Bionanosci. 11(2): 136–140.

Namasivayam, S.K.R., Avimanyu. 2011. Silver nanoparticle synthesis from *Lecanicillium lecanii* and evalutionary treatment on cotton fabrics by measuring their improved antibacterial activity with antibiotics against *Staphylococcus aureus* (ATCC 29213) and *E. coli* (ATCC 25922) strains. Int. J. Pharm. Pharm. Sci. 3(4): 190–195.

Nayak, B. K., Nanda, A., and Prabhakar, V. 2018. Biogenic synthesis of silver nanoparticle from wasp nest soil fungus, *Penicillium italicum* and its analysis against multi drug resistance pathogens. Biocatal. Agric. Biotechnol. 16(0): 412–418.

Netala, V.R., Bethu, M.S., Pushpalatah, B., Baki, V.B., Aishwarya, S., Rao, J.V., and Tartte, V. 2016. Biogenesis of silver nanoparticles using endophytic fungus *Pestalotiopsis microspora* and evaluation of their antioxidant and anticancer activities. Int. J. Nanomed. 11(0): 5683–5696.

Niraimathi, K.L., Sudha, V., Lavanya, R., and Brindha, P. 2013. Biosynthesis of silver nanoparticles using *Alternanthera sessilis* (Linn.) extract and their antimicrobial, antioxidant activities. Colloids Surf B Biointerfaces 102(0): 288–291.

Noor, S., Shah, Z., Javed, A., Ali, A., Hussain, S.B., Zafar,DS., Ali, H., and Muhammad, S.A. 2020. A fungal based synthesis method for copper nanoparticles with the determination of anticancer, antidiabetic and antibacterial activities. J. Microbiol. Meth. 174(0): 105966.

Noorafsha, Gade, J., Kashyap, A., Kashyap, A., Vishwakarma, D., and Azizi, S. 2020. A critical analysis of the biogenic synthesis of transition metal nanoparticles along with its application and stability. Eur J Mol Clin Med. 7(11): 6368–6397.

Omran, B.A., Nassar, H.N., Younis, S.A., El-Salamony, R.A., Fatthallah, N.A., Hamdy, A. et al. 2020. Novel mycosynthesis of cobalt oxide nanoparticles using *Aspergillus brasiliensis* ATCC 16404 – optimization, characterization and antimicrobial activity. J. Appl. Microbiol. 128(2): 438–57.

Othman, A.M., Elsayed, M.A., Al-Balakocy, N.G., Hassan, M.M., and Elshafei, A.M. 2019. Biosynthesis and characterization of silver nanoparticles induced by fungal proteins and its application in different biological activities. J. Genet. Eng. Biotechnol. 17(0): 8.

Ottoni, C.A., Simões, M.F., Fernandes, S., Santos, J.G., Silva, E.S., Souza, R.F.B., and Maiorano, A.E. 2017. Screening of filamentous fungi for antimicrobial silver nanoparticles synthesis. AMB Express. 7(0): 31.

Owaid, M.N., Raman, J., Lakshmanan, H., Al-Saeedi, S.S.S., Sabaratnam, V., and Abed, I.A. 2015. Mycosynthesis of silver nanoparticles by *Pleurotus cornucopiae* var. *citrinopileatus* and its inhibitory effects against *Candida* sp. Mater Lett. 153(0): 186–190.

Pantidos, N., and Horsfall, L.E. 2014. Biological synthesis of metallic nanoparticles by bacteria, fungi and plants. Journal of Nanomedicine & Nanotechnology 5: 0–0.

Parikh, R.Y., Singh, S., Prasad, B.L.V., Patole, M.S., Sastry, M., and Shouche, Y.S. 2008. Extracellular synthesis of crystalline silver nanoparticles and molecular evidence of silver resistance from *Morganella* sp.: towards understanding biochemical synthesis mechanism. Chembiochem. 9(9): 1415–1422

Pereira, L., Dias, N., Carvalho, J., Fernandes, S., Santos, C., and Lima, N. 2014. Synthesis, characterization and antifungal activity of chemically and fungal produced silver nanoparticles against *Trichophyton rubrum*. J. Appl. Microbiol. 117(6): 1601–1613.

Pimprikar, P., Joshi, S., Kumar, A., Zinjarde, S., and Kulkarni, S. 2009. Influence of biomass and gold salt concentration on nanoparticle synthesis by the tropical marine yeast *Yarrowia lipolytica* NCIM 3589. Colloids Surf B: Biointerf. 74(1): 309–316.

Poblete, H., Agarwal, A., Thomas, S.S., Bohne, C., Ravichandran, R., Phopase, J.B., Alarcon, E.I., and Comer, J. 2013. Understanding the interaction between biomolecules and silver nanoparticles. Biophys. J. 110(3): 341a.

Prakash, P., Gnanaprakasam, P., Emmanuel, R., Arokiyaraj, S., and Saravanan, M. 2013. Green synthesis of silver nanoparticles from leaf extract of *Mimusops elengi*, Linn. for enhanced antibacterial activity against multi drug resistant clinical isolates. Colloids Surf B Biointerf. 108(0): 255–259.

Qin, W., Wang, C.-Y., Ma, Y.-X., Shen, M.-J., Li, J., Jiao, K., Tay, F.R., and Niu, L.-N. 2020. Microbe-mediated extracellular and intracellular mineralization: Environmental, industrial, and biotechnological applications. Adv. Mater. 32(22): 1907833.

Rai, M., Yadav, A., and Gade, A. 2009. Silver nanoparticles as a new generation of antimicrobials. Biotechnol Adv. 27(1): 76–83.

Rai, M., Kon, K., Ingle, A., Durán, N., Galdiero, S., and Galdiero, M. 2014. Broad-spectrum bioactivities of silver nanoparticles: The emerging trends and future prospects. Appl. Microbiol. Biotechnol. 98(5): 1951–1961.

Rai, M., Ingle, A.P., Gade, A., and Durán, N. 2015a. Synthesis of silver nanoparticles by *Phoma gardenia* and *in vitro* evaluation of their efficacy against human disease-causing bacteria and fungi. IET Nanobiotechnol. 9(2): 71–75.

Rai, M.,, Ingle, A.P., Gade, A., Duarte, M., and Durán, N. 2015b. Three-Phoma spp. synthesized novel silver nanoparticles that possess excellent antimicrobial efficacy. IET Nanobiotechnol. 9(5): 280–297.

Rai, M., Bonde, S., Golinska, P., Trzcinska-Wencel, J., Gade, A., Abd-Elsalam, K.A., Shende, S., Gaikwad, S., and Ingle, A.P. 2021. *Fusarium* as a novel fungus for the synthesis of nanoparticles: mechanism and applications. J. Fungi. 7(2): 139.

Rajput, S., Werezuk, R., Lange, R.M., and McDermott, M.T. 2016. Fungal isolate optimized for biogenesis of silver nanoparticles with enhanced colloidal stability. Langmuir 32(34): 8688–8697.

Raliya, R., and Tarafdar, J.C. 2013. ZnO nanoparticle biosynthesis and its effect on phosphorous-mobilizing enzyme secretion and gum contents in clusterbean (*Cyamopsis tetragonoloba* L.). Agric. Res. 2(1): 48–57.

Raman, J., Reddy, G.R., Lakshmanan, H., Selvaraj, V., Gajendran, B., Nanjian, R., Chinnasamy, A., and Sabaratnam, V. 2015. Mycosynthesis and characterization of silver nanoparticles from *Pleurotus djamor* var. *roseus* and their *in vitro* cytotoxicity effect on PC3 cells. Process Biochem. 50(1): 140–147.

Rani, R., Sharma, D., Chaturvedi, M., and Yadav, J.P. 2017. Green synthesis, characterization and antibacterial activity of silver nanoparticles of endophytic fungi *Aspergillus terreus*. J. Nanomed. Nanotechnol. 8(0): 4.

Ray, S., Sarkar, S., and Kundu, S. 2011. Extracellular biosynthesis of silver nanoparticles using the mycorrhizal mushroom *Tricholoma crassum* (BERK.) Sacc.: its antimicrobial activity against pathogenic bacteria and fungus, including multidrug resistant plant and human bacteria. Dig. J. Nanomater. Bios. 6(3): 1289–1299.

Riddin, T.L., Gericke, M., and Whiteley, C.G. 2006. Analysis of the inter- and extracellular formation of platinum nanoparticles by *Fusarium oxysporum* f. sp. *Lycopersici* using response surface methodology. Nanotechnology 17(4): 3482–3489.

Rodrigues, A.G., Ping, L.Y., Marcato, P.D., Alves, O.L., Silva, M.C.P., Ruiz, R.C., Melo, I.DS., Tasic. L., and De Souza, A.O. 2013. Biogenic antimicrobial silver nanoparticles produced by fungi. Appl. Microbiol. Biotechnol. 97(2): 775–782.

Rodrigues, A.G., Ruiz, RdC., Selari, P.J.R.G., de Araujo,W.L., and de Souza, A.O. 2021. Anti-biofilm action of biological silver nanoparticles produced by *Aspergillus tubingensis* and antimicrobial activity of fabrics carrying it. Biointerface Res. Appl. Chem. 11(6): 14764–14774.

Rodriguez-Serrano, C., Guzman-Moreno, J., Angelez-Chavez, C., Rodriguez-Gonzalez, V., Ortega-Sigala, J.J., Ramírez-Santoyo, R.M., and Vidales-Rodríguez, L.E. 2020. *In vitro* and *in vivo* evaluation of biologically synthesized silver nanoparticles for topical applications: effect of surface coating and loading into hydrogels PloS One 15(13): e0230275.

Rossi-Bergmann, B., Pacienza-Lima, W., Marcato, P.D., De Conti, R., and Durán, N. 2012. Therapeutic potential of biogenic silver nanoparticles in murine cutaneous leishmaniasis. J. Nano Res. 20(0): 89–97.

Roy, S., Mukherjee, T., Chakraborty, S., and Das, T.K. 2013. Biosynthesis, characterisation & antifungal activity of silver nanoparticles synthesized by the fungus *Aspergillus foetidus* MTCC8876. Dig. J. Nanomater. Biostruct. 8(1): 197–205.

Rubilar, O., Rai, M., Tortella, G., Diez, M.C., Seabra, A.B., and Durán, N. 2013. Biogenic Nanoparticles: Copper, copper oxides, copper sulfides, complex copper nanostructures and their applications. Biotechnol. Lett. 35(9): 1365–1375.

Saied, E., Eid, A.M., Hassan, S.E.-D., Salem, S.S., Radwan, A.A., Halawa, M., Saleh, F.M., Saad, H.A., Saied, E.M., and Fouda, A. 2021. The catalytic activity of biosynthesized magnesium oxide nanoparticles (MgO-NPs) for inhibiting the growth of pathogenic microbes, tanning effluent treatment, and chromium ion removal. Catalysts. 11(7): 821.

Saifuddin, N., Wong, C.W., and Yasumira, A.A. 2009. Rapid biosynthesis of silver nanoparticles using culture supernatant of bacteria with microwave irradiation. J. Chem. 6(1): 61–70.

Saitawadekar, A., and Kakde U.B.. 2020. Green synthesis of copper nanoparticles using *Aspergillus flavus*. J. Crit. Rev. 7(16): 1083–1090.

Sandhu, S.S., Shukla, H., and Shukla, S. 2017. Biosynthesis of silver nanoparticles by endophytic fungi: Its mechanism, characterization techniques and antimicrobial potential. African J. Biotechnol. 16(14): 683–698.

Santos, L.M., Stanisic, D., Menezes, U.J., Mendonça, M.A., Barral, T.D., Seyffert, N., Azevedo, V., Durán, N., Meyer, R., Tasic, L., and Portela, R.W. 2019. Biogenic Silver nanoparticles as a post-surgical treatment for *Corynebacterium pseudotuberculosis* infection in small ruminants. Front Microbiol. 10(0): 824.

Saravanan, M., and Nanda, A. 2010. Extracellular synthesis of silver bionanoparticles from Aspergillus clavatus and its antimicrobial activity against MRSA and MRSE. Colloids Surf B: Biointerf. 77(2): 214–218.

Saravanakumar, K., MubarakAli, D., and Kandasamy, K. 2016. Biogenic metallic nanoparticles as catalyst for bioelectricityproduction: a novel approach in microbial fuel cells. Mater Sci. Eng. 203(0): 27–34.

Saravanakumar, K., and Wang, M.-H. 2018. *Trichoderma* based synthesis of anti-pathogenic silver nanoparticles and their characterization, antioxidant and cytotoxicity properties. Microb. Pathogen. 114(0): 269–273.

Saravanakumar, K., Shanmugam, S., and Babu, N. 2019. Biosynthesis and characterization of copper oxide nanoparticles from indigenous fungi and its effect of photothermolysis on human lung carcinoma. J. Photochem. Photobiol B: Biol. 190(0): 103–109.

Saravanakumar, K., Jeevithan, E., Hu, X., Chelliah, R., Ho, D.-H., and Wang, M-H. 2020. Enhanced anti-lung carcinoma and anti-biofilm activity of fungal molecules mediated biogenic zinc oxide nanoparticles conjugated with β-D-glucan from barley. J. Photochem. Photobiol. B: Biol. 203(0): 111728.

Sarsar, V., Selwal, M.K., and Selwal, K.K. 2016. Biogenic synthesis, optimisation and antibacterial efficacy of extracellular silver nanoparticles using novel fungal isolate *Aspergillus fumigatus* MA. IET Nanobiotechnol. 10(4): 215–221.

Saxena, J., Sharma, P.K., Sharma, M.M., and Singh, A. 2016. Process optimization for green synthesis of silver nanoparticles by *Sclerotinia sclerotiorum* MTCC 8785 and evaluation of its antibacterial properties. Springerplus 5(1): 861.

Schröfel, A., Gabriela Kratošová, Šafařík, I., Šafaříková, M., Raška, I., and Shor, L.M. 2014. Applications of biosynthesized metallic nanoparticles – A review. Acta Biomater. 10(0): 4023–4042.

Senapati, S., Ahmad, A., Khan, M.I., Sastry, M., and Kumar, R. 2005. Extracellular biosynthesis of bimetallic Au–Ag alloy nanoparticles. Small 1(5): 517–520.

Shamim, A., Abid, M.B., and Mahmood, T. 2019. Biogenic synthesis of zinc oxide (ZnO) nanoparticles using a fungus (*Aspergillus niger*) and their characterization. Inter. J. Chem. 11(2): 119–126.

Siddiqi, K.S., and Husen, A. 2016. Fabrication of metal nanoparticles from fungi and metal salts: scope and application . Nanoscale Res. Lett. 11(1): 98.

Silva, T.A., Andrade, P.F., Segala, K., Silva, L.S.C., Silva, L.P., Nista, S.V.G., Mei, L.H.I., Durán, N., and Teixeira, M.F.S. 2017. Silver nanoparticles biosynthesis and impregnation in cellulose acetate membrane for anti-yeast therapy. Afr. J. Biotechnol. 16(27): 1490–1500.

Singh, D., Rathod, V., Ninganagouda, S., Hiremath, J., Singh, A.K., and Mathew, J. 2014. Optimization and characterization of silver nanoparticle by endophytic fungi *Penicillium* sp. isolated from *Curcuma longa* (Turmeric) and application studies against MDR *E. coli* and *S. aureus*. Bioinorg Chem. App. 2014(0): 408021.

Singh, P., and Raja, R.B. 2011. Biological synthesis and characterization of silver nanoparticles using the fungus *Trichoderma harzianum*. Asian J. Exp. Biol. Sci. 2(4): 600–605.

Singh, T., Jyoti, K., Patnaik, A., Singh, A., Chauhan, R., and Chandel, S.S. 2017. Biosynthesis, characterization and antibacterial activity of silver nanoparticles using an endophytic fungal supernatant of *Raphanus sativus*. J. Genet. Eng. Biotechnol. 15(1): 31–39.

Soni, N., and Prakash, S. 2012. Synthesis of gold nanoparticles by the fungus *Aspergillus niger* and its efficacy against mosquito larvae. Rep. Parasitolol. 2(0): 1–7.

Spadaro, D., and Gullino, M.L. 2005. Improving the efficacy of biocontrol agents against soilborne pathogens. Crop Protection. 24(7): 601–613.

Srivastava, S., Bhargava, A., Pathak, N., and Srivastava, P. 2019. Production, characterization and antibacterial activity of silver nanoparticles produced by *Fusarium oxysporum* and monitoring of protein-ligand interaction through *in silico* approaches. Microb. Pathog. 129(0): 136–145.

Stanisic, D., Fregonesi, N.L., Barros, C.H.N., Pontes, J.G.M., Fulaz, S., Menezes, U.J., Nicoleti, J.L., Castro, T.L.P., Seyffert, N., Azevedo, V., Duran, N., Portela, R.W., and Tasic, L. 2018. NMR insights on nano silver post-surgical treatment of superficial caseous lymphadenitis in small ruminants. RSC Adv. 8(71): 40778–40786.

Sulaiman, G.M., Hussein, T.H., and Saleem, M.M. 2015. Biosynthesis of silver nanoparticles synthesized by *Aspergillus flavus* and their antioxidant, antimicrobial and cytotoxicity properties. Bull Mater Sci. 38(0): 639–644.

Suryavanshi, P., Pandit, R., Gade, A., Derita, M., Zachino, S., and Rai, M. 2017. *Colletotrichum* sp.-mediated synthesis of sulphur and aluminium oxide nanoparticles and its *in vitro* activity against selected food-borne pathogens. LWT–Food Sci Technol. 81(0): 188–194.

Syed, A., and Ahmad, A. 2012 Extracellular biosynthesis of platinum nanoparticles using the fungus *Fusarium oxysporum*. Colloids Surf. B 97(0): 27–31.

Thakker, J.N., Dalwadi, P., and Dhandhukia, P.C. 2013. Biosynthesis of gold nanoparticles using *Fusarium oxysporum* f. sp. *cubense* JT1, a plant pathogenic fungus. ISRN Biotechnol. 2013(0): 515091.

Thakkar, K.N., Mhatre, S.S., and Parikh, R.Y. 2010. Biological synthesis of metallic nanoparticles. Nanomedicine. 6(2): 257–262.

Thirumurugan, G., Shaheedha, S.M., and Dhanaraju, M.D. 2009. *In vitro* evaluation of antibacterial activity of silver nanoparticles synthesised by using *Phytophthora infestans*. Int. J. Chem. Tech. Res. 1(3): 714–716.

Tyagi, S., Tyagi, P.K., Gola, D., Chauhan, N., Randhir, K., and Bharti, R.K. 2019. Extracellular synthesis of silver nanoparticles using entomopathogenic fungus: characterization and antibacterial potential. SN Appl. Sci. 1(12): 1545.

Usher, A., McPhail, D.C., and Brugger, J.A. 2009. Spectrophotometric study of aqueous Au(III) halide-hydroxide complexes at 25–80°C. Geochim. Cosmochim. Acta. 73(11): 3359–3380.

Vahabi, K., Mansoori, G.A., and Karimi, S. 2011. Biosynthesis of silver nanoparticles by fungus *Trichoderma reesei* (a route for large-scale production of AgNPs). Insciences J. 1(1): 65–79.

Vigneshwaran, N., Ashtaputre, N.M., Varadarajan, P.V., Nachane, R.P., Paralikar, K.M., and Balasubramanya, R.H. 2007. Biological synthesis of silver nanoparticles using the fungus *Aspergillus flavus*. Mater Lett. 61(6): 1413–1418.

Vijayanandan, A.S., and Balakrishnan, R.M. 2018. Biosynthesis of cobalt oxide nanoparticles using endophytic fungus *Aspergillus nidulans*. J. Environ. Manag. 218(0): 442–450.

Wang, D., Xue, B., Wang, L., Zhang, Y., Liu, L., and Zhou, Y. 2021. Fungus-mediated green synthesis of nano-silver using Aspergillus sydowii and its antifungal/antiproliferative activities. Sci. Rep. 11(0): 10356.

Yadav, A., Kon, K., Kratosova, G., Duran, N., Ingle, A.P., and Rai, M. 2015. Fungi as an efficient mycosystem for the synthesis of metal nanoparticles: progress and key aspects of research. Biotechnol. Lett. 37: 2099–2120.

Yusof, H.M., Mohamad, R., and Zaidan, U. 2019. Microbial synthesis of zinc oxide nanoparticles and their potential application as an antimicrobial agent and a feed supplement in animal industry: a review. J. Animal. Sci. Biotechnol. 10(1): 57.

Xue, B., He, D., Gao, S., Wang, D., Yokoyama, K., and Wang, L. 2016. Biosynthesis of silver nanoparticles by the fungus *Arthroderma fulvum* and its antifungal activity against genera of *Candida, Aspergillus* and *Fusarium*. Int. J. Nanomed. 11(0): 1899–1906.

Zhang, S.Y., Zhang, J., Wang, H.Y., and Chen, H.Y. 2004. Synthesis of selenium nanoparticles in the presence of polysaccharides. Mat. Lett. 58(21): 2590–2594.

Zhang, X., Yan, S., Tyagi, R.D., and Surampalli, R.Y. 2011. Synthesis of nanoparticles by microorganisms and their application in enhancing microbiological reaction rates. Chemosphere. 82(4): 489–494.

Zhao, W. 2010. Nucleotide-mediated Strategy for the Synthesis of Bio-functionalized Gold-nanoparticles, Thesis (Ph.D.)--Hong Kong University of Science and Technology, 2010, Hong Kong, 2010. http://hdl.handle.net/1783.1/7001.

Section III
Toxicity of Nanoparticles to Human and Environment

CHAPTER 14

Toxicity of Mycosynthesised Nanoparticles

Indarchand Gupta,[1,*] *Arun Ghuge,*[2] *Hanna Dahm*[3] and *Mahendra Rai*[3]

Introduction

Nanotechnology is continuously gaining interest as it has promising effects for human welfare (Dos Santos et al. 2014). Hence, for meeting the supply of nanoparticles for applications, there is a need to work on several methods for their synthesis which can produce the nanoparticles of desired characteristics. Currently, the nanoparticle synthesis methods are broadly categorised into three methods viz. chemical, physical and biological methods (Rai et al. 2021). The physical methods of synthesis are widely used but they are labour intensive and time consuming. However, these methods produce the highly monodisperse and stable nanoparticles (Srivastava and Bhargava 2022). Chemical methods are also being preferred for the synthesis of many nanoparticles. Many types of chemical methods are already in practice. In general, these methods involve the use of chemicals, most of which are toxic. The use of toxic chemicals in the synthesis of nanoparticles adds toxicity to the nanoparticles (Arshad et al. 2021). However, it is almost difficult to remove the toxic chemicals from synthesised nanoparticle suspension. Though synthesised nanoparticles are highly monodispersive and stable, they are considered to be biologically incompatible (Zafar et al. 2020). Hence, before any biological applications, it is necessary to consider their toxic effects on humans and environment. The biologically synthesised nanoparticles are good alternative for the production of biocompatible nanoparticles.

---

[1] Department of Biotechnology, Government Institute of Science, Nipatniranjan Nagar, Caves Road, Aurangabad-431 004, Maharashtra, INDIA.
[2] Department of Forensic Biology, Government Institute of Forensic Science, Nipatniranjan Nagar, Caves Road, Aurangabad-431 004, Maharashtra, INDIA.
[3] Department of Microbiology, Nicolas Copernicus University, Torun, POLAND.
* Corresponding author: indarchandgupta@gmail.com

The biosynthesis methods make use of various organisms (Saravanan et al. 2021). Herein, plants, bacteria, actinomycetes, algae and fungi are being used (Alavi 2022). In general, biosynthesis of nanoparticles does not require any harmful material. Moreover, as organisms are synthesising nanoparticles, these methods are called to be biocompatible and hence, also called as green synthesis methods (Salem and Fouda 2021). Among biosynthesis methods the plant and bacteria mediated synthesis do have some drawbacks (Behzad et al. 2021). The plant mediated synthesis methods require the extraction of plants whereas in the bacterial method of synthesis the bacterial cells need its rupturing so that the biologically active components for nanoparticle synthesis come out of the cell (Paiva-Santos et al. 2021; Huq et al. 2022). Both these methods are tedious and time consuming. However, the fungi could be a better alternative to these. The synthesis of nanoparticles by using fungi has been called as mycosynthesis (Rai et al. 2021). In this process, fungi act as reducing and stabilizing agents for the nanoparticle synthesis. In recent era fungi are appearing to be attractive agents for nanoparticle synthesis due to some advantages associated with their use. The fungi can be easily grown and do not require the sophisticated instrumentation facilities, they can be easily handled (Gade et al. 2022). They have high potential for production of many compounds. It is estimated that microscopic filamentous fungi such as ascomycetes and imperfect fungi can produce approximately 6,400 bioactive substances (Guilger-Casagrande and de Lima 2019). They give high yield of proteins. During mycosynthesis the fungal derived biomolecules get coated on nanoparticles. It provides better stability, biocompatibility and better biological activity (Paul and Roychoudhury 2021). Additionally, mycosynthesis can produce nanoparticles of various shapes, sizes and surface charge. The synthesis mechanism is yet to be fully elucidated, however it is believed that fungal biomolecules play a major role in the nanoparticle synthesis (Rai et al. 2021). The duration of fungal growth, temperature and pH of cultivation media plays a significant role in the nanoparticle synthesis (Priyadarshani et al. 2021). Mycosynthesised nanoparticles are showing promising prospective applications (Priyadarshini et al. 2021). But while studying the beneficial effects, the harmful effects also need to be considered. In recent era many reports suggest that nanoparticles also show the toxicity, which is called as nanotoxicity. There is a need to focus on the nanotoxicological aspects of myconsythesised nanoparticles. Hence, the present chapter aims to majorly focus on the toxicity of fungal derived nanoparticles.

Nanotoxicity

Nanotoxicology is subdiscipline of toxicology dealing specifically with the toxicity of nanoparticles in humans and environment. It majorly encompasses the study of harmful effect of nanoparticles *in vitro* and *in vivo*. The main aim of nanotoxicity study is to find the list of factors which contribute to their toxicity, indirectly helping the nanotoxicologist to consider them for the safe application of nanobased products. Nanoparticles once released into the environment after their designated application, can sustain there and may enter the human body through various routes, i.e., dermal exposure, respiration and oral route.

Routes of nanoparticle entry

Nanoparticles used for any type of applications, are finally released into the environment. After getting released in the environment, it can enter into the human body through absorption into the skin, ingestion through oral route and inhalation through aerial route (See Figure 1). The nanoparticles entered through these routes can accumulate in the human body and cause the harmful effects (Gupta et al. 2012).

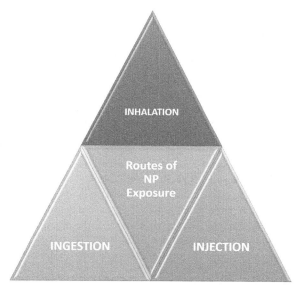

Figure 1. Different routes of nanoparticle exposure to humans.

Inhalation

It is the major route of nanoparticle entry into the human body (Sonwani et al. 2021). Once nanoparticles are released in the air, they remain suspended as particulate matter (Souza et al. 2021). Those nanoparticles inhaled through airway settle down in the lungs and then come into the lung tissues, alveoli and capillaries. During long term exposure, the nanoparticles can get accumulated in the lung tissues, where they can obstruct the process of respiration (Lin et al. 2022). Later on, through blood circulation they can be translocated into the highly active organs such as liver, heart, kidney and nervous tissues (Nikzamir et al. 2021). The nanoparticles are cleared out from the body depending upon their concentration and duration of exposure. Smaller nanoparticles are highly active and hence, they exert more toxic effects (Rashid et al. 2021).

Injection

Skin is the protective cover for human body. But still there are many modes for its damage including physical damage, damage due to skin infections and ultraviolet

286 *Mycosynthesis of Nanomaterials: Perspectives and Challenges*

rays coming from sunlight (Sun et al. 2021). Hence, many nanobased products are already available in the market which act as antimicrobials and sunscreen (Farjadian et al. 2019) protecting the skin from any damage. Prolonged use of these products results into their accumulation in the dermal layer. Many studies suggest that the nanoparticles smaller than 100 nm in size are able to penetrate into the skin (Hongbo et al. 2013). There is no mechanism of nanoparticle translocation from the skin and hence, they remain in the skin for long duration, interfering with the skin function.

Ingestion

Nanoparticles are being used to increase the food shelf life by coating its thin layer on it (Jafarzadeh et al. 2021). Moreover, they are also being used in water filter for making water potable, personal care products, toothpaste and sunscreens as mentioned above (Kidd et al. 2021). After the consumption of NP coated fruits or water from nano–based water filter, there is risk of nanoparticle entry into the human alimentary canal (Aguilar-Pérez et al. 2021). The main role of digestive system is absorbing the nutrients from the food and hence, with nutrients, nanoparticles also get absorbed into the blood circulatory system. Through blood, the nanoparticles reach the liver, kidney, spleen, lungs thus resulting into the hepatoxicity, nephrotoxicity, splenotoxicity and lung damage (Kreyling et al. 2017). However, as far as exposure to nanoparticles through these routes are concerned, many reports suggest the nanotoxicity of physically and chemically synthesised nanoparticles (Kapoor and Singh 2021) and there are a few reports covering the toxicity of biosynthesised nanoparticles, especially mycosynthesised nanoparticles. Therefore, there is need to focus on the toxic impacts of mycosynthesised nanoparticles on human and environmental components.

Toxicity of mycosynthesised nanoparticles

As stated above, mycosynthesised NPs have been used in many areas, but on the contrary there is a need to assess their cytotoxic effects on humans and environment. For example, silver nanoparticles (AgNPs) synthesised through fungus *Pestalotiopsis microspore* presented the cytotoxic effects on mouse melanoma cells (B16F10) with IC50¼26.43 ± 3.4108 g/mL. Moreover, it also showed IC50¼16.24 ± 2.4808 g/mL, IC50¼39.83 ± 3.7408 g/mL and IC50¼27.71 ± 2.8908 g/mL on the human ovarian carcinoma (SKOV3), human lung adenocarcinoma (A549) and human prostate carcinoma (PC3) cell lines respectively (Netala et al. 2016). The toxicity of nanoparticles against normal and cancer lines differs (AshaRani et al. 2009). It is generally found that the nanoparticles are more toxic to cancer lines as compared to normal cell line. This is because the cancer cells are metabolically more active as compared to normal cells and thus they assimilate higher amount of nanoparticles into their cytoplasm as compared to normal cells. Hence, higher interaction with nanoparticles leads to the higher chance of showing harmful effects. The study lead by Potara et al. (2015) also corroborates the findings of the other study emphasising toxicity of mycosynthesised nanoparticles. The AgNPs synthesised by using *Pestalotiopsis microspora* didn't show any significant harmful effects on viability of HaCaT cells at the concentration ranging from 2–4 µg/ml. The study further mentioned that 98% of the exposed HaCaT cells remained viable after 24-hour exposure time. However, as compared to the said fungal derived AgNPs, the plant (*Azadirachta indica*) derived AgNPs at the concentration of 4 µg/ml were found to be more toxic, resulting into the 60% of viable cells during same exposure time. The fungi *Guignardia mangiferae* have been also exploited for the synthesis of AgNPs (Balakumaran et al. 2015). Further, their toxicity was assessed on Vero cell, HeLa and MCF-7 cell lines. The IC50 values were found to be 63.37 µg/ml, 27.54 µg/ml and 19 µg/ml for mycosynthesised AgNPs. It is significant to note here that the mycosynthesised AgNPs have been found to be more active against cancer cells viz. MCF-7, followed by HeLa cells, showing higher cell growth inhibition at lower concentration. El-Sonbaty (2013) performed the toxicity study of AgNPs synthesised by using the fungus *Agaricus bisporus*. For *in vitro* toxicity study AgNPs were exposed to MCF-7 cell line and *in vivo* toxicity effects were studied against Ehrlich carcinoma in mice. The study shows the concentration dependent toxicity against the breast carcinoma cells with LC50 value to be 3.5 ng. *In vivo* study found the decrease of blood vessels and increase of apoptotic cells in mice. Moreover, the toxic effects were intensified when application of AgNPs was combined with the gamma radiation. Mycosynthesised AgNPs have been also exploited as the antifungal agents against some phytopathogen. For this purpose, Guilger et al. (2017) synthesised AgNPs by using the fungal filtrate of *Trichoderma harzianum* as a reducing agent and stabilizer. Along with their phytopathogenic effects, the group also studied their toxic effects on animal lines viz. 3T3, HeLa, HaCat, and V79. It is reported that the higher sensitivity in 3T3 cell line is with lowest IC50 value of 0.21×10^{12} NPs/ml. Whereas, the HeLa and A549 cells showed higher resistance with the IC50 value of 0.91×10^{12} NPs/ml. The study hence showed that the different cell lines respond differently to the exposed AgNPs (Mukherjee et al. 2012).

Larvicides are required for controlling mosquitoes. Various chemical agents are being used and there is continuous interest in searching novel larvicides. Salunkhe et al. (2011) have tested the larvicidal activity of mycosynthesised AgNPs against the larvae of *Aedes egypti* and *Anopheles stephensi*. The group synthesised AgNPs by using filamentous fungus *Cochliobolus lunatus*. It was found to show the dose dependant toxic effects on larvae with LC50 at 1.29, 1.48, and 1.58 and LC90 at 3.08, 3.33, and 3.41 ppm against second, third, and fourth instar larvae of *A. aegypti*. Similarly, mycosynthesised AgNPs showed LC50 at 1.17, 1.30, and 1.41; and LC90 at 2.99, 3.13, and 3.29 ppm second, third, and fourth instar larvae of *A. stephensi*. However, the same nanoparticles didn't show any toxicity at LC50 and LC90 doses to the non-target fish species *Poecilia reticulate*. Hence, this study indicates that mycosynthesised nanoparticles are eco-friendly and nontoxic, which could favour their use vector control without causing any harm to predatory fish.

Endophytic fungi synthesise bioactive compounds which have medicinal value and hence they were also tested for the synthesis of nanoparticles which could be used for biomedical applications. Ranjani et al. (2020) has synthesised AgNps by using *Eupatorium triplinerve*, an endophytic fungus. For the toxicity evaluation of mycosynthesised AgNPs *Caenorhabditis elegans* was used as an animal model. The study revealed that mycosynthesised AgNPs were non-toxic to *C. elegans*. Moreover,

288   *Mycosynthesis of Nanomaterials: Perspectives and Challenges*

it showed increase in the survival rate of *C. elegans* by decreasing the mortality rate when treated with *Staphylococcus aureus*. When *C. elegans* were treated with *S. aureus*, the worms were able to survive only for 7 days. On the contrary mycosynthesised AgNPs treated against *S. aureus* has increased the survival rate of nematode up to 9 days. It is known that the soil contains microflora. Most of the organisms present therein are beneficial for plant growth. One of those bacteria is *Pseudomonas putida*. Gupta et al. (2015) synthesised AgNPs by using *F. moniliforme, F. culmorum, P. glomerata and F. tricinctum* and tested their toxicity effects on *P. putida*. They revealed that the mycosynthesised AgNPs have toxicity to *P. putida* in the concentration range from 0.4 µg/ml to 0.8 µg/ml. The study hence demonstrated the similar toxicity of similar sized AgNPs regardless of their origin from where they have been synthesised. Gaikwad et al. (2013) made use of *Alternaria* sp., *Fusarium oxysporum, Curvularia* sp., *Chaetomium indicum*, and *Phoma* sp. for the synthesis of AgNPs. These nanoparticles showed potential antiviral effects against HSV-1, HSV-2 and HPIV-3 in cell cultures. Moreover, the AgNPs synthesised by using *F. oxysporum, Curvularia* sp., and *C. indicum* didn't show significant cytotoxicity. Moreover, the interaction of fungal synthesised AgNPs is concerned, it showed significant alteration in the plant metabolism. Lima et al. (2013) reported the *F. oxysporum* synthesised AgNPs to show dose and duration dependent toxic effects on *Allium cepa* by inducing significantly higher alteration rates as compared to the negative control. Mycosynthesised AgNPs have been also exploited as the antifungal agents against some phytopathogen. In this regards, Guilger et al. (2017) have also studied their toxicity to animal cell lines viz. 3T3, HeLa, HaCat, and V79. It is reported the higher sensitivity in 3T3 cell line with the lowest IC50 value of 0.21×10^{12} NPs/ml. Whereas, the HeLa and A549 cells showed higher resistance with the IC50 value of 0.91×10^{12} NPs/ml. The study hence showed that the different cells lines respond differently to the exposed AgNPs (Mukherjee et al. 2012). All of these studies signified the toxicity of fungal generated nanoparticles. As compared to physical and chemical synthesised nanoparticles, they appear to be safe and biocompatible and hence, more attention needs to be given to the biomedical applications of fungal generated nanoparticles. However, still there is need to generate the toxicity database for fungal generated nanoparticles and more specifically the mechanism of nanoparticle toxicity at cellular level needs to be focussed upon. Such study will surely help us to develop nanotechnology for safe use.

Ecotoxicity of NPs

Nanoparticles are considered to be environment friendly options for many types of applications. They are also helpful for plants to increase their production and growth. However, if their concentration crosses a certain limit, then they may pose substantial threat to the components of ecosystem. As far as studies published in this regard, there are many knowledge gaps in our understanding of the ecotoxicity of mycosynthesised NPs. The phytotoxicity of NPs needs to be given due consideration wherein the harmful effects of NPs to plant root length, seed germination and their uptake needs to be studied. Park et al. (2006) has made use of nanosized silica-silver particles to control the powdery mildew diseases of cucurbits. The said NPs have been highly effective in controlling the disease. However, at higher concentration viz. 3200 ppm, the NPs were found to exert harmful effects on cucumber and pansy plants. Similarly, Corredor et al. (2009) have demonstrated the plant species dependent nanotoxicity. The NPs showed negative effect on root elongation in cabbage, carrot, corn, cucumber, soybean and tomato. ZnONPs and TiO_2NPs have also been employed for many applications. After their targeted application, if they reach the soil, then they can show huge impact on soil microbial flora. When ZnONPs and TiO_2NPs were used to help for soybean cultivation, they were shown to induce the direct harmful effects on soil microflora including the N2-fixing symbiosis. Impact of nanoparticles on soil components are always a major concern as it can affect plants and the surrounding microflora. AgNPs synthesised by *Trichoderma* species has no detrimental effect on soybean germination and its development, but prevents the growth of white mold, a phytopathogen (Guilger et al. 2017). As far as interaction of NPs with plants is concerned, there is need of efforts to be made to get clear idea about interaction of NPs with plants, their beneficial and pathogenic microflora. Additionally, there is also need to give attention on nanotoxicity of NPs on plant and their interaction with microflora present in their vicinity.

Role of shape, size, dose and surface capping of mycosynthesised nanoparticle in toxicity

In general, toxicity of nanoparticles depends upon their shape, size, concentration and surface capping. Nanoparticles are small in size and hence they have larger surface area which allows them to interact more with the external environment. Therefore, the information regarding these factors playing role in their toxicity needs to be considered.

Effect of NP size on nanotoxicity

Nanoparticle size has been demonstrated to determine their toxicity. It is a widely accepted fact that smaller nanoparticles have a larger surface area to volume ratio. Hence, smaller nanoparticles show more toxicity and they can easily permeate in cell membrane as compared to their larger counterpart (Tao 2018). The blood kinetics of biosynthesised AgNPs were studied in rats (Kakakhel et al. 2021). The size of AgNPs was in the range of 20, 80 and 100 nm. After entry into the blood they were found to get disappeared quickly and get dispersed to all organs. After repeated intravenous injections, they were found to exert toxic effects depending upon their size and concentration. Santos-Rasera et al. (2022) have analysed toxic effects of ZnONPs on *Daphnia magna*. The NPs used were of 20 nm and 40 nm size. They found that smaller NPs were more toxic as compared to their counterparts. Moreover, the 300 nm ZnONPs were found to be least toxic in the same study.

Effect of NPs shape on nanotoxicity

Shape of nanoparticle also plays an important role in determining their toxicity. Synthesised nanoparticles have various shapes viz. spherical, rod, flower, triangular,

290   *Mycosynthesis of Nanomaterials: Perspectives and Challenges*

hexagonal, irregular and star shaped (Tao 2018; Santos et al. 2021). Spherical nanoparticles are more toxic as compared to others. In a study, ZnONPs were synthesised by using *Fusarium keratoplasticum* and *Aspergillus niger*. The method produced ZnONPs of two different shapes viz. hexagonal and nanorod having similar size and they were tested for their toxic effects. The study claimed that the nanoparticles were safe at concentration less than 20.1 and 57.66 ppm respectively, suggesting that rod shaped ZnONPs are safer as compared to hexagonal shaped NPs (Mohamed et al. 2019). In another study *F. oxysporum* mediated synthesised AgNPs were found to have shapes such as cube, sphere, rice, rod and blunt rod (Soleimani et al. 2018). The study concluded that all of the above mentioned AgNPs are inherently nontoxic to human cells at concentration lower than 10 µg/ml. However, at higher concentration the toxicity increases depending on shape and dose of AgNPs. Among them nanocubes, nanorice and nonorods appeared to be more toxic as compared to the spheres and blunt nanorods. The stronger vertex in sharp ends of AgNPs structures and their geometries, especially cubic structures can interact more with cell membrane, thus causing greater toxicity (Soleimani et al. 2018). Awad et al. (2022) have synthesised AgNPs by using *A. niger* and studied their larvicidal and pupicidal activity against *A. aegypti* at the concentrations ranging from 5, 10, 15, 20, 25 and 30 ppm. The investigation revealed that the AgNPs showed toxic effects in dose-dependent manner.

Effect of NP concentration on nanotoxicity

Guilger-Casagrande et al. (2021) has performed the biogenic synthesis of AgNPs. They have synthesised AgNPs by using filtrate of *Trichoderma harzianum*. After studying their anti-phytopathogenic, they have performed their toxicity assessment on *Allium cepa*. The study reported that AgNP has genotoxic effect on *Allium cepa* at higher concentration only. However, at lower concentrations, it appeared to be non-toxic. Similar results were obtained by Skladanowski et al. (2016). They have synthesised AgNPs by using *Streptomyces* sp. and evaluated their cytotoxicity to L929 mouse fibroblast cell line. The study found no cytotoxicity at lowest concentration, whereas IC50 reached while using the cells at higher concentrations. *Trichoderma atroviride* synthesised AgNPs were studied for possible toxic effects on MDB-MB-231, a human breast cancer cells, IMR-90, U251 and A549 cells (Saravanakumar et al. 2018). The AgNPs were found to reduce the viability of MDA-MB-231 cells at the concentration of 16.5 µg/ml. Further analysis of data revealed that AgNPs caused cell damage, cytoplasmic compression depending upon the increasing concentration.

Effect of capping

Capping of nanoparticles plays an important role in controlling size of nanoparticles and their biological activity (Guilger-Casagrande et al. 2021). Capping agent remain bound to nanoparticle surface and hence it also contributes to the nanoparticle interaction with any other entity. While studying the genotoxicity of capped and uncapped AgNPs synthesised by *Trichoderma harzianum*, it was found that the uncapped AgNPs caused higher genotoxicity as compared to the capped nanoparticles. The comet assays also showed that the uncapped AgNPs cause higher damage to the exposed cells as compared to the capped AgNPs. The toxicity difference between capped and uncapped nanoparticles may be due to the fact that in aqueous environment, the uncapped nanoparticles get more readily ionised as compared to capped nanoparticles (Zhang et al. 2010; Jain et al. 2015). Therefore, while comparing the toxicity of nanoparticles synthesised by various sources, their capping needs to be considered.

Mechanism of Nanotoxicity

As discussed in the earlier section, nanoparticles are very fine particles. When they come in contact with the target cell, they disrupt their morphology and mostly result into unrepairable damage. At cellular level, nanoparticles tend to interact with the cell membrane, organelles and nuclear DNA (Asghar and Asghar 2020) (see Figure 2). Initially nanoparticles come in contact with the cell membrane. The interaction leads to damage to plasma membrane components, making it unstable

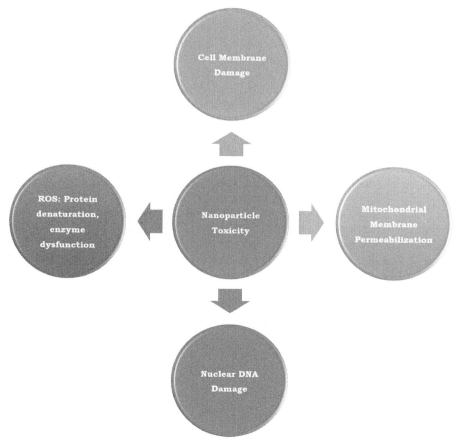

Figure 2. Mechanism of Nanoparticle toxicity at cellular level.

(Mandic et al. 2019). Finally, it results into the formation of pores in cell membrane, thus causing the leakage of cytoplasmic fluid out of the cell (Wang et al. 2020). The nanoparticles get entry inside the cell cytoplasm via endocytosis into an endosome and later on into the lysosomes (Rennick et al. 2021). Through the pore formed in cell membrane, nanoparticles interfere in the metabolic process of cell. It reacts with the proteins, more specifically with the enzymes and hence it affects the normal metabolic pathways. Moreover, nanoparticles induce the accumulation of reactive oxygen species (ROS) (Yu et al. 2020). ROS affects the cell membrane integrity and initiate apoptosis. ROS are highly reactive and they also react with the cellular protein and nuclear DNA (Kowalska et al. 2020). The nanoparticles thus cause the DNA damage (Attia et al. 2018; El-Demerdash et al. 2021). Moreover, nanoparticle exposure also induces the expression of oxidative stress related enzymes (Nayek et al. 2022). It includes the induction of genes coding for superoxide dismutase, catalase and glutathione peroxidase (Koner et al. 2021). Overall, nanoparticles cause the damage to cell at every location, which force the cell to undergo programmed cell death.

Conclusions

Fungal-mediated synthesis of nanoparticles is showing promising potential in terms of their applications for the betterment of humans. The mycosynthesis of nanoparticles is considered to be eco-friendly and biocompatible as compared to physical and chemical mode of nanosynthesis. In spite of their usefulness, the concerns about their toxicity are rising. Therefore, there is utmost need to focus on this aspect. The harmful effects of mycosynthesised nanoparticles has been studied by many research groups across the globe. In general, these studies show that mycosynthesised nanoparticles are although toxic, they are less toxic as compared to physically and chemically synthesised nanoparticles. It indicates that the mycosynthesised nanoparticles are highly biocompatible, thus making them a preferred candidate for any biological applications. The toxicity of these nanoparticles depends upon their size, shape, surface capping and their concentration. It is important to note that negligible reports are available on the *in vitro* toxicity and ecotoxicity of mycosynthesised nanoparticles, thus pressing the need to perform more studies to get in depth idea about their harmful effects. Moreover, *in vivo* toxicity study of these nanoparticles also needs to focussed on, knowing which it will help to develop and deliver nanobased products safely for human and environmental applications.

Acknowledgements

MR thankfully acknowledges the financial support rendered by the Polish National Agency for Academic Exchange (NAWA) (Project No. PPN/ULM/ 2019/1/00117/U/00001) to visit the Department of Microbiology, Nicolaus Copernicus University, Toruń, Poland.

References

Aguilar-Pérez, K.M., Ruiz-Pulido, G., Medina, D.I., Parra-Saldivar, R., and Iqbal, H.M. 2021. Insight of nanotechnological processing for nano-fortified functional foods and nutraceutical—opportunities, challenges, and future scope in food for better health. Crit. Rev. Food Sci. Nutri. 1–18.

Alavi, M. 2022. Bacteria and fungi as major bio-sources to fabricate silver nanoparticles with antibacterial activities, Expert Rev. Anti Infect. Ther. 1–10 DOI: 10.1080/14787210.2022.2045194.

Arshad, H., Sami, M.A., Sadaf, S., and Hassan, U. 2021. *Salvadora persica* mediated synthesis of silver nanoparticles and their antimicrobial efficacy. Sci. Rep. 11(1): 1–11.

Asghar, M.A., and Asghar, M.A. 2020. Green synthesized and characterized copper nanoparticles using various new plants extracts aggravate microbial cell membrane damage after interaction with lipopolysaccharide. Intl. J. Biol. Macromolecules 160: 1168–1176.

AshaRani, P.V., Low Kah Mun, G., Hande, M.P., and Valiyaveettil, S. 2009. Cytotoxicity and genotoxicity of silver nanoparticles in human cells. ACS Nano 3(2): 279–290.

Attia, H., Nounou, H., and Shalaby, M. 2018. Zinc oxide nanoparticles induced oxidative DNA damage, inflammation and apoptosis in rat's brain after oral exposure. Toxics, 6: 29. DOI: 10.3390/toxics6020029.

Awad, M.A., Eid, A.M., Elsheikh, T.M.Y., Al-Faifi, Z.E., Saad, N., Sultan, M.H., Selim, S., Al-Khalaf, A.A., and Fouda, A. 2022. Mycosynthesis, characterization, and mosquitocidal activity of silver nanoparticles fabricated by *Aspergillus niger* Strain. J. Fungi. 8: 396. https://doi.org/ 10.3390/jof8040396.

Balakumaran, M.D., Ramachandran, R., and Kalaichelvan, P.T. 2015. Exploitation of endophytic fungus, *Guignardia mangiferae* for extracellular synthesis of silver nanoparticles and their *in vitro* biological activities. Microbiol. Res. 178: 9–17.

Behzad, F., Naghib, S.M., Tabatabaei, S.N., Zare, Y., and Rhee, K.Y. 2021. An overview of the plant-mediated green synthesis of noble metal nanoparticles for antibacterial applications. J. Ind. Eng. Chem. 94: 92–104.

Corredor, E., Testillano, P.S., Coronado, M.-J., González-Melendi, P., Fernández-Pacheco, R., Marquina, C., Ibarra, M.R., de la Fuente, J.M., Rubiales, D., Pérez-de-Luque, A., and Risueño, M.C. 2009. Nanoparticle penetration and transport in living pumpkin plants: *in situ* subcellular identification. BMC Plant. Biol. 9: 45.

Dos Santos, C.A., Seckler, M.M., Ingle, A.P., Gupta, I., Galdiero, S., Galdiero, M., Gade, A., and Rai, M. 2014. Silver nanoparticles: therapeutical uses, toxicity, and safety issues. J. Pharm. Sci. 103(7): 1931–1944.

El-Demerdash, F.M., El-Magd, M.A., and El-Sayed, R.A. 2021. Panax ginseng modulates oxidative stress, DNA damage, apoptosis, and inflammations induced by silicon dioxide nanoparticles in rats. Env. Toxicol. 36(7): 1362–1374.

El-Sonbaty, S.M. 2013. Fungus-mediated synthesis of silver nanoparticles and evaluation of antitumor activity. Cancer Nanotechnol. 4: 73–79.

Farjadian, F., Ghasemi, A., Gohari, O., Roointan, A., Karimi, M., and Hamblin, M.R. 2019. Nanopharmaceuticals and nanomedicines currently on the market: challenges and opportunities. Nanomedicine. 14(1): 93–126.

Gade, A., Shende, S., and Rai, M. 2022. Potential role of *Phoma* spp. for mycogenic synthesis of silver nanoparticles. *In*: Rai, M., Zimowska, B., and Kövics, G.J. (eds.). Phoma: Diversity, Taxonomy, Bioactivities, and Nanotechnology. Springer, Cham. DOI: 10.1007/978-3-030-81218-8_17.

Gaikwad, S., Birla, S., Ingle, A., Gade, A.,Marcato, P., Rai, M., and Duran, N. 2013. Screening of different *Fusarium* species to select potential species for the synthesis of silver nanoparticles. J. Braz. Chem. Soc. 24(2): 1974–1982.

Guilger, M., Pasquoto-Stigliani, T., Bilesky-Jose, N., Grillo, R., Abhilash, P.C., Fraceto, L.F., and De Lima, R. 2017. Biogenic silver nanoparticles based on *Trichoderma harzianum*: synthesis, characterization, toxicity evaluation and biological activity. Sci. Rep. 7: 44421.

294 *Mycosynthesis of Nanomaterials: Perspectives and Challenges*

Guilger-Casagrande, M., and de Lima, R. 2019. Synthesis of silver nanoparticles mediated by fungi: A review. Frontiers Bioeng. Biotechnol. 7(287). DOI: 10.3389/fbioe.2019.00287.

Guilger-Casagrande, M., Germano-Costa, T., Bilesky-José, N., Pasquoto-Stigliani, T., Carvalho, L., Fraceto, L.F., and de Lima, R. 2021. Influence of the capping of biogenic silver nanoparticles on their toxicity and mechanism of action towards *Sclerotinia sclerotiorum*. J. Nanobiotechnol. 19,53. https://doi.org/10.1186/s12951-021-00797-5.

Gupta, I., Duran, N., and Rai, M. 2012. Nano-silver toxicity: emerging concerns and consequences in human health. pp. 525–548. *In*: Rai, M., and Cioffi, N. (eds.). Nanoantimicrobials: Progress and Prospects. Springer Verlag, Germany.

Gupta, I.R., Anderson, A.J., and Rai, M. 2015. Toxicity of fungal-generated silver nanoparticles to soil-inhabiting *Pseudomonas putida* KT2440, a rhizospheric bacterium responsible for plant protection and bioremediation. J. Hazard. Mater. 286: 48–54.

Hongbo, S., Magaye, R., Castranova, V., and Zhao, J. 2013. Titanium dioxide nanoparticles: a review of current toxicological data. Part. Fibre Toxicol. 10: 1–33.

Huq, M.A., Ashrafudoulla, M., Rahman, M.M., Balusamy, S.R., and Akter, S. 2022. Green synthesis and potential antibacterial applications of bioactive silver nanoparticles: a review. Polymers (Basel). 4(4): 742 DOI: 10.3390/polym14040742.

Jain, N., Bhargava, A., Rathi, M., Dilip, V., and Panwar, J. 2015. Removal of protein capping enhanced the antibacterial efficiency of biosynthesized silver nanoparticles. PloS One. 10: 1–19.

Jafarzadeh, S., Nafchi, A.M., Salehabadi, A., Oladzad-Abbasabadi, N., and Jafari, S.M. 2021. Application of bio-nanocomposite films and edible coatings for extending the shelf life of fresh fruits and vegetables. Adv. Coll. Interf. Sci. 291: 102405.

Lima, R., Feitosa, L.O., Ballottin, D., Marcato, P.D., Tasic, L., and Durán, N. 2013. Cytotoxicity and genotoxicity of biogenic silver nanoparticles. J. Phy.: Conf. Series. 429(1): 012020.

Kakakhel, M.A., Sajjad, W., Wu, F., Bibi, N., Shah, K., Yali, Z., and Wang, W. 2021. Green synthesis of silver nanoparticles and their shortcomings, animal blood a potential source for silver nanoparticles: a review. J. Hazard. Mater. Adv. 1: 100005.

Kapoor, D., and Singh, M.P. 2021. Nanoparticles: sources and toxicity. *In*: Singh, V.P., Singh, S., Tripathi, D.K., Prasad, S.M., Chauhan, D.K. (eds.). Plant Responses to Nanomaterials. Nanotechnology in the Life Sciences. Springer, Cham. DOI: 10.1007/978-3-030-36740-4_9.

Kidd, J., Westerhoff, P., and Maynard, A. 2021. Survey of industrial perceptions for the use of nanomaterials for in-home drinking water purification devices. NanoImpact. 22: 100320.

Koner, D., Banerjee, B., Kumari, A., Lanong, A.S., Snaitang, R., and Saha, N. 2021. Molecular characterization of superoxide dismutase and catalase genes, and the induction of antioxidant genes under the zinc oxide nanoparticle-induced oxidative stress in air-breathing magur catfish (*Clarias magur*). Fish Physiol. Biochem. 47(6): 1909–1932.

Kowalska, M., Piekut, T., Prendecki, M., Sodel, A., Kozubski, W., and Dorszewska, J. 2020. Mitochondrial and nuclear DNA oxidative damage in physiological and pathological aging. DNA Cell Biol. 39(8): 1410–1420.

Kreyling, W.G., Holzwarth, U., Schleh, C., Kozempel, J., Wenk, A., Haberl, N., Hirn, S., Schäffler, M., Lipka, J., Semmler-Behnke, M., and Gobson, N. 2017. Quantitative biokinetics of titanium dioxide nanoparticles after oral application in rats: Part 2. Nanotoxicology. 11: 443–453.

Lin, K.C., Yen, C.Z., Yang, J.W., Chung, J.H.Y., and Chen, G.Y. 2022. Airborne toxicological assessment: The potential of lung-on-a-chip as an alternative to animal testing, Mat. Today Adv. 14: 100216. DOI: 10.1016/j.mtadv.2022.100216.

Mandic, L., Sadzak, A., Strasser, V., Baranovic, G., Domazet Jurasin, D., Dutour Sikiric, M., and Šegota, S. 2019. Enhanced protection of biological membranes during lipid peroxidation: study of the interactions between flavonoid loaded mesoporous silica nanoparticles and model cell membranes. Intl. J. Mol. Sci. 20(11): 2709.

Mohamed, A.A., Fouda, A., Abdel-Rahman, M.A., Hassan, S.E.-D., El-Gamal, M.S., Salem, S.S., and Shaheen, T.I. 2019. Fungal strain impacts the shape, bioactivity and multifunctional properties of green synthesized zinc oxide nanoparticles. Biocatal. Agric. Biotechnol. 19: 101103.

Mukherjee, S.G., O'claonadh, N., Casey, A., and Chambers, G. 2012. Comparative *in vitro* cytotoxicity study of silver nanoparticle on two mammalian cell lines. Toxicol. *In Vitro* 26: 238–251.

Nayek, S., Lund, A.K., and Verbeck, G.F. 2022. Inhalation exposure to silver nanoparticles induces hepatic inflammation and oxidative stress, associated with altered renin-angiotensin system signaling in Wistar rats. Env. Toxicol. 37(3): 457–467.

Netala, V.R., Bethu, M.S., Pushpalatha, B., Baki, V.B., Aishwarya, S., Rao, J.V., and Tartte, V. 2016. Biogenesis of silver nanoparticles using endophytic fungus *Pestalotiopsis microspora* and evaluation of their antioxidant and anticancer activities. Int. J. Nanomedicine. 11: 5683–5696.

Nikzamir, M., Akbarzadeh, A., and Panahi, Y. 2021. An overview on nanoparticles used in biomedicine and their cytotoxicity. J. Drug Deliv. Sci. Technol. 61: 102316. DOI:10.1016/j.jddst.2020.102316.

Paiva-Santos, A.C., Herdade, A.M., Guerra, C., Peixoto, D., Pereira-Silva, M., Zeinali, M., Mascarenhas-Melo, F., Paranhos, A., and Veiga, F. 2021. Plant-mediated green synthesis of metal-based nanoparticles for dermopharmaceutical and cosmetic applications. Int. J. Pharm. 597: 120311. DOI: 10.1016/j.ijpharm.2021.120311.

Park, H.-J., Kim, S.H., Kim, H.J., and Choi, S.-H. 2006. A new composition of nanosized silica-silver for control of various plant diseases plant. Pathol. J. 22(3): 295–302.

Paul, A., and Roychoudhury, A. 2021. Go green to protect plants: repurposing the antimicrobial activity of biosynthesized silver nanoparticles to combat phytopathogens. Nanotechnol. Environ. Eng. 6: 10. DOI: 10.1007/s41204-021-00103-6.

Potara, M., Bawaskar, M., Simon, T., Gaikwad, S., Licarete, E., Ingle, A., Banciu, M., Vulpoi, A., Astilean, S., and Rai, M. 2015. Biosynthesized silver nanoparticles performing as biogenic SERS-nanotags for investigation of C26 colon carcinoma cells. Colloids Surf. B Biointerf. 133: 296–303.

Priyadarshini, E., Priyadarshini, S.S., Cousins, B.G., and Pradhan, N. 2021. Metal-fungus interaction: review on cellular processes underlying heavy metal detoxification and synthesis of metal nanoparticles. Chemosphere, 274: 129976. DOI: 10.1016/j.chemosphere.2021.129976.

Rai, M., Bonde, S., Golinska, P., Trzcińska-Wencel, J., Gade, A., Abd-Elsalam, K.A., Shende, S., Gaikwad, S., and Ingle, A.P. 2021. *Fusarium* as a novel fungus for the synthesis of nanoparticles: mechanism and applications. J. Fungi (Basel). 7(2): 139. DOI: 10.3390/jof7020139.

Ranjani, S., Shariq Ahmed, M., MubarakAli, D., Ramachandran, C., SenthilKumar, N., and Hemalatha, S. 2020. Toxicity assessment of silver nanoparticles synthesized using endophytic fungi against nosacomial infection, Inorganic Nano-Metal Chem. DOI: 10.1080/24701556.2020.1814332.

Rashid, M.M., Forte Tavcer, P., and Tomšič, B. 2021. Influence of titanium dioxide nanoparticles on human health and the environment. Nanomaterials (Basel, Switzerland). 11(9): 2354. DOI :10.3390/nano11092354.

Rehman, S., Ansari, M.A., Al-Dossary, H.A., Fatima, Z., Hameed, S., Ahmad, W., and Ali, A. 2021. Current perspectives on mycosynthesis of nanoparticles and their biomedical application. *In* Modeling and control of drug delivery systems, Academic Press, pp: 301–311.

Rennick, J.J., Johnston, A.P., and Parton, R.G. 2021. Key principles and methods for studying the endocytosis of biological and nanoparticle therapeutics. Nat. Nanotechnol. 16(3): 266–276.

Saravanakumar, K., and Wang, M.H. 2018. *Trichoderma* based synthesis of anti-pathogenic silver nanoparticles and their characterization, antioxidant and cytotoxicity properties. Microb. Pathog. 114: 269–273.

Salem, S.S., and Fouda, A. 2021. Green synthesis of metallic nanoparticles and their prospective biotechnological applications: an overview. Biol. Trace Elem. Res. 199: 344–370.

Salunkhe, R.B., Patil, S.V., Patil, C.D., and Salunke, B.K. 2011. Larvicidal potential of silver nanoparticles synthesized using fungus *Cochliobolus lunatus* against *Aedes aegypti* (Linnaeus, 1762) and *Anopheles stephensi* Liston (Diptera; Culicidae). Parasitol. Res. 109(3): 823–831.

Santos-Rasera, J.R., Monteiro, R.T.R., and de Carvalho, H.W.P. 2022. Investigation of acute toxicity, accumulation, and depuration of ZnO nanoparticles in *Daphnia magna*. Sci. Total Environ. 15;821:153307. doi: 10.1016/j.scitotenv.2022.153307.

Saravanan, A., Kumar, P.S., Karishma, S., Vo, D.N., Jeevanantham, S., Yaashikaa, P.R., and George, C.S. 2021. A review on biosynthesis of metal nanoparticles and its environmental applications. Chemosphere. 264(Pt2): 128580. DOI:10.1016/j.chemosphere.2020.128580.

Santos, T.S., Silva, T.M., Cardoso, J.C., Albuquerque-Júnior, R.L.C., Zielinska, A., Souto, E.B., Severino, P., and Mendonça, M.D.C. 2021. Biosynthesis of silver nanoparticles mediated by entomopathogenic fungi: antimicrobial resistance, nanopesticides, and toxicity. Antibiotics (Basel). 10(7): 852. doi: 10.3390/antibiotics10070852.

Skladanowski, M., Golinska, P., Rudnicka, K., Dahm, H., and Rai, M. 2016. Evaluation of cytotoxicity, immune compatibility and antibacterial activity of biogenic silver nanoparticles. Med. Microbiol. Immunol. 205: 603–613.

Soleimani, F.F., Saleh, T., Shojaosadati, S.A., and Poursalehi, R. 2018. Green synthesis of different shapes of silver nanostructures and evaluation of their antibacterial and cytotoxic activity. BioNanoScience 8: 72–80.

Sonwani, S., Madaan, S., Arora, J., Suryanarayan, S., Rangra, D., Mongia, N., and Saxena, P. 2021. Inhalation exposure to atmospheric nanoparticles and its associated impacts on human health: a review. Frontiers Sust. Cities, 3: 690444. DOI:10.3389/frsc.2021.690444.

Souza, I.D.C., Morozesk, M., Mansano, A.S., Mendes, V.A.S., Azevedo, V.C., Matsumoto, S.T., Elliott, M., Monferrán, M.V., Wunderlin, D.A., and Fernandes, M.N. 2021. Atmospheric particulate matter from an industrial area as a source of metal nanoparticle contamination in aquatic ecosystems. Sci. Total Environ. 753: 141976. DOI: 10.1016/j.scitotenv.2020.141976.

Srivastava, S., and Bhargava, A. 2022. Green synthesis of nanoparticles. *In* Green nanoparticles: The future of nanobiotechnology. Springer, Singapore. DOI:10.1007/978-981-16-7106-7_4.

Sun, K., Song, Y., He, F., Jing, M., Tang, J., and Liu, R. 2021. A review of human and animals exposure to polycyclic aromatic hydrocarbons: health risk and adverse effects, photo-induced toxicity and regulating effect of microplastics. Sci. Total Environ. 773: 145403. DOI: 10.1016/j.scitotenv.2021.145403.

Tao, C. 2018 Antimicrobial activity and toxicity of gold nanoparticles: research progress, challenges and prospects. Lett. Appl. Microbiol. 67(6): 537–543.

Wang, L., Hartel, N., Ren, K., Graham, N.A., and Malmstadt, N. 2020. Effect of protein corona on nanoparticle–plasma membrane and nanoparticle–biomimetic membrane interactions. Env. Sci. Nano. 7(3): 963–974.

Yu, Z., Li, Q., Wang, J., Yu, Y., Wang, Y., Zhou, Q., and Li, P. 2020. Reactive oxygen species-related nanoparticle toxicity in the biomedical field. Nanoscale Res. Lett. 15(1): 1–14.

Zafar, N., Madni, A., Khalid, A., Khan, T., Kousar, R., Naz, S.S., and Wahid, F. 2020. Pharmaceutical and biomedical applications of green synthesized metal and metal oxide nanoparticles. Curr. Pharm. Des. 26(45): 5844–5865.

Zhang, H., Chen, B., and Banfield, J.F. 2010. Particle size and pH effects on nanoparticle dissolution. J. Phys. Chem. C. 114: 14876–14884.

Index

A

AFM 244, 247, 248
AgNO$_3$ 127, 135, 137, 138
Antibacterial activity 121
anticancer 158–160, 165
Antifungal activity 121
antimicrobial 159, 161, 162, 165, 173, 183, 187–190
Applications 3, 10, 15–18
AuNPs 146–161, 164

B

Biogenic Nanoparticles 236
Biogenic Production 4, 5
biogenic synthesis 254, 255, 260–262, 264–266, 269
biological activity 67, 72
biological synthesis 56, 72, 83, 88, 89
bio-recovery 38
biosynthesis 255, 260, 261, 263–265, 267–270

C

Capping agent 83, 85, 89, 92
Carbon Nanomaterials (CNMs) 195, 199, 200, 203, 205, 206
Characterization 85
Characterization techniques 200, 231, 235, 236, 247, 248
Chitosan nanoparticles 99–103, 105–109
Copper nanoparticles 112–115, 119, 122
Copper oxide nanoparticles 112–115, 122

D

downstream processing 220, 224, 225

E

enzyme-mediated 107, 109
extracellular enzymes 216
Extracellular Synthesis 173, 175–178, 190

F

fungi 81–83, 86–90, 92, 99, 103, 107–109, 126–135, 139, 140, 215–217, 219–225, 254–256, 260, 263–265, 267–269, 284, 287

G

gold 254, 255, 259, 260, 266–268
green synthesis 42, 148, 151, 152, 156, 164

I

Intracellular 194
Intracellular Synthesis 173, 174, 176–178, 189, 190
Iron oxide nanoparticles 81–90

M

Mechanism 7, 8, 14, 17
mechanistic aspects 255, 260, 264, 268–270
metal-based nanoparticles 51
Metal Nanomaterials 5
Metallic nanoparticles 254, 255, 260, 267
Myconanotechnology 5, 9, 17, 165
Mycosynthesis 99, 103, 109, 180, 189, 284, 292

N

Nanoparticle 1, 3, 5, 6, 8, 29, 30, 33, 38, 40–43, 45, 281, 283–292
nanotechnology 103
Nanotoxicity 284, 286, 289–291

O

optimization parameters 127

P

Penicillium family 75

S

Selenium 27–46
SEM 236, 241, 243–246
silver 254–258, 260, 262, 264–269

T

TEM 236, 240, 243–246
Toxicity 281, 283, 284, 286–292

U

upstream processing 220–222

X

XRD 233, 236, 239, 240, 246

Y

yeast 27, 28, 30, 38–45, 128, 133

Z

ZnO Nanoparticles 177, 184

Editors' Biography

Dr. Mahendra Rai is a Senior Professor and UGC-Basic Science Research Faculty at the Department of Biotechnology, Sant Gadge Baba Amravati University, Maharashtra, India. Presently, he is a Visiting Professor at the Department of Microbiology, Nicolaus Copernicus University, Poland. He was Visiting Scientist at the University of Geneva, Debrecen University, Hungary; the University of Campinas, Brazil; Nicolaus Copernicus University, Poland, VSB Technical University of Ostrava, Czech Republic, National University of Rosario, Argentina, and the Federal University of Piaui, Teresina, Brazil. Dr. Rai has published more than 400 research papers in national and international journals with 71 index. In addition, he has edited/authored more than 60 books and 6 patents. The main focus of his research is Myconanotechnology and its applications in medicine and sustainable agriculture.

Dr. Patrycja Golińska, currently works at the Department of Microbiology at Nicolaus Copernicus University (NCU) as an Associate Professor. She received a fellowship from the Marshal of Kuyavian-Pomeranian Voivodeship for the best doctoral students in 2006 and from the NCU in 2011 to complete one-year postdoctoral internship at the School of Biology of Newcastle University, UK. She has fourteen years of teaching and eighteen years of research experience. Dr. Golińska scientific contribution encompasses 46 original and 13 review articles, 14 book chapters, and editor of one book. She has received the grant within the Horizon Europe programme of European Commission. The main focus of Dr. Golińska research is the biosynthesis of metal nanoparticles, mainly by actinobacteria and fungi, and their use to combat bacterial and fungal pathogens, the study of actinobacterial diversity in extreme biomes and the use of actinobacteria as biocontrol agent of fungal phytopathogens.